INTRODUCTORY PHYSICS WITH ALGEBRA

Mastering Problem-Solving

INTRODUCTORY PHYSICS WITH ALGEBRA

Mastering Problem-Solving

STUART E. LOUCKS
American River College

WILEY
JOHN WILEY & SONS, INC.

Executive Publisher *Kaye Pace*
Acquisitions Editor *Stuart Johnson*
Production Editor *Nicole Repasky*
Marketing Manager *Amanda Wygal*
Creative Director *Harry Nolan*
Designer *Michael St. Martine*
Production Management Services *Ingrao Associates*
Editorial Assistant *Krista Jarmas, Alyson Rentrop*
Senior Media Editor *Thomas Kulesa*

This book was set in Times by Techbooks

To order books or for customer service, please call 1-800-CALL WILEY (225-5945).

Library of Congress Cataloging-in-Publication Data
Loucks, Stuart E.
Introductory physics with algebra : mastering problem-solving / Stuart E. Loucks.
p. cm.
ISBN-13: 978-0-471-76250-8 (pbk. : acid-free paper)
ISBN-10: 0-471-76250-4 (pbk. : acid-free paper)
1. Mathematical physics—Textbooks. 2. Algebraic logic—Textbooks. 3. Problem solving—Mathematics—Textbooks. I. Title.
QC20.7.A4L68 2007
530.15'2—dc22

2006007384

10 9 8 7 6 5 4

To Jennifer Bei-Yu, my precious wife,
and Jason Stuart Ming-Yen,
Jared Brian Ming-Sun,
Jessie Ann Ming-Jun,
and Joanna Lee Ming-Mei,
my wonderful children.
I could not have done it without you!

And

In memory of Sue Ann Reynolds, M.D.,
my dear mother who was so loving,
and so loved by so many.
I miss you very much, and know you
would have been proud!

PREFACE

Why can one student sit down, read a physics problem, and immediately start making progress toward a solution, but another student reading the same problem stares in frustration, unable to make any headway? Both students basically understand the physics concepts from the textbook and lecture. The difference is that the successful problem solver knows how to use these concepts to set up the problem, which gets the right values into the right equations, which in turn lead to the correct solution. The frustrated nonsolver does not see how to apply the physics ideas to set up the problem, and so has learned to plug in numbers without really being sure whether the equations apply or not.

Skill in problem solving is almost always gained by good training and practice rather than by raw talent. The aim of this book is to give you good training. You will have to work hard to practice. If you do, problem solving will soon come together for you. Of course, you *will* experience frustration. That is part of the process and is often necessary and even useful to bring about learning. If you stick with it, you can come out of your physics class with the level of understanding you want and with the grade you want.

Competitive high divers train their bodies to do things that don't come naturally. Once they leave the diving board, they have to move their body parts into the right positions. If they trust in their training, not in their instincts, then they go into the pool painlessly and head first. *Knowing why* is never complete until they have dived successfully. With each good dive, they see the fruit of their training, and their confidence and understanding grow.

As you go through this book, you will often have to trust in the training we give you and practice thinking in a way that is *not automatic* for you. When you apply what you learn here to solve new problems, your confidence and understanding will grow. Thinking in this way will gradually become more natural to you.

We will not cover every single physics problem you will encounter. Our goal is to coach you in the fundamental concepts and approaches needed to set up and solve the major problem types. As you learn how to deal with these kinds of problems, you will be better equipped to tackle problems you have never seen before.

Rather than being overwhelmed when you read a new problem, you will recognize details that tell you how to begin. You might not see the entire solution at first glance, but you will know how to effectively organize the information, decide on the correct equations, and ultimately solve the problem.

Does this book replace your textbook and class lectures? No. These important resources will familiarize you with aspects of physics not included in this book. We will help you make sense of your textbook and class notes so that you can use them more effectively. Our goal is to help you break through barriers you face when trying to solve a new physics problem, to bridge the gap between *reading the problem statement* and *setting up a successful solution*.

Work hard, learn, and enjoy your success!

ACKNOWLEDGMENTS

I would never have dreamed of trying to write a book completely on my own. Were it not for the help provided by so many others, it would never have happened.

My first "thank you" goes to Eric Yager of Wiley, who initiated this process for me two years ago, and who has provided continual encouragement along the way. I am also grateful to David Klein, for getting this ball rolling with his excellent book, *Organic Chemistry as a Second Language,* the origin of this *As a Second Language* series.

From the beginning of this project, the invaluable support of many at Wiley has helped bring it to completion. To editor Stuart Johnson I am thankful for pulling me in and making this book a reality. Somehow combining patience, frank and upfront criticism, and sincere encouragement, Stuart helped me become a much better writer than I otherwise would have been. Senior editorial assistant, Krista Jarmas, with help from Aly Rentrop, has provided fantastic, patient support in our frequent communications throughout the writing, editing, and production processes. For their outstanding work in each step of production, I thank Jeanine Furino and Nicole Repasky. Jane Shifflet at Techbooks has been very helpful in working out all of the little details with the art and composition. I would also like to mention the fine work of Amanda Wygal in marketing, and Harry Nolan and Michael St. Martine in design.

Along the way, many have given indispensable help in reviewing and editing. "Excellence" is the word that comes to mind when I think of Suzanne Ingrao of Ingrao Associates, and Karen Osborne, whose attention to detail made up for my many little mistakes. Professor Russ Poch of Howard Community College provided very thoughtful and detailed feedback which helped make the book more effective. I so appreciate the detailed accuracy check and many helpful suggestions of my colleague, Professor Mike Anderson of American River College.

Finally, I am grateful to the many students who gave encouraging feedback and constructive criticism. Many participated in formal reviews, including students organized by Professor Heath Hatch of the University of Massachusetts, and those in the Physics 350 class taught by my colleague, Professor Bill Simpson of American River College. My own wonderful Physics 311 and Physics 410 students at American River College have also been extremely helpful in fine-tuning the details of this book.

CONTENTS

CHAPTER 1

THE BOTTOM LINE FOR SOLVING PHYSICS PROBLEMS

IN AN EVERYDAY math problem, you get an equation with x as the unknown, and you solve for x. Maybe there are two equations and two unknowns, and you solve for both x and y. Then there are word problems. Before you can solve the equations, you have to set up the problem in order to *figure out what the equations are*.

That is the exact challenge in physics problems, which are just word problems related to certain physical situations. Your focus when facing a physics problem should be on the physics: *setting up the problem*. Once set up, it becomes simply a math problem. At this point, hypothetically, you could give the problem to a friend who knows no physics but only knows the math (and likes you enough to finish it for you).

You might be tempted to try a shortcut instead of taking time to set up the problem. Or maybe you want to set it up effectively but don't see how. In either situation, the results are usually wrong answers and frustration.

For example, look at this physics problem:

A car travels for 8.0 seconds in a straight line, traveling at an average velocity of 40 m/s east. The first half of the displacement was traveled at an average velocity of 35 m/s east. What is the average velocity for the last half of the trip?

Here are some failed attempts to answer this question without a good setup:

- **Answering too quickly:** "*Obviously*, the correct answer is $45\frac{\text{m}}{\text{s}}$!" This answer is WRONG!
- **Plugging into an equation:** Find an equation with the right kinds of terms, plug in known values, and solve:

$$\text{Use } v_{\text{avg}} = \frac{v_1 + v_2}{2}, \text{ with } v_1 = 35\tfrac{\text{m}}{\text{s}} \text{ and } v_{\text{avg}} = 40\tfrac{\text{m}}{\text{s}}, \text{ and then solve for } v_2 = 45\tfrac{\text{m}}{\text{s}}.$$

Same answer, still WRONG! This is a more "conscientious" version of the quick answer above. Why is it wrong? We will discuss that in the next chapter.

The main problem in these attempts is that the information is not organized in a useful way. If it were, the thinking errors could be avoided. How should we sort out the information? If we recognize the need, we might try the following:

- **Not-very-effective organizing**

  Known: $t = 8.0$ s, $v_{\text{avg}} = 40\frac{\text{m}}{\text{s}}$ for whole trip, $v_{\text{avg}} = 35\frac{\text{m}}{\text{s}}$ for first half
  Unknown: $v_{\text{avg}} = ?$ for second half

This is better, but it still does not give us any guidance on what to do next. What equations should we use? And in the equations, which values go where?

Here are some other not-so-productive ways to deal with not knowing how to solve the problem:

- Looking it up in a solutions manual
- Finding one like it in a physics problem-solving book
- Finding someone to show you how to do it

In each of these cases, you are relying on another source that might not be available or reliable. Worse, you will be no better off when you come across a new and different problem that you can't solve. These sources often leave out crucial *mental* steps, assuming they are obvious, and do little to help you learn the thinking pattern you need.

The best way to deal with a physics problem like this is to learn the *mental* and *written* steps to set it up. The benefits are many:

- *You* know where to start! You do not depend on an outside source of help.
- It is more efficient in the long run. It takes time to learn, but once you do, you will spend much less time on failed attempts.
- When facing a problem you have never seen before (like on a test), you will be equipped to solve it successfully.
- Once a problem is set up, you just do the math, which you already know.

Don't make the mistake of thinking that *problem solving* is the same as *doing algebra*. The algebra is the last step, which we do only after the physics part is already accomplished.

Focus on the process of *setting up* problems and you will succeed in your physics class. We will also help with the more difficult math, but the heart of this book is to help you to know what to have *in your mind* and *on your paper* when setting up physics problems!

CHAPTER 2

LINEAR VELOCITY AND ACCELERATION

AS YOU LEARN to solve linear motion problems, you will lay a foundation for your entire physics course. The first step is to understand how to use the motion equations. Then you can set up a problem to get the right quantities into the right equations. That is the physics, and you are left with only the math to finish. When you succeed here, your confidence and skill will grow for working out problems in later topics.

The algebra in motion problems is often more involved than what we will see in many other topics. To help with this, we will complete the math in some exercises with more difficult algebra.

Your own solutions to problems will be much shorter on paper than those given here. In addition to the *written* steps, we detail the *mental* steps that you would not normally put on paper. This will help you bridge the gap between reading a problem and solving it successfully.

2.1 LINEAR MOTION EQUATIONS

In your lecture and textbook, you have been or will be introduced to several linear motion equations. Here we group the main equations into two sets:

Equations for motion with constant or average speed or velocity		Equations for motion with constant or average acceleration	
		$x = v_0t + \frac{1}{2}at^2$	(2.3a)
Speed: speed $= \frac{d}{t}$	(2.1)	$x = \left(\frac{v_0 + v}{2}\right)t$	(2.3b)
Velocity: $v = \frac{x}{t}$	(2.2)	$v = v_0 + at$	(2.3c)
		$v^2 = v_0^2 + 2ax$	(2.3d)

Your textbook or class notes might have them with slightly different symbols. Equation 2.3a, for example, is sometimes written as $\Delta x = v_0 t + \frac{1}{2}at^2$, or $x - x_0 = v_0 t + \frac{1}{2}at^2$, or $x = x_0 + v_0 t + \frac{1}{2}at^2$. These all mean essentially the same thing. After going through this chapter, you will easily recognize each of these equations and know how to use them.

2.2 *THE* IDEA BEHIND HOW TO USE MOTION EQUATIONS

A time *interval* has "length" or duration, while a time *instant* is a single moment with no duration. Each time *interval* has a beginning or *initial instant* (I) and an ending or *final instant* (F). We can *see* it like this:

In the motion equations, each quantity corresponds to one of the specific parts[1] of an interval: the *initial instant,* the *entire interval,* or the *final instant.* For example, look at the quantities in Equation 2.3c:

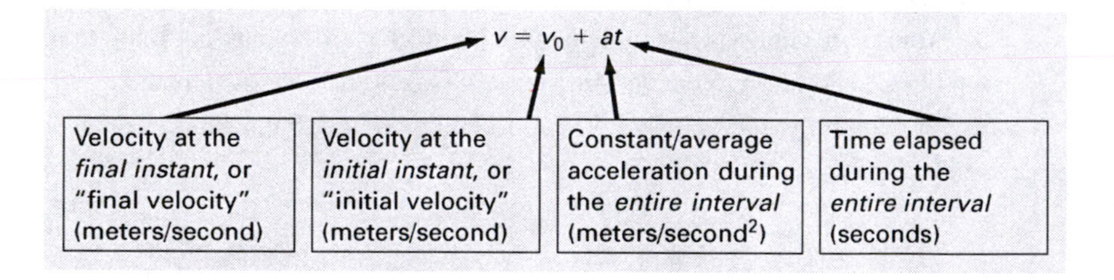

When solving a motion problem, especially a complicated one, we sketch a simple diagram to help *see* these connections:

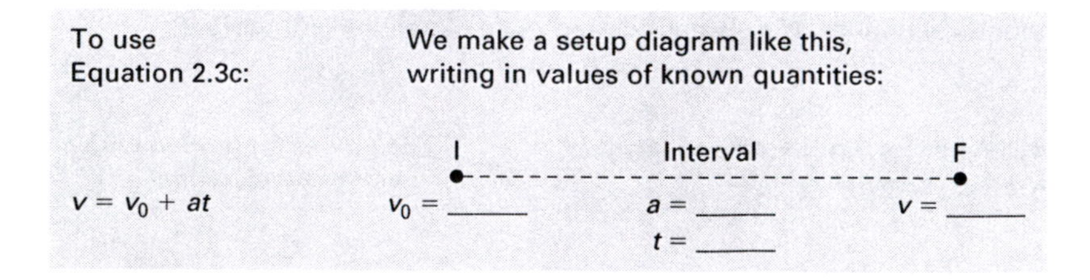

This is THE idea behind how to use motion equations! If you understand this idea, then you can *set up* and solve even challenging problems. We will start by looking at how to use Equations 2.1 and 2.2.

[1]In equations using integral calculus, there is one more option: Quantities within an integral correspond to an arbitrary instant within the interval.

2.3 CONSTANT/AVERAGE SPEED OR VELOCITY PROBLEMS

Before solving any problems, we need to understand the quantities in Equations 2.1 and 2.2—specifically, *where each quantity fits in an interval*. First, Equation 2.1:

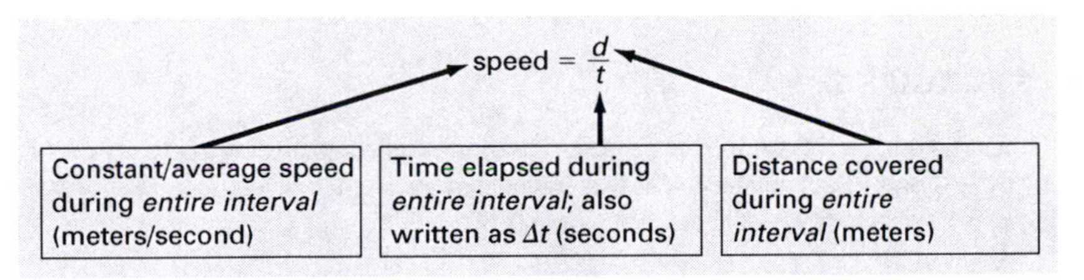

And Equation 2.2:

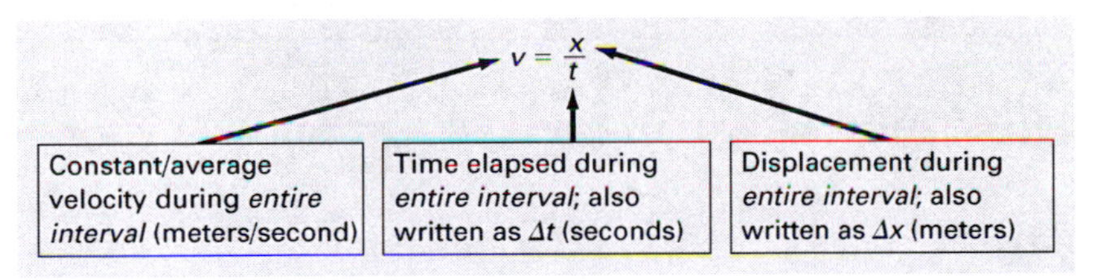

In both of these equations, each quantity corresponds to an *entire interval*. There are *no initial or final values*. We have written these equations for *constant* speed and velocity; for *average* values, we put a bar on top like this: $\overline{\text{speed}}$ or $\bar{v}$.

The main difference between Equations 2.1 and 2.2 is in the use of positives and negatives. In Equation 2.1, "speed" and d are *always positive*. In Equation 2.2, v and x can be *positive or negative*; the *sign* represents the *direction*. For horizontal motion, for example, we usually have *right* as *positive* and *left* as *negative*.

PAY ATTENTION TO DETAILS IN DEFINITIONS!

In physics, we take words like *speed* and *velocity,* which mean the same thing in everyday speech, and give them very specific definitions, so that *speed* means one thing and *velocity* means something slightly different.

Though not necessary for the simplest problems, it is very helpful for more difficult problems to make setup diagrams like one of the following:

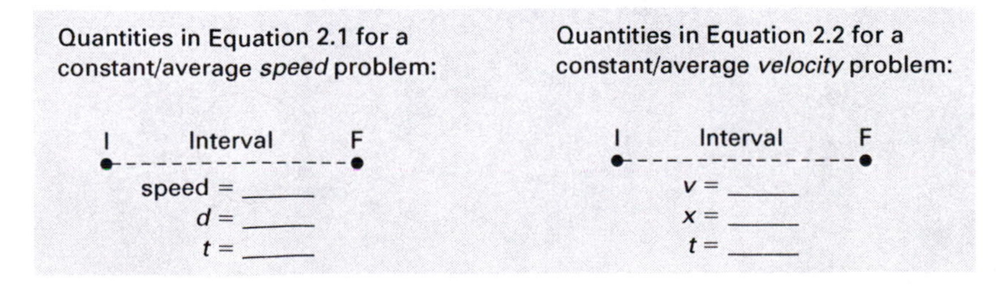

2.4 CONSTANT/AVERAGE SPEED OR VELOCITY—TWO INTERVALS, SAME DIRECTION

We have already considered some unproductive approaches to the motion problem discussed in Chapter 1. Here is a successful way to solve it.

EXERCISE 2.1

A car travels for 8.0 seconds in a straight line, traveling at an average velocity of 40 m/s east. The first half of the displacement was traveled at an average velocity of 35 m/s east. What is the average velocity for the last half of the trip?

BE PATIENT HERE, AND IT WILL SAVE YOU TIME LATER!

Learn the *written* and *mental* steps that we outline throughout this book. Let them be a guide to your thinking, not a restrictive recipe. In this first exercise, we include even more detail than usual.

Solution

(1) Type of problem

Here we *mentally* identify the type of problem and thus which equation(s) to use. The problem mentions "average velocity" in a few places, so we will use Equation 2.2.

(2) Sort by interval

This step is both mental and written. Before calculating anything, we organize, or sort, the information from the problem *by interval*. There are three intervals mentioned: the *entire trip,* the *first half of the trip,* and the *second half of the trip*.

First, we can mentally picture how the intervals are related:

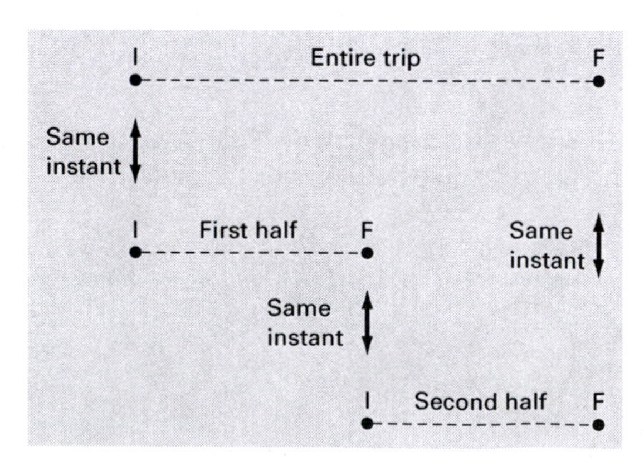

Beginning with the interval for the *entire trip,* we sort out the quantities for Equation 2.2:

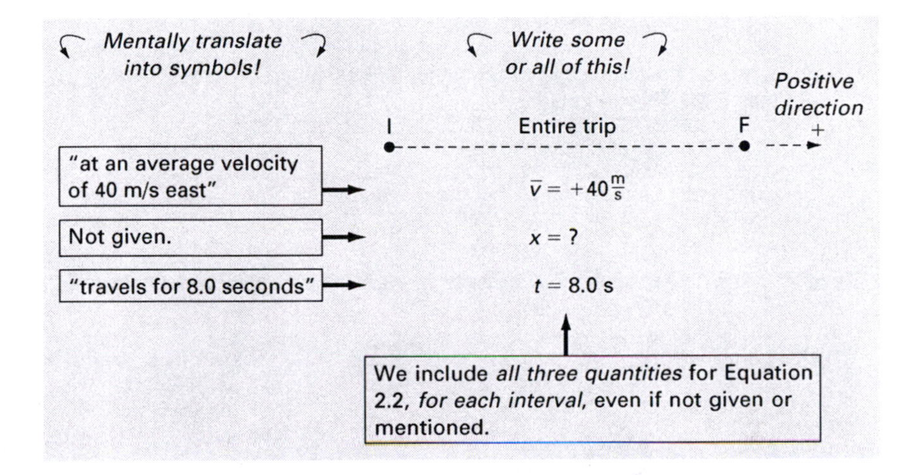

We made *east* the *positive* direction, so all displacements and velocities in this problem are *positive*. Now for the other two intervals:

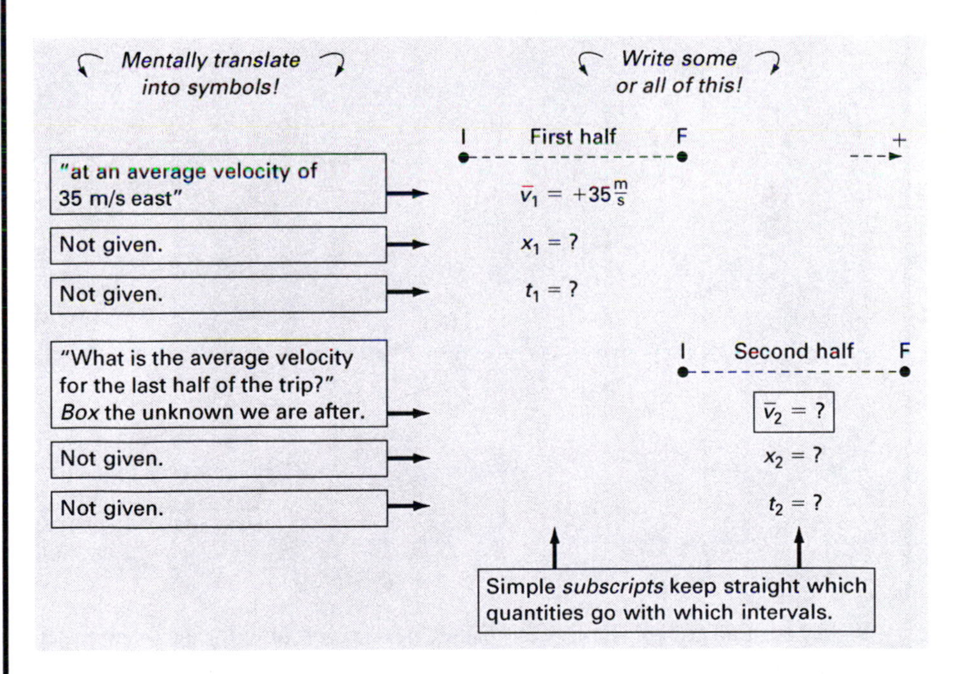

(3) Equations & unknowns

This mental and written step is amazingly helpful in leading us directly to an outline of the solution. Don't skip this step! If you do, you will be plagued with frustration, especially in more complicated problems.

Because there are multiple intervals, we need equations *relating the intervals* and equations *for each interval.*

First, we *relate the intervals*:

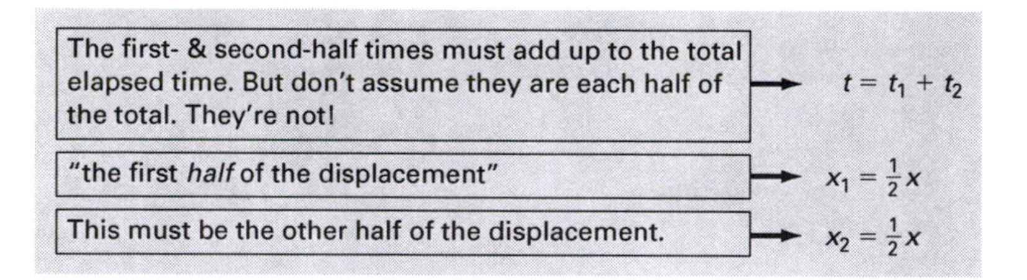

Then we just write Equation 2.2, separately *for each interval*. Here is our completed setup:

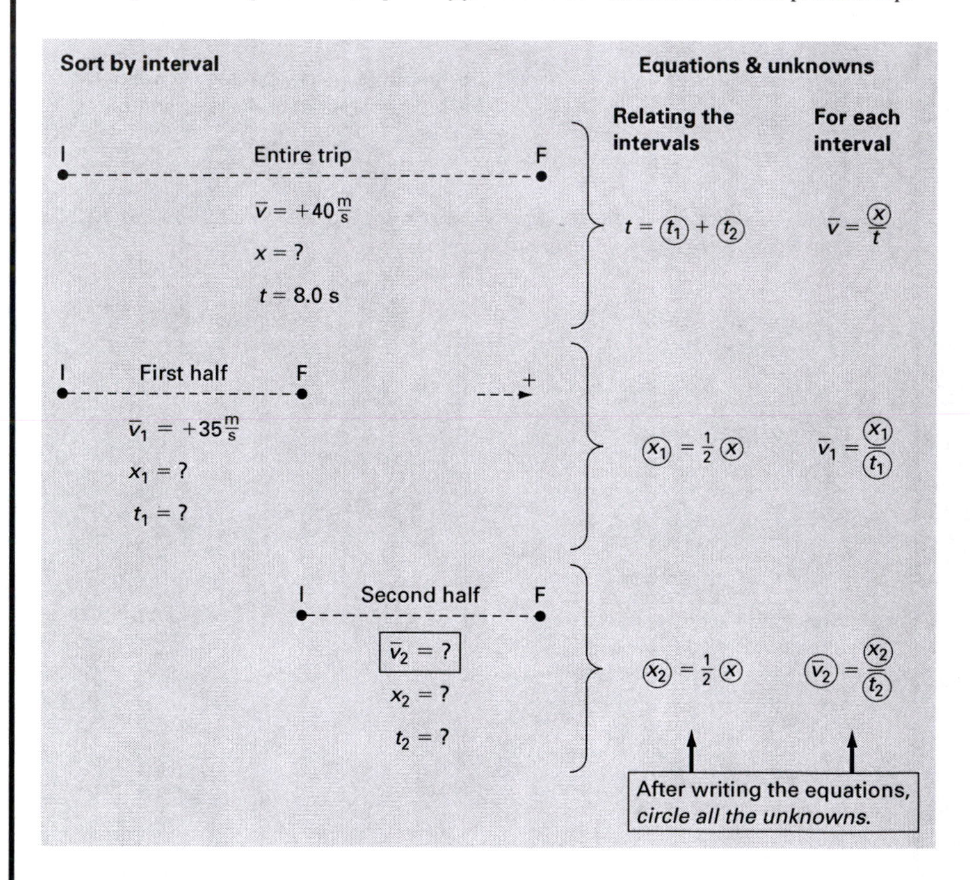

We may not need every single equation, but we include them all so you know how to get them if you need them.

USE CONSISTENT NOTATION!

Use the *exact same symbols* in your *setup diagram* and *equations*. Make your notation as simple as possible, but add subscripts if necessary to keep from mixing up quantities.

You don't want a quantity for one interval ending up in another interval's equation! For example:

For this interval:

I Second half F

$\boxed{\bar{v}_2 = ?}$

$x_2 = ?$

$t_2 = ?$

Write this

$$\bar{v}_2 = \frac{x_2}{t_2}$$

but NOT this!

$$\bar{v} = \frac{x}{t}$$

(4) Outline solution

Now we look for *any equation* with only *one unknown*. We solve it, then substitute the result into another equation, solve it, and so on until we reach our final solution. Do you see where to start? Look for an equation with one unknown:

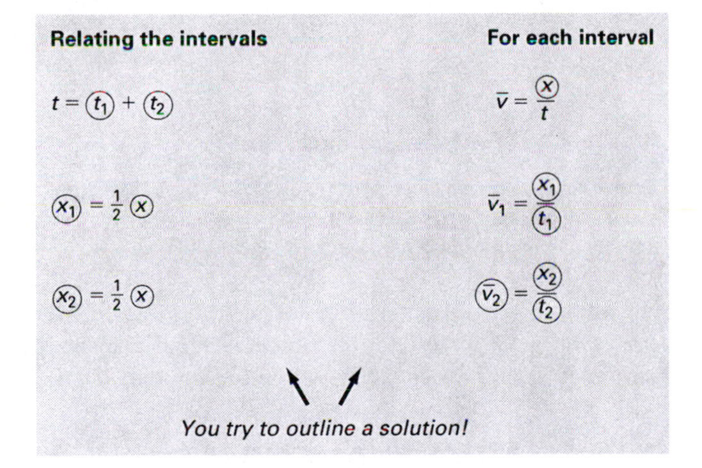

You can check our outline and the math in the answers at the end of this exercise.

Answers for (4) Outline solution Here is one possible outline. Start with (1), then (2), and so on:

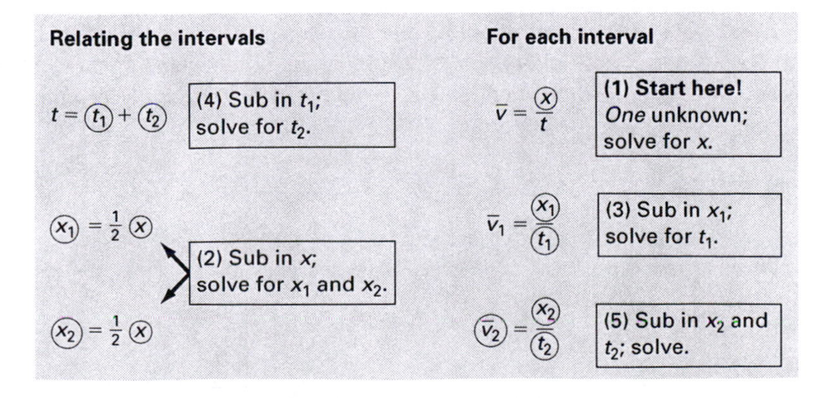

Answers for the math

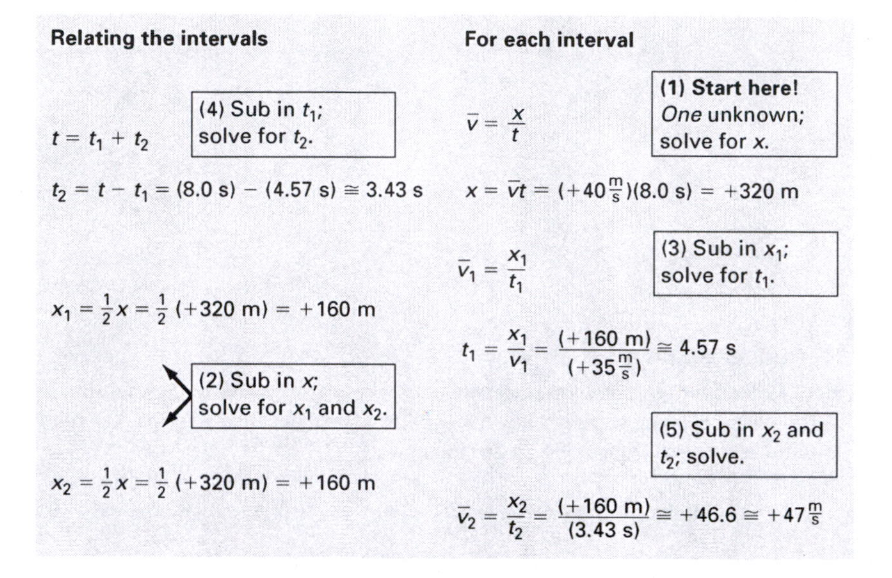

Answer $\bar{v}_2 \cong +46.6 \cong +47\frac{\text{m}}{\text{s}}$ (*positive* means *east*).

As promised in Chapter 1, an explanation of why $45\frac{\text{m}}{\text{s}}$ is wrong: The second half of the trip does not take *as much time* as the first half. That means the second-half velocity must be even *faster* than $45\frac{\text{m}}{\text{s}}$ in order to end up with an entire trip average of $40\frac{\text{m}}{\text{s}}$. ■

Each quantity given in the previous exercise had *two* sig. figs. (significant figures). We keep *one extra* sig. fig., or *three*, in the *intermediate* calculations. When rounding off our *final answer*, it is generally correct to *two*, or the *original number* of sig. figs.

2.5 CONSTANT/AVERAGE SPEED OR VELOCITY—TWO INTERVALS, DIRECTION CHANGE

In the previous exercise, all velocities and displacements were in one direction (to the east, or positive), so we did not really need to worry about signs. In the next exercise, the motion changes from one direction to the opposite direction; this makes positive and negative signs more important.

EXERCISE 2.2

You travel east 50.0 m for 10.0 s, change directions, and then travel west 15.0 m for 3.00 s. (a) For the trip west, what is your average velocity? (b) For the entire trip, what is your average velocity?

Solution

(1) Type of problem

"Average velocity" is mentioned, so we use Equation 2.2.

(2) Sort by interval and/or object

There are three intervals: the entire trip, the trip east, and the trip west. We let Equation 2.2 guide us: List *all three quantities* ($\bar{v}$, x, and t) *for each interval.*

We make *east* the *positive* direction, so all displacements and velocities will be *positive* if *east, negative* if *west*. We leave a few things for you to fill in, with answers at the end of the exercise:

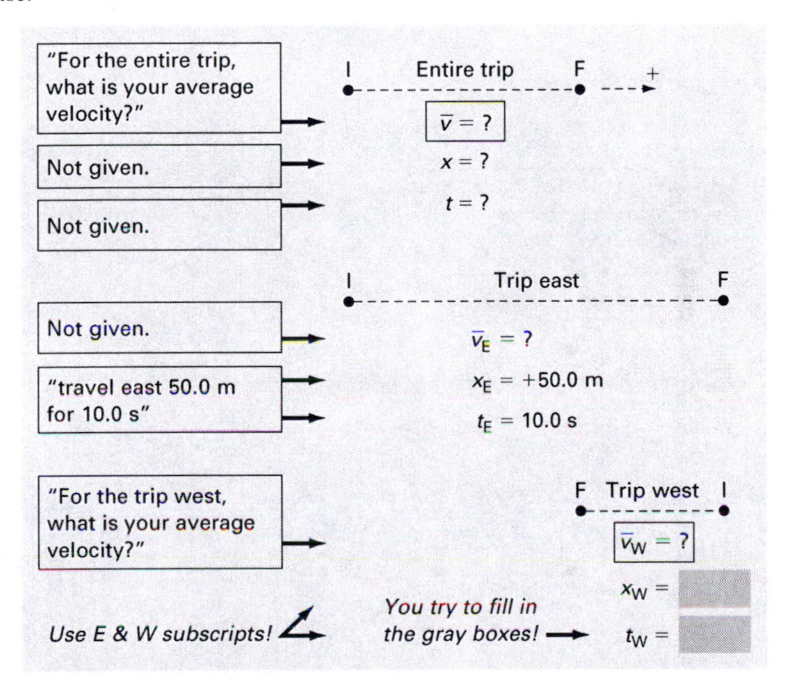

(3) Equations & unknowns

Again, we need equations *relating the intervals* and equations *for each interval*. To *relate the intervals,* the displacement for the entire trip is the sum of the other two displacements: $x = x_E + x_W$. The elapsed time also adds: $t = t_E + t_W$. Then, *for each interval,* we use Equation 2.2. This gives us:

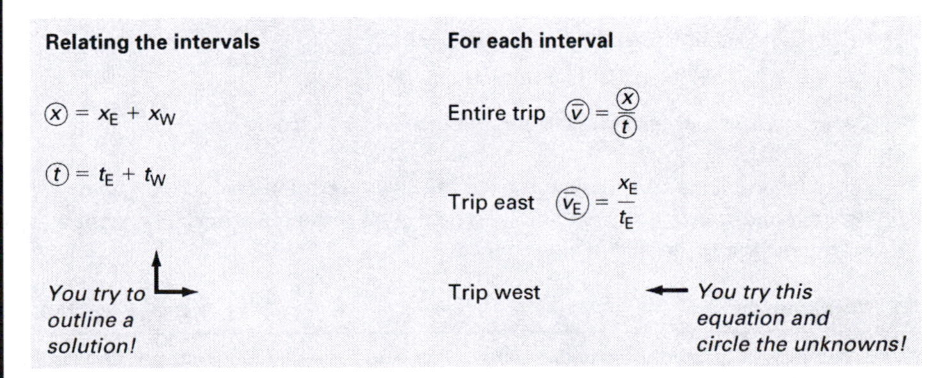

(4) Outline solution

Do you see any equations with one unknown? Try to outline a solution on the previous figure, and then check the answers.

DON'T WORRY ABOUT THE ORDER OF THE QUESTIONS!

Regardless of which unknowns are asked for, or the order in which they are asked for, set up the problem in this way. Once set up, solve for *anything you can*. Then look for a "chain reaction" to lead to the final solution.

Answers for gray boxes

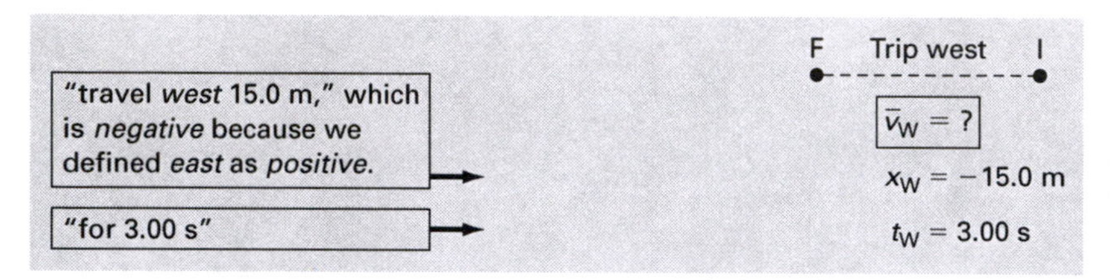

Answers for (3) Equations & unknowns and (4) Outline solution

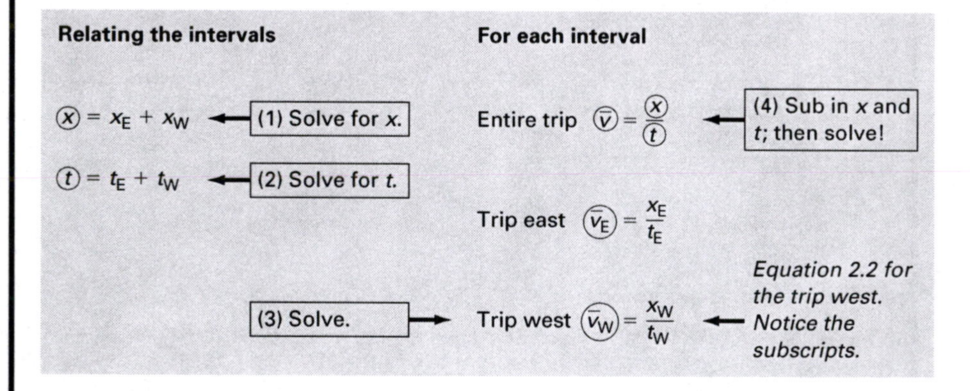

Answers

- $\bar{v}_W = -5.00\frac{\text{m}}{\text{s}}$ (*negative* means to the *west*).
- $\bar{v} \cong +2.692 \cong +2.69\frac{\text{m}}{\text{s}}$ (*positive* means to the *east*).

There are also some unknowns that we never solved for but could have. ■

In problems like the one we just did, we are often also asked about *distance covered* and *speed*. To explain how to deal with these quantities, we begin by comparing *displacement* and *distance covered*:

Displacement	**Distance covered**
• Depends only on *starting* and *ending points*, and ignores the path taken in between.	• Depends on the *path* taken.
• Sign gives direction.	• No direction.

For the previous problem, this results in the following:

Interval	Displacement	Distance covered	Relationship
Trip east	$x_E = +50.0$ m	$d_E = 50.0$ m	Straight-line path, so $d_E = \|x_E\|$.
Trip west	$x_W = -15.0$ m	$d_W = 15.0$ m	Straight-line path, so $d_W = \|x_W\|$.
Entire trip	$x = x_E + x_W$ $x = +35.0$ m	$d = d_E + d_W$ $d = 65.0$ m	*Direction change* in path, so $d \neq \|x\|$. *No relationship.*

The same thing happens when we compare average *velocity* (from Equation 2.2) and *speed* (from Equation 2.1):

Interval	Average velocity	Average speed	Relationship
Trip east	$\bar{v}_E = \frac{x_E}{t_E} = +5.00\frac{m}{s}$	$\overline{speed}_E = \frac{d_E}{t_E} = 5.00\frac{m}{s}$	Straight-line path, so $\overline{speed}_E = \|\bar{v}_E\|$.
Trip west	$\bar{v}_W = \frac{x_W}{t_W} = -5.00\frac{m}{s}$	$\overline{speed}_W = \frac{d_W}{t_W} = 5.00\frac{m}{s}$	Straight-line path, so $\overline{speed}_W = \|\bar{v}_W\|$.
Entire trip	$\bar{v} = \frac{x}{t} \cong +2.69\frac{m}{s}$	$\overline{speed} = \frac{d}{t} = 5.00\frac{m}{s}$	*Direction change* in path, so $\overline{speed} \neq \|\bar{v}\|$. *No relationship.*

2.6 CONSTANT/AVERAGE SPEED OR VELOCITY—TWO OBJECTS

When there are two or more objects in a problem, we sort the information *by object*.

EXERCISE 2.3

A woman runs with a constant speed of 4.0 m/s toward a cable car, which moves with a constant speed of 3.0 m/s toward the woman. The woman begins at a distance of 13 m from the front of the cable car. (a) How much time until she reaches the front of the cable car? (b) How far will she have to run?

Solution

(1) Type of problem

Both objects move with constant speed, so we use Equation 2.1 (no signs) *for each object*. We could also use Equation 2.2 with positive and negative signs, but it is more difficult.

Why can we ignore signs here? Even though the woman and cable car move in opposite directions from each other, all quantities *for the woman* are in one direction, and all quantities *for the cable car* are in one (different) direction. So we don't necessarily need to bother with signs for either object. However, in the previous exercise, for *the one object*, some quantities were east and others were west. This meant we could not ignore signs for that object.

(2) Sort by interval and/or object

There is just one interval, but there are *two objects*. We sort out *all three quantities* in Equation 2.1 (speed, d, and t) *for each object,* using subscripts 1 for the woman and 2 for the cable car:

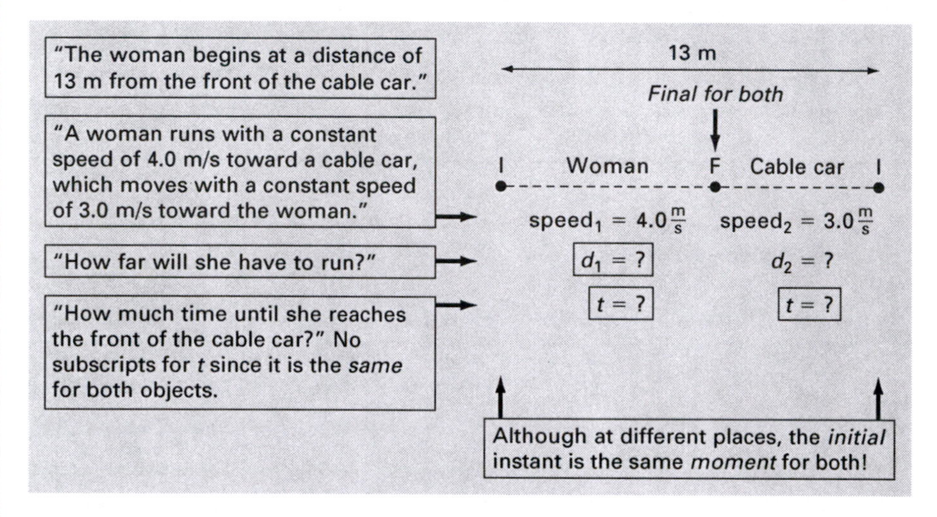

(3) Equations & unknowns and (4) Outline solution

To get an equation *relating the objects,* we use the fact that together they travel a total distance of 13 m: $d_1 + d_2 = 13$ (temporarily ignoring units). We write this with Equation 2.1 *for each object*:

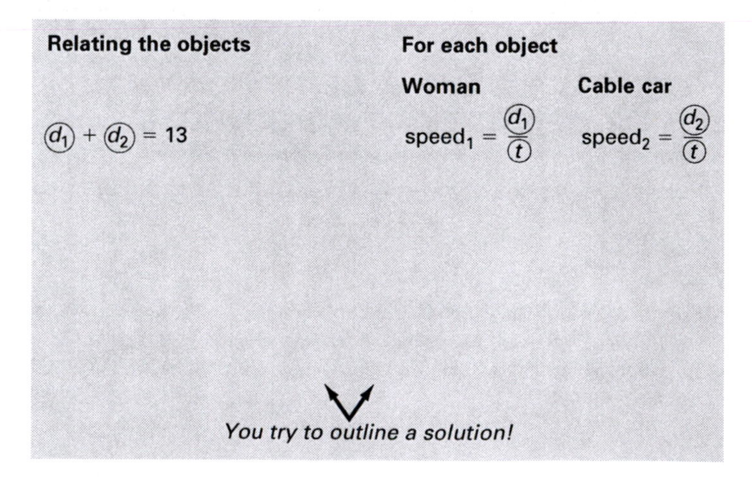

Can we solve it with just these equations?

COUNT EQUATIONS AND UNKNOWNS!

If we have as many *equations* as *unknowns,* we can solve for the unknowns:

- *Three* equations and *three* unknowns. Result: Happiness! We can solve it!
- *Two* equations and *three* unknowns. Result: Misery. We cannot solve it.

Exception: If we have more unknowns than equations, one or more unknowns may *cancel out* in the algebra. Then we can still solve it. Happiness! We will see this in later chapters.

We have *three equations* and *three unknowns*. Count them! THIS means we have enough to solve it. But the outline is a bit tricky.

Answers for (4) Outline solution One possible outline:

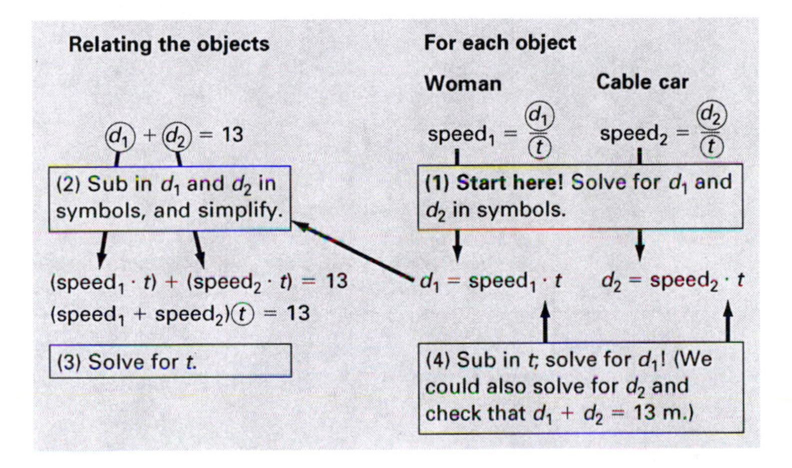

Answers $t \cong 1.86 \cong 1.9$ s, $d_1 \cong 7.43 \cong 7.4$ m ■

There is a more clever solution to the previous exercise, using the fact that *relative to the cable car,* the woman is running at 7.0 m/s. The math is simpler, but *the concept is trickier*. Clever is nice but *useless* if you don't see the "trick." If you do see it, use it, and write down what you did. Otherwise, set up the problem step by step. It always works!

What if t is different for the two objects? For example, if object 1 moves for 3.0 seconds longer (i.e., starts 3.0 seconds earlier) than object 2, then the equation *relating the objects* would be: $t_1 = t_2 + 3$.

2.7 HOW TO SET UP CONSTANT/AVERAGE SPEED OR VELOCITY PROBLEMS

Here we summarize what we have done in the previous few exercises. The setup process is meant to help you, not confine you. If you see one equation with one unknown before writing down the other equations, go ahead and solve it. Then, if needed, go back to these setup steps.

Constant/Average Speed or Velocity Problems—Mental and Written Steps

Mental →	**(1) Type of problem** The type of problem tells us the main equation(s) to use: • Constant or average speed: Use Equation 2.1. • Constant or average velocity: Use Equation 2.2.
Mental and written →	**(2) Sort by interval and/or object** First, identify intervals and objects. Then Equation 2.1 or 2.2 tells us which quantities to sort *for each interval* and/or *for each object.*

Mental and written →	**(3) Equations & unknowns**
	• For one object and one interval, write Equation 2.1 or 2.2.
	• For two or more *intervals,* write:
	  • Equation(s) *relating the intervals*
	  • Equation 2.1 or 2.2 *for each interval*
	• For two or more *objects,* write:
	  • Equation(s) *relating the objects*
	  • Equation 2.1 or 2.2 *for each object*
	Circle each unknown in each equation.

Mental →	**(4) Outline solution** Look for *one equation* with *one unknown.* Or look for *two equations* with the same *two unknowns,* and so on. Then look for a *chain reaction* to lead to the final answer.

Think through the problem piece by piece, and write as much as you need to, but don't write too much.

WHAT ABOUT CONSTANT OR AVERAGE ACCELERATION PROBLEMS?

We can also solve the simplest constant/average acceleration problems with the methods we have shown up to now, using this equation instead:

$$a = \frac{\Delta v}{t} \quad \text{or} \quad a = \frac{\Delta v}{\Delta t}$$

However, the rest of this chapter shows how to use Equations 2.3a–d to solve more involved problems with constant/average acceleration.

2.8 CONSTANT/AVERAGE ACCELERATION PROBLEMS

When a problem involves constant or average *acceleration,* Equations 2.3a–d all apply. Unlike problems with constant or average *velocity,* here we have (1) *initial* and *final* values for each interval and (2) more equations to choose from.

There are five total quantities in Equations 2.3a–d. Be consistent with units! The most common set of units for these equations are shown here:

Quantity	x	t	v_0 and v	a
Units	meters	seconds	$\frac{\text{meters}}{\text{second}}$	$\frac{\text{meters}}{\text{second}^2}$

When a problem involves motion, always think in terms of *intervals*. Each of these quantities fits in an interval as shown here:

Part of interval	Initial instant (I)	Entire interval	Final instant (F)
Quantities that correspond	Initial velocity, v_0	• Displacement, x • Acceleration, a • Time elapsed, t	Final velocity, v

Regardless of the question(s) being asked, when we set up a constant or average acceleration problem, we organize these quantities *for each interval* like this:

I Interval F

v_0 = ____ x = ____ v = ____

a = ____

t = ____

We also need to pick a *positive direction*. Displacement, x, velocities, v_0 and v, and acceleration, a, all have *direction,* and so can be positive or negative. Time elapsed, t, is always positive.

2.9 CONSTANT/AVERAGE ACCELERATION—ONE INTERVAL

Here we show two basic constant/average acceleration exercises, one with an object gaining speed and the other with an object losing speed.

EXERCISE 2.4

A car, initially at rest, accelerates at a constant rate for 6.0 s and covers a distance of 32 m along a straight line. What is the car's (a) acceleration and (b) final velocity?

Solution

(1) Type of problem

The car *accelerates at a constant rate*: Use Equations 2.3a–d (any or all of them).

(2) Sort by interval and/or object

There is one interval (while the car is accelerating) and one object (the car). We make the positive direction to the right:

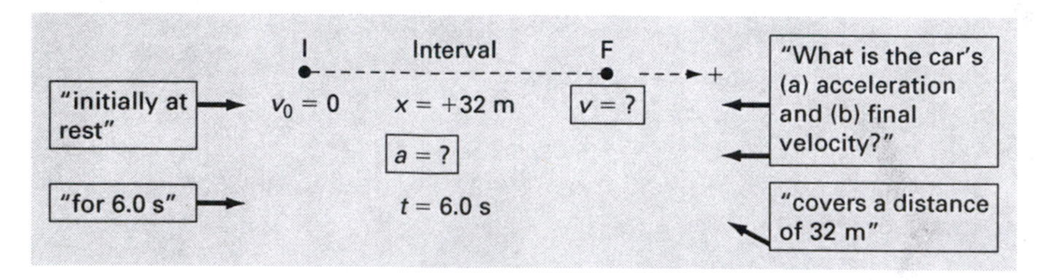

(3) Equations & unknowns and (4) Outline solution

We list Equations 2.3a–d, circle the unknowns, and outline a simple solution:

$$x = v_0 t + \tfrac{1}{2} a t^2 \quad \text{(1) One unknown; solve for } a.$$

$$x = \left(\frac{v_0 + v}{2}\right) t \quad \text{(2) One unknown; solve for } v.$$

$$v = v_0 + at$$

$$v^2 = v_0^2 + 2ax$$

If you see this solution before writing *all* the equations, *go ahead and solve it*! You can also use the other equations to check yourself.

Answers $v \cong +10.7 \cong +11\frac{\text{m}}{\text{s}}$, $a \cong +1.78 \cong +1.8\frac{\text{m}}{\text{s}^2}$ (*positive* means to the *right*). ■

EXERCISE 2.5

A car traveling in a straight line, initially at 20 m/s, slows at an average rate of 3.5 m/s^2 until it is moving at 14 m/s in the same direction. While slowing, (a) how far does the car travel and (b) how much time passes?

Solution

(1) Type of problem

The car has an *average* (treat just like *constant*) acceleration of 3.5 m/s^2. The units, m/s^2, tell us it is acceleration. So we use Equations 2.3a–d.

(2) Sort by interval and/or object

There is one interval (while the car is slowing) and one object (the car). *Positive* is to the *right*, and we sort the quantities in Equations 2.3a–d:

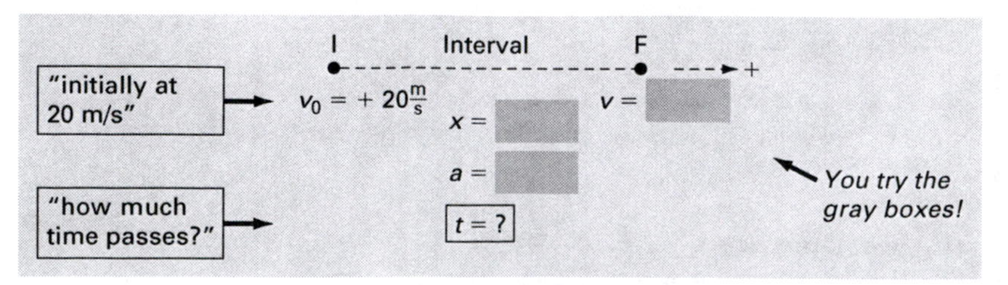

Check your answers with those at the end of the exercise.

(3) Equations & unknowns and (4) Outline solution

We write Equations 2.3a–d:

$$x = v_0 t + \tfrac{1}{2}at^2$$
$$x = \left(\frac{v_0 + v}{2}\right)t$$
$$v = v_0 + at$$
$$v^2 = v_0^2 + 2ax$$

You circle the unknowns and try an outline!

Answers for gray boxes

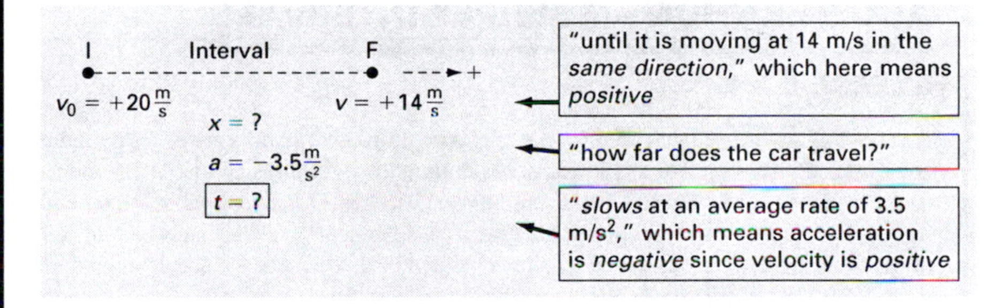

Answers for (3) Equations & unknowns and (4) Outline solution

$$\textcircled{x} = v_0\textcircled{t} + \tfrac{1}{2}a\textcircled{t}^2$$
$$\textcircled{x} = \left(\frac{v_0 + v}{2}\right)\textcircled{t}$$
$$v = v_0 + a\textcircled{t}$$
$$v^2 = v_0^2 + 2a\textcircled{x}$$

You can use these to check your answers!

(1) One unknown. Solve for t.

(2) One unknown. Solve for x.

Answers for the math

$$v^2 = v_0^2 + 2ax \quad \Rightarrow \quad x = \frac{v^2 - v_0^2}{2a} = \frac{\left(+14\frac{m}{s}\right)^2 - \left(+20\frac{m}{s}\right)^2}{2\left(-3.5\frac{m}{s^2}\right)} \cong 29.1 \cong +29\ \text{m}$$

$$v = v_0 + at \quad \Rightarrow \quad t = \frac{v - v_0}{a} = \frac{\left(+14\frac{m}{s}\right) - \left(+20\frac{m}{s}\right)}{\left(-3.5\frac{m}{s^2}\right)} \cong 1.71 \cong 1.7\ \text{s}$$

Watch the *negative signs!*

Answers

- $x \cong +29.1 \cong +29$ m (*positive* means to the *right*).
- $t \cong 1.71 \cong 1.7$ s (elapsed time should always be positive). ■

HOW ARE THE SIGNS OF VELOCITY AND ACCELERATION RELATED?

Look back at the previous two exercises to see the following:

- When the object is *gaining* speed, the signs of velocity and acceleration are always *the same* (+ and +, OR − and −).
- When the object is *losing* speed, the signs of velocity and acceleration are always *opposite* (+ and −, OR − and +).

2.10 CONSTANT/AVERAGE ACCELERATION—MULTIPLE INTERVALS

EXERCISE 2.6

A car is initially traveling at 15 m/s on a long straight road. The driver sees a stop light ahead and eventually slows down to 12 m/s, at which time the light turns green. At this moment the driver begins to accelerate at 2.5 m/s^2 until the car reaches 17 m/s. A total of 8.0 seconds pass from the moment the driver begins to slow down to the moment the car is finally moving at 17 m/s. (a) For how much time was the driver slowing down? (b) What was the acceleration during the slowing? (c) What was the displacement for the entire 8.0-second trip?

Solution

(1) Type of problem

There are two consecutive intervals, (A) *losing speed* and (B) *gaining speed,* each with its own constant acceleration: Use Equations 2.3a–d *for each interval separately.*

The *entire trip* does NOT have constant acceleration, so we do NOT set up Equations 2.3a–d for this interval!

(2) Sort by interval and/or object

The quantities in Equations 2.3a–d, sorted for each interval:

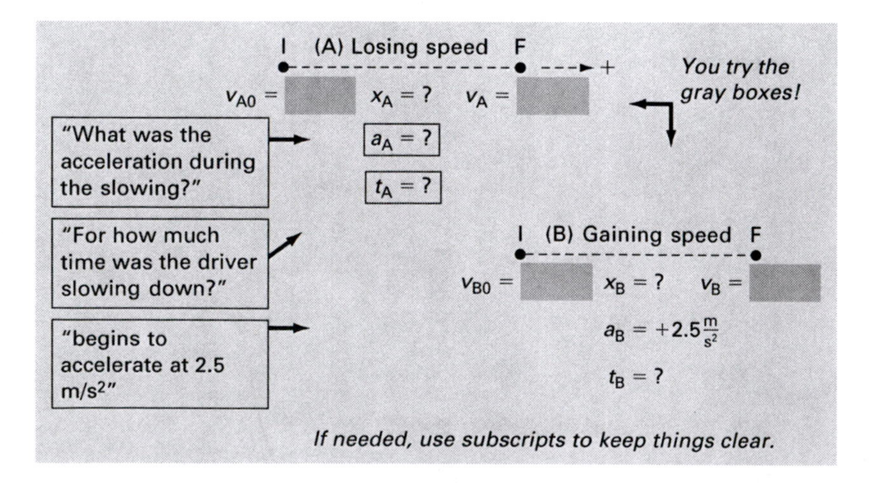

(3) Equations & unknowns and (4) Outline solution

First, the equations *relating the intervals*:

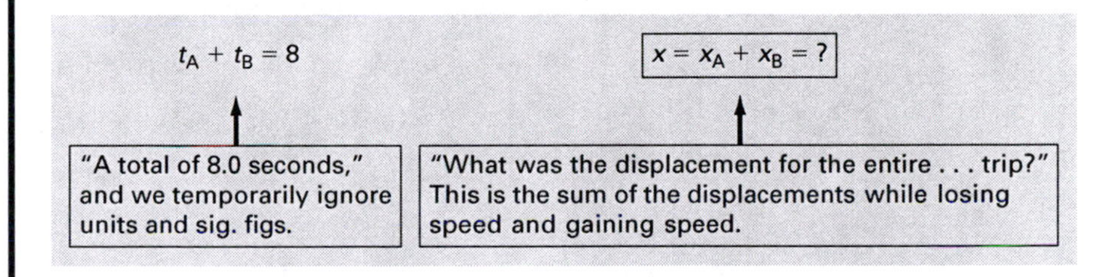

Then, we put these together with Equations 2.3a–d *for each interval*:

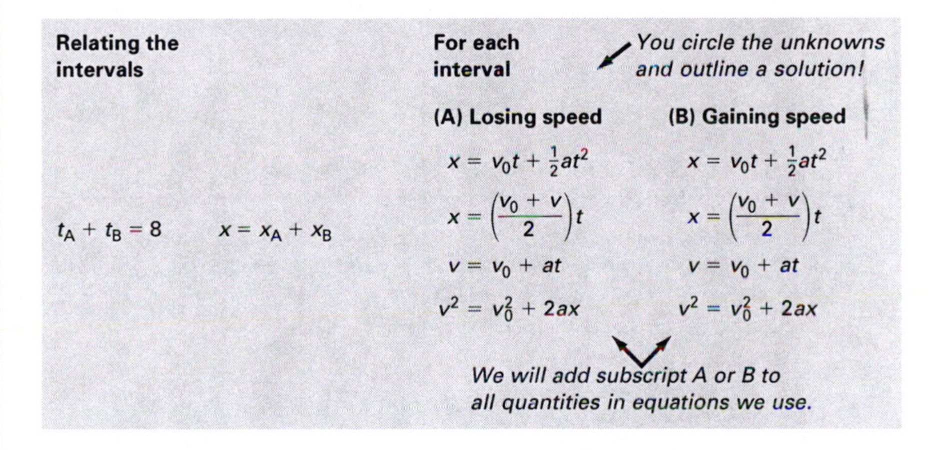

Check the answers if you get stuck.

Answers for gray boxes

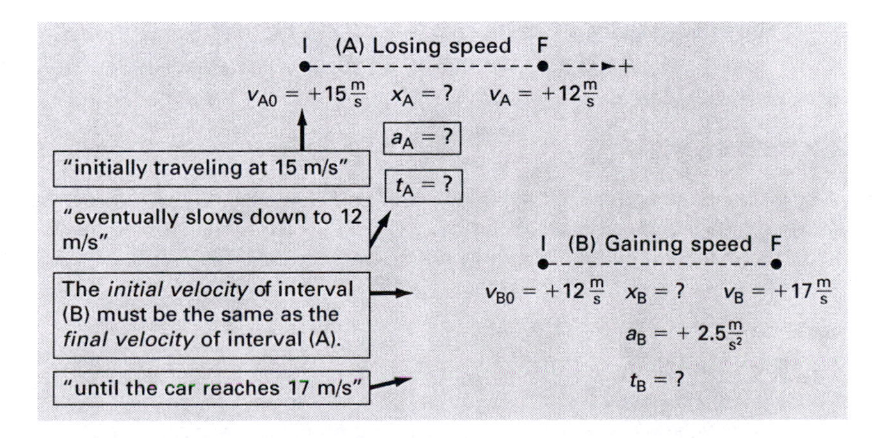

Answers for (3) Equations & unknowns and (4) Outline solution

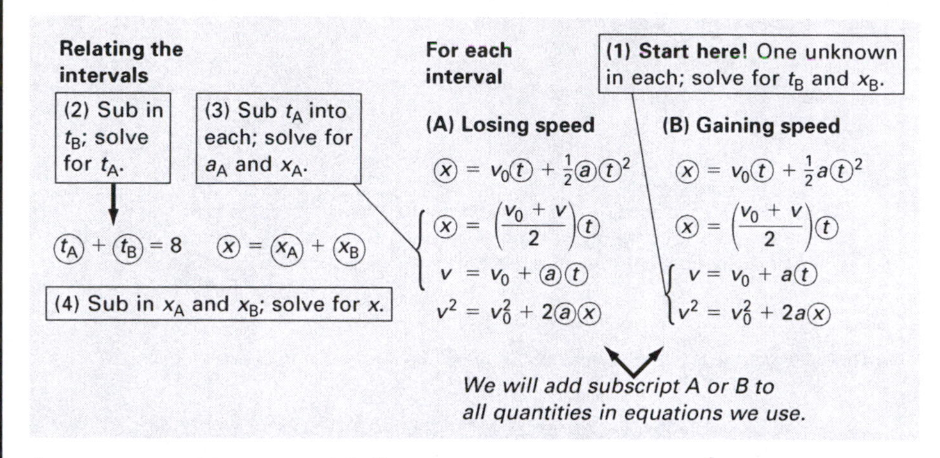

Answers $t_A = 6.0\ \text{s}, a_A = -0.50\frac{\text{m}}{\text{s}^2}, x = +110\ \text{m}$

Intermediate answers: $t_B = 2.0\ \text{s}, x_B = +29\ \text{m}, x_A = +81\ \text{m}$ ■

THE BEST WAY TO LEARN PHYSICS IS TO DO IT!

Even if you don't *perfectly* understand all the physics, go ahead and try to set up and solve a problem. You will get frustrated at times and may need help along the way. But after you finally get through it, many concepts that once were confusing will have become clearer in your mind. You can learn physics *by doing it*!

2.11 CONSTANT/AVERAGE ACCELERATION—"FREE-FALL"

The term *free-fall* sounds like it is only for objects going down, but in physics it means *any motion under the influence of gravity alone* (no air resistance). Free-fall motion can have *upward* velocity, *downward* velocity, or even *zero* velocity (like at the highest point of an object's path).

Near the earth's surface, the acceleration in free-fall has a constant value of $a = -g = -9.8\ \text{m/s}^2$. We put a *negative* sign on it because the acceleration is *down*, and we usually define *up* as *positive*. The quantity $g = 9.8\ \text{m/s}^2$ is positive.

EXERCISE 2.7

A ball is launched vertically upward from ground level. It reaches its highest point, and then on the way down, it is caught by a person when the ball's speed is 3.0 m/s, at a height of 1.2 meters above the ground. (a) What is the maximum height reached? (b) What is the launch velocity? (c) How much time passes from the launch to when the ball is caught?

Solution

(1) Type of problem

This entire motion is free-fall: Use Equations 2.3a–d. The problem does not explicitly tell us to take air resistance into account, which means we should *ignore air resistance,* which again means free-fall.

Important point: We are looking at *only* the free-fall motion, which begins at the moment *just after* launch and ends at the moment *just before* the catch. We are NOT dealing with the actual processes of launching and catching the ball, but *only the motion in between*.

(2) Sort by interval and/or object and (3) Equations & unknowns

There are three intervals to think about, all of which have constant acceleration:

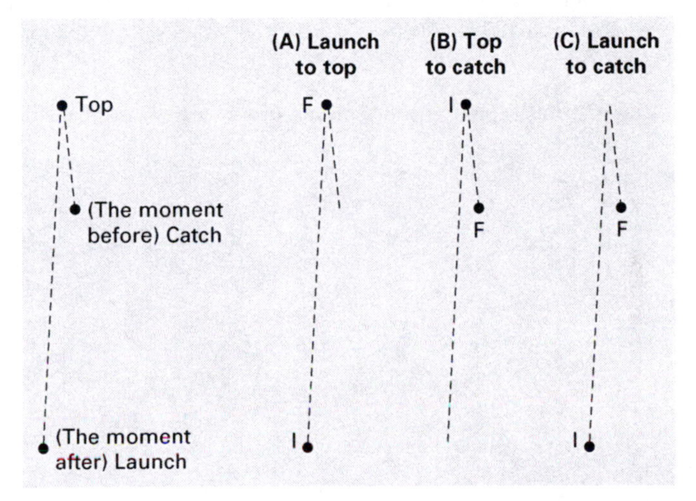

Most people believe the acceleration is constant for (A) and (B). Harder to swallow is the fact that the acceleration is constant for the *entire* interval (C), *even at the highest point*.

WHAT IS ZERO AT THE TOP?

NOT acceleration! *Velocity* is zero at the top of the path. If acceleration were also zero there, the velocity *would not change*—remember, acceleration is the rate of *change* of velocity—and the ball would *stay at the top*. That does NOT happen, so the acceleration is NOT ZERO at the top.

We define *up* as *positive*, sort quantities in Equations 2.3a–d, and write the equations, one interval at a time. First, interval (A):

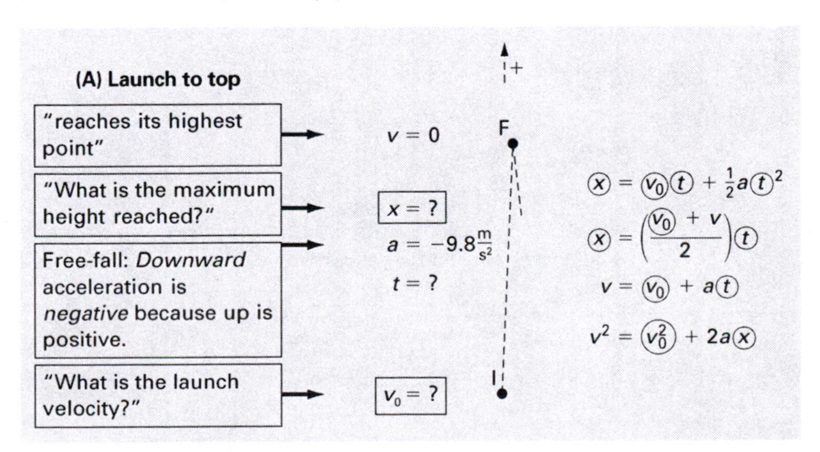

We will try it without subscripts. We could add subscripts for each interval, but it is a pain when there are too many.

THIS IS THE MOST IMPORTANT STEP!

Problem solving does NOT equal *doing algebra*. Make sure you understand the details in the setup diagrams. Good setup → good equations → good solution!

For interval (A), there are too many unknowns, but we have two more intervals:

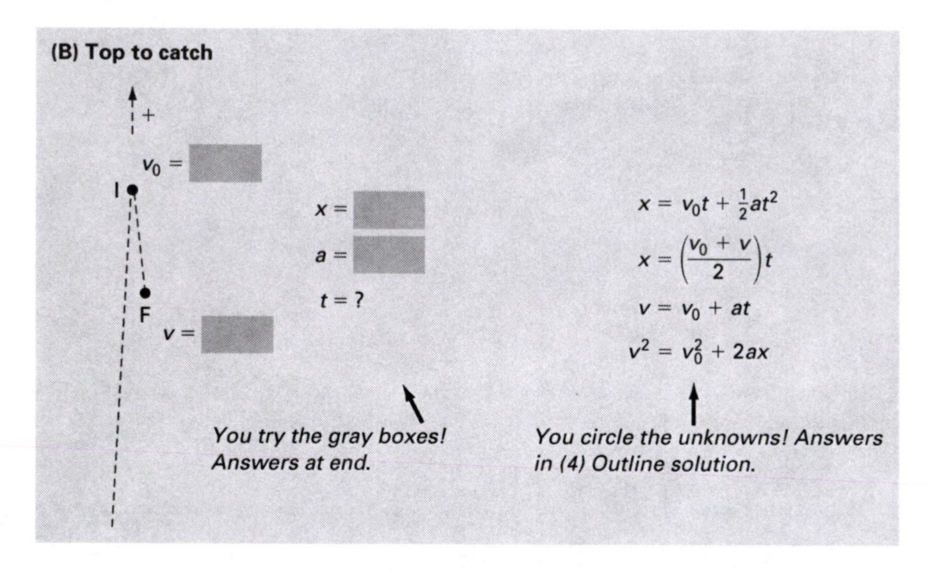

And:

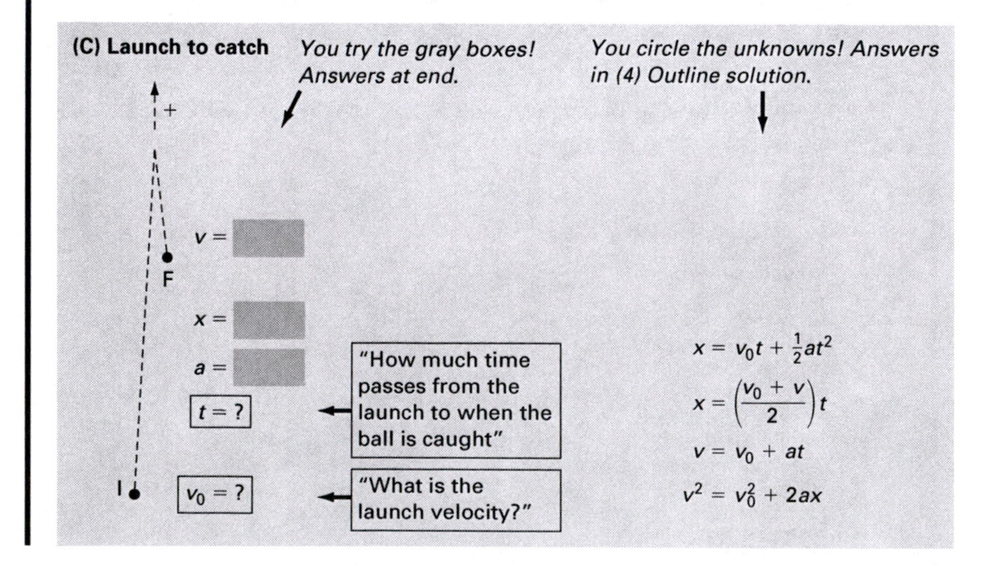

BE CONSISTENT WITH SIGNS!

If up is positive, then make sure all downward displacements, velocities, and accelerations are negative (and vice versa).

We could also write equations *relating the intervals,* but we don't need them here.

(4) Outline solution

We put together the equations *for each interval,* and circle the unknowns:

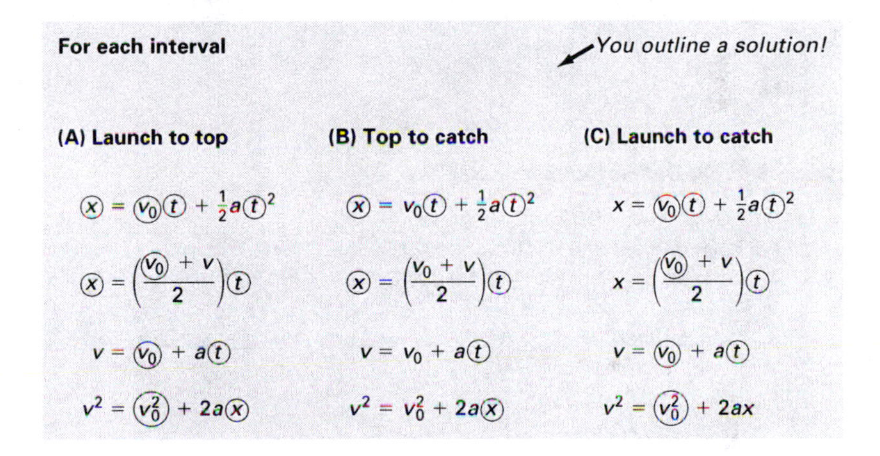

Answers for gray boxes

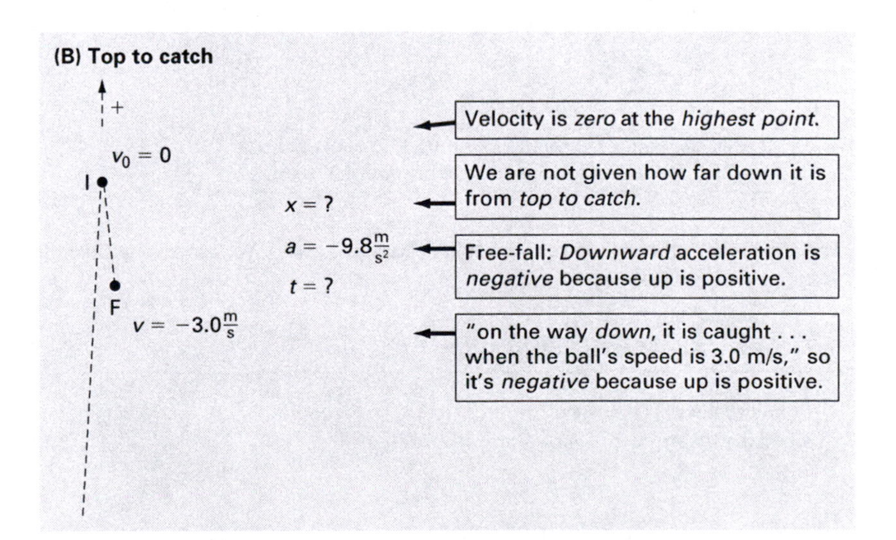

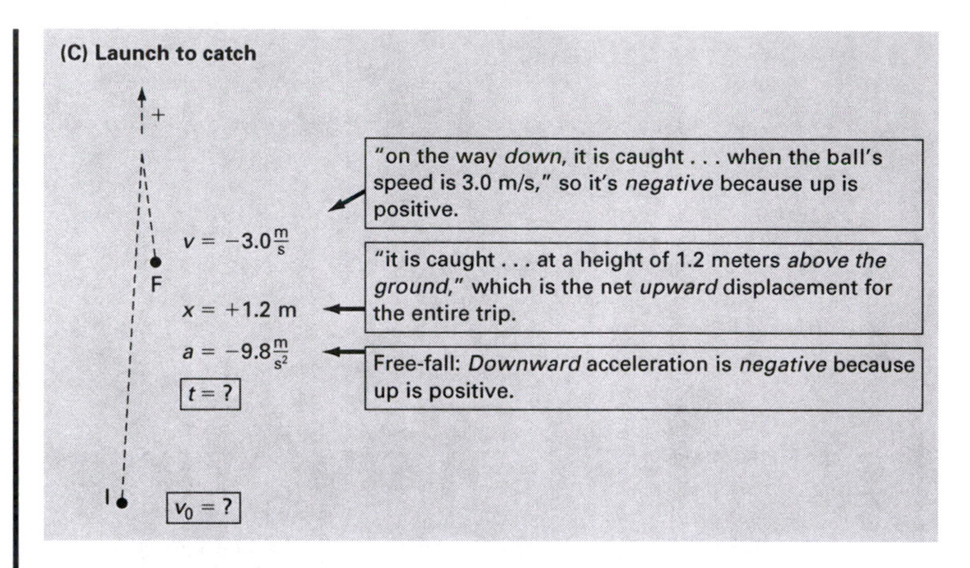

Answers for (4) Outline solution

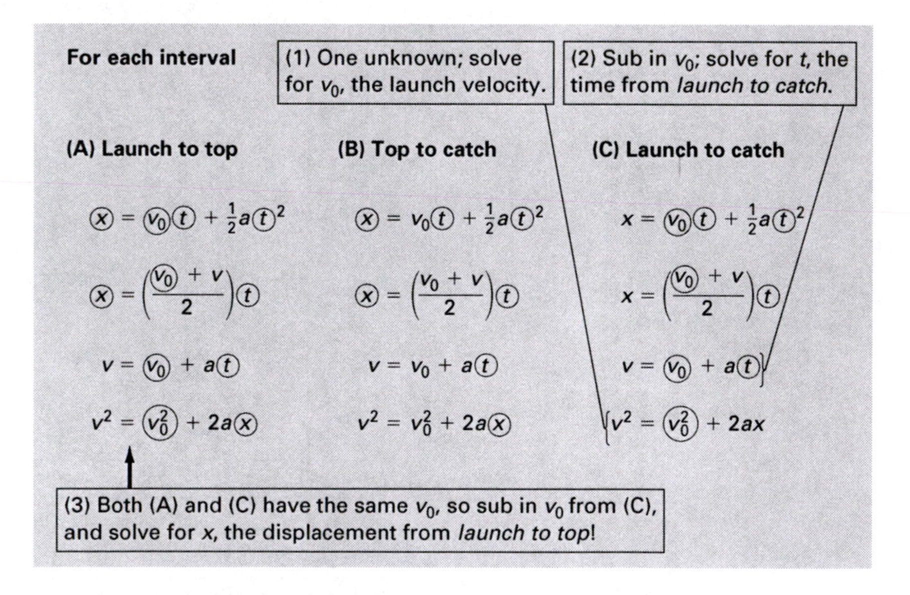

There are many other unknowns we could solve for if asked. We can also use the other equations to check our answers.

CONSIDER ALL POSSIBLE INTERVALS!

In this problem it is easiest to start calculations with interval (C), which is the one interval that many people *don't think of using*!

Answers

- How high? This is x for interval (A): $x \cong +1.66 \cong +1.7$ m (*positive* means *up*).

- Launch velocity? This is v_0 for both intervals (A) and (C): $v_0 \cong +5.70 \cong +5.7\frac{\text{m}}{\text{s}}$ (*positive* means *up*).
- Time elapsed from launch to catch? This is t for interval (C): $t \cong 0.888 \cong 0.89$ s (elapsed time should always come out positive). ■

2.12 CONSTANT/AVERAGE ACCELERATION—TWO OBJECTS

In the next exercise, two objects move simultaneously during an interval, one with constant *acceleration* and the other with constant *velocity*.

EXERCISE 2.8

A truck travels at a constant velocity of 18 m/s. The car in the next lane is initially at rest. At the instant the truck is 20 m ahead of the car, the car begins accelerating at a constant rate of 2.0 m/s^2 in the same direction the truck is moving. (a) How much time does it take for the car to pass the truck? (b) How far has the car traveled by then? (c) How fast is the car traveling at the instant it passes the truck?

Solution

(1) Type of problem

- The truck has constant *velocity*: Use Equation 2.2 for the truck.
- The car has constant *acceleration*: Use Equations 2.3a–d for the car.

(2) Sort by interval and/or object

There are *one interval* and *two objects*. Our setup has quantities from Equation 2.2 for the truck and quantities from Equations 2.3a–d for the car:

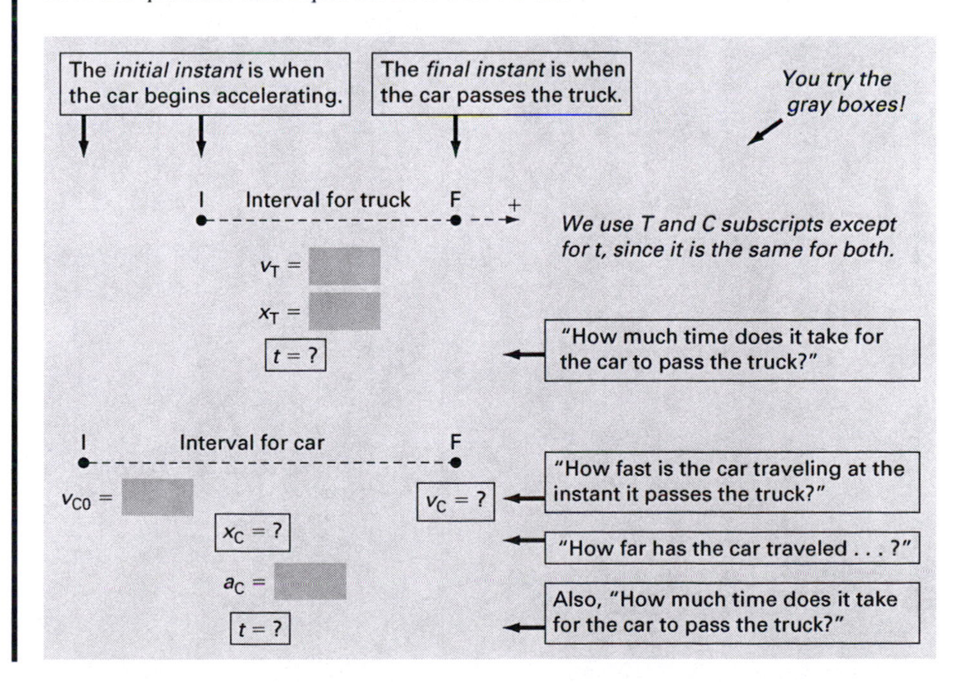

(3) Equations & unknowns and (4) Outline solution

One common mistake is to assume that the final velocities are equal. NOT TRUE! If they were, how could the car *pass* the truck?

To get the correct equation *relating the objects,* we use the fact that at the initial instant "the truck is 20 m ahead of the car." This means that when the car passes (is even with) the truck at the final instant, it has traveled 20 m farther than the truck: $x_C = x_T + 20$.

We put this together with the equations *for each object*:

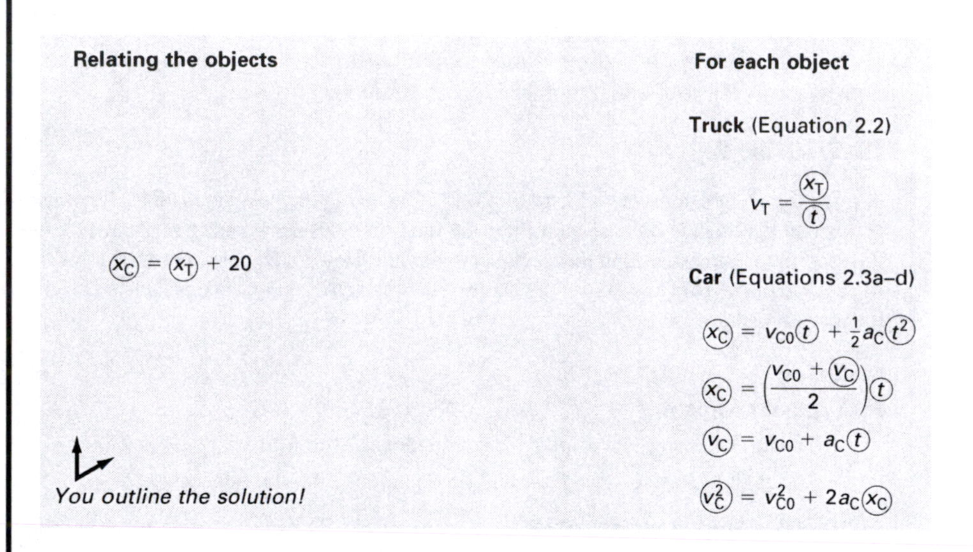

Answers for gray boxes

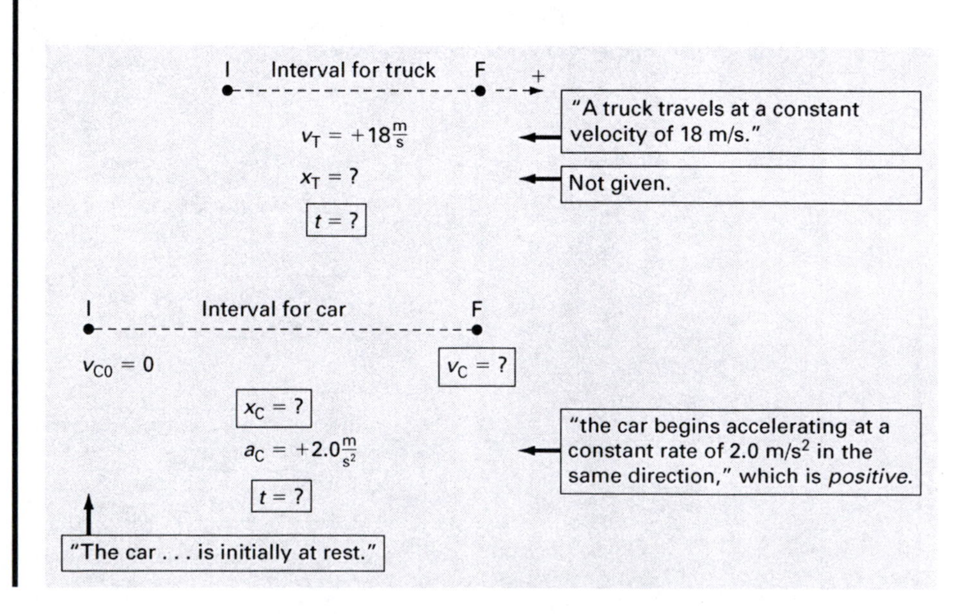

Answers for (4) Outline solution One possible outline:

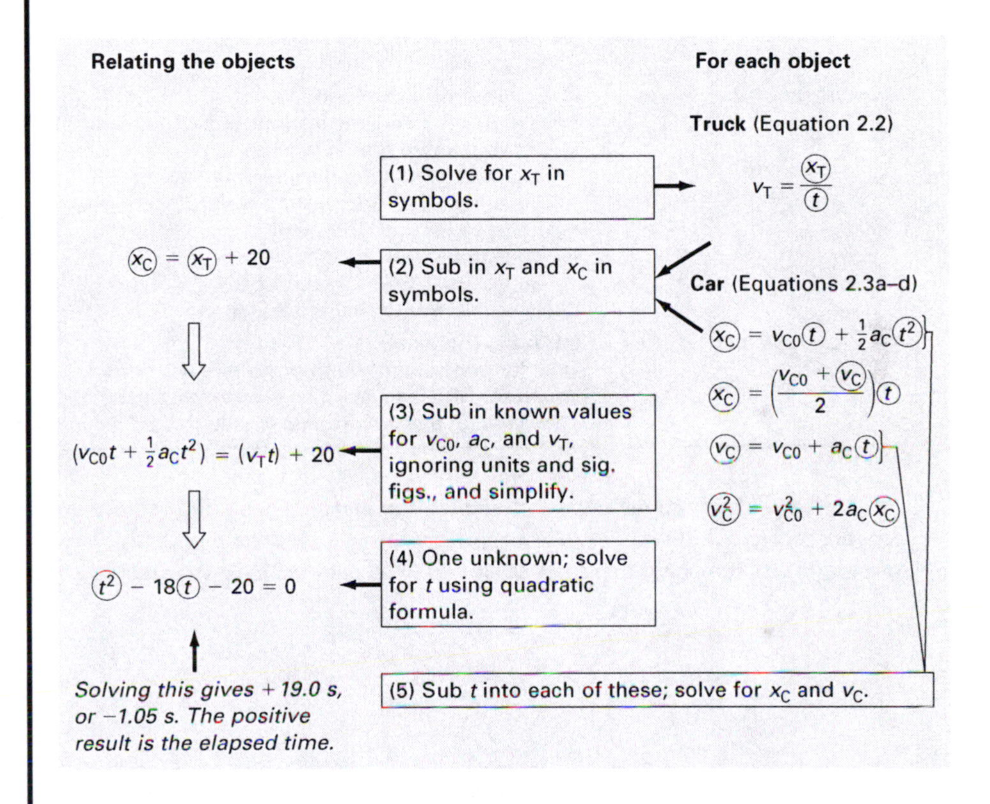

Answers $t \cong 19.0 \cong 19$ s, $x_C \cong +363 \cong +360$ m, $v_C \cong +38.0 \cong +38\frac{\text{m}}{\text{s}}$ ■

2.13 HOW TO SET UP CONSTANT/AVERAGE ACCELERATION PROBLEMS

Here we summarize the setup steps from the previous several problems. We might combine these with the steps for constant/average speed or velocity if there are both kinds of motion (like in the previous exercise). Write as much as you need to, but practice doing some of the steps mentally.

Constant/Average Acceleration Problems—Mental and Written Steps

Mental →	**(1) Type of problem** Constant/average *acceleration*: Use Equations 2.3a–d.
Mental and written →	**(2) Sort by interval and/or object** Identify intervals and objects. Sort quantities for Equations 2.3a–d *by interval* and/or *by object*.

Mental and written →	**(3) Equations & unknowns**
	• For one object and one interval, write Equations 2.3a–d.
	• For two or more *intervals,* write:
	  • Equation(s) *relating the intervals*
	  • Equations 2.3a–d *for each interval*
	• For two or more *objects,* write:
	  • Equation(s) *relating the objects*
	  • Equations 2.3a–d *for each object*
	Circle each unknown in each equation!

Mental →	**(4) Outline solution** Look for *one equation* with *one unknown,* or *two equations* with the same *two unknowns,* and so on. Then look for a *chain reaction* to lead to the final answer.

You will not likely see the entire solution at one glance when you first face a challenging problem. But *practice thinking in this way* and it will help you solve any kind of motion problem, even those not exactly like the ones we have shown here.

CHAPTER 3

VECTORS

MANY OF THE diagrams in your physics book have something in common: *arrows.* In most cases, they represent quantities *with direction,* called *vectors.* Displacement, velocity, acceleration, and many quantities that we will see later are all vectors. Quantities like temperature or volume, which have *no direction,* are called *scalars.* Vectors show up when *direction* is relevant, which is often the case in physics problems. That means you need to learn to handle vectors confidently in order to do well in your physics class.

From your study of linear motion, you already know something about vectors. For a car moving with a displacement *east* or a displacement *west,* the *direction* in that case was represented by a *positive* or *negative sign.* Velocities and accelerations also had signs that told us their directions.

Using the sign alone to represent the direction works well for linear (1D, or one-dimensional) situations like the car moving along an east–west line. But in two-dimensional (2D) situations, it is more involved.

3.1 MAGNITUDE AND DIRECTION, AND x- AND y-COMPONENTS

Imagine that you are at airport 1 and a pilot asks you how to get to airport 2. You could tell the pilot the *displacement vector* from airport 1 to airport 2, which includes *two pieces of information:* how far (the *magnitude*) and the angle relative to a map direction (the *direction*). Let's say this displacement vector (call it $\vec{A}$) is 25 km at 70 degrees north of east. We write this as:

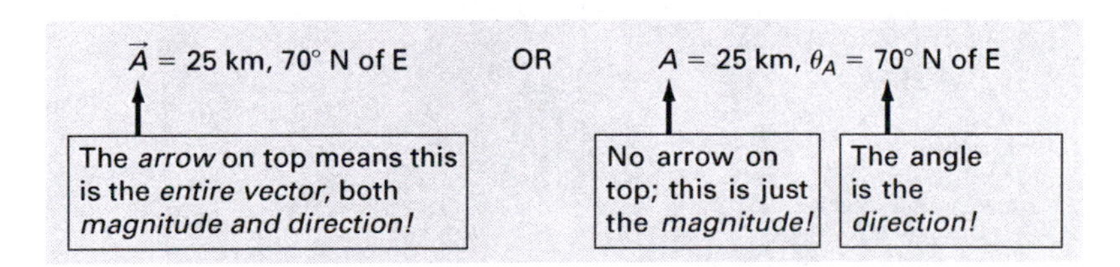

Textbooks often use **bold** for vectors and might write ***A*** instead of $\vec{A}$.

USE GOOD NOTATION FOR VECTORS!

This will help keep all the parts of the vector clear in your mind. The $\vec{A}$ symbol stands for the *entire vector.* Do *not* write $\vec{A} = 25$ km (bad notation). The magnitude alone can be written as $A = 25$ km (good notation) with the angle written separately, as in the previous figure.

Instead of *magnitude* and *direction,* A and θ_A, it would also be enough to tell the pilot *how far east/west* and *how far north/south* to go. These *two pieces* of information are the *x*- and *y-components,* A_x and A_y, of the vector. If we know A and θ_A, we can easily calculate A_x and A_y by using a right triangle and trigonometry.

The following exercise is a very simple problem and has a very quick solution. But as usual, we discuss the mental and written steps in detail, and so our solutions are much longer than yours will be on paper. Don't focus on getting the "right answer" but instead on the *process* of effective problem solving. Then many right answers will follow!

EXERCISE 3.1

Calculate the *x*- and *y*-components of $\vec{A}$ ($A = 25$ km, $\theta_A = 70°$ N of E). Assume east is the positive *x*-direction and north is the positive *y*-direction.

Solution

(1) Type of problem

We *mentally* note that this is a vector problem. As we will see later in this exercise, this means we should use *right triangle trigonometry equations.*

(2) Sort by vector

Before calculating anything, we sort the information. Here we sort *by vector,* diagramming and labeling each vector with its own information.

For the one vector in this problem, we mentally picture it on a map, going from airport 1 to airport 2. We translate this to a coordinate system with positive *x*- and *y*-directions, as shown, and write what we know about A, θ_A, A_x, and A_y:

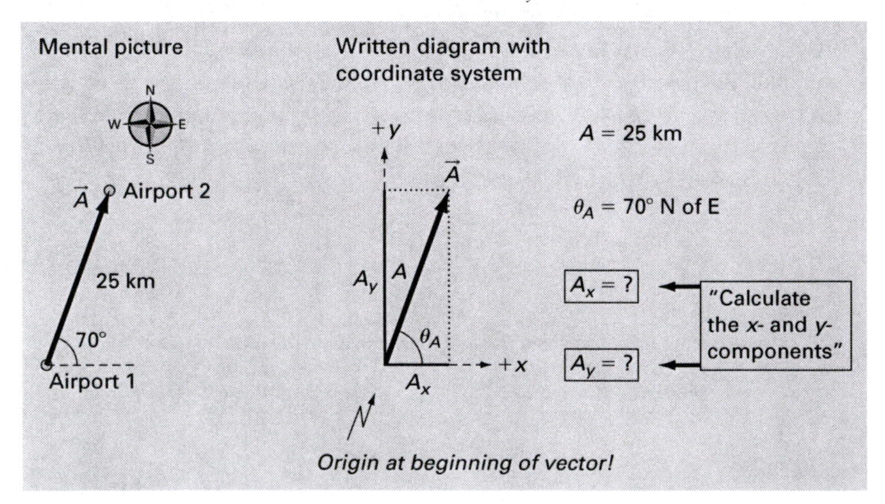

WHAT DOES "70 DEGREES NORTH OF EAST" MEAN?

Physics textbooks often use language like this to describe a direction. Make sure you understand its meaning: *Starting from east, measure an angle 70 degrees toward north.* This is exactly the same direction as "20 degrees east of north," which means: Starting from north, measure an angle 20 degrees toward east.

(3) Equations & unknowns and (4) Outline solution

In studying the previous figure, we can see two right triangles for which we can write down the trig equations: cosine, sine, tangent, and the Pythagorean theorem. We then circle the unknowns and outline a solution:

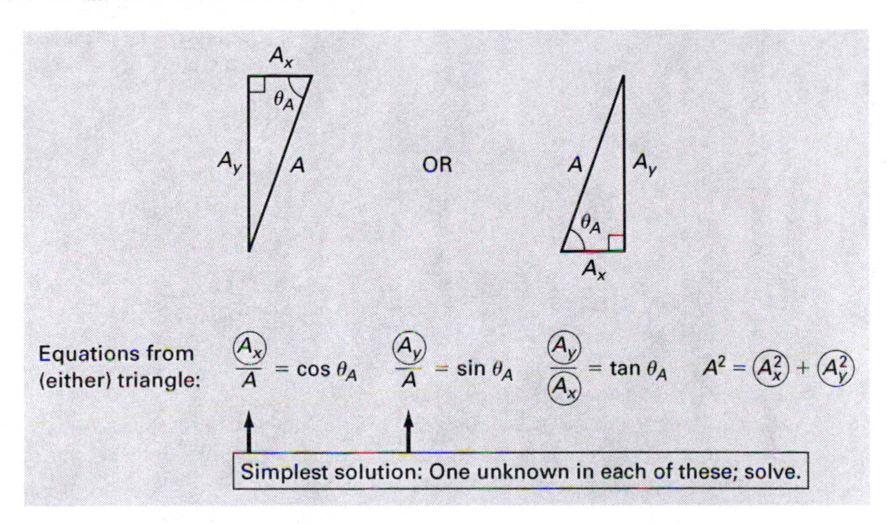

THINK EQUATIONS AND UNKNOWNS!

In this problem we have two unknowns. We can solve this problem with *any two* of the four trig equations.

Answers

Here we show the math:

Calculator in *degree* mode!

Positive because *east.*

$$\frac{A_x}{A} = \cos\theta_A \Longrightarrow A_x = A\cos\theta_A = (25\text{ km})\cos(70°) \cong +8.55 \cong +8.6\text{ km}$$

$$\frac{A_y}{A} = \sin\theta_A \Longrightarrow A_y = A\sin\theta_A = (25\text{ km})\sin(70°) \cong +23.49 \cong +23\text{ km}$$

Positive because *north.*

Having drawn a diagram with approximately the correct angle helps us to see that these components are about the right size in comparison to the magnitude of 25 km.

Another way to solve it
The problem is now solved, but now we show another way using a different but equivalent angle, 20 degrees east of north. We can make either of the two right triangles in the next figure, where everything is the same as before, except now $\theta_A' = 20°$ E of N.

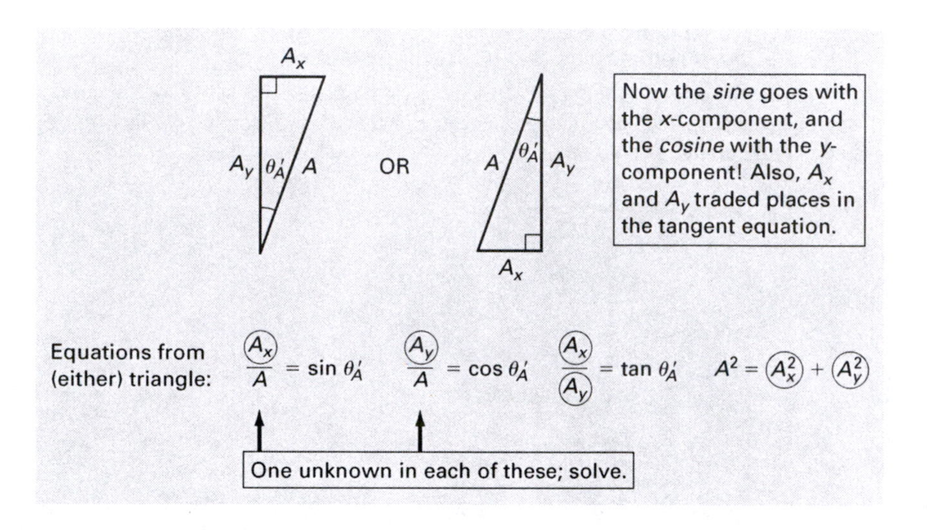

In the end, we get the same answers as before!

Don't make the mistake of assuming that the x-component always goes with the cosine. It depends on the axis from which we measure the angle. If in doubt, draw a right triangle to figure out which component gets the sine and which gets the cosine. ■

The setup steps we just followed are very similar to those for motion problems. Here is a preliminary summary:

Vector Problems—Mental and Written Steps

Mental→	**(1) Type of problem**
	Vector problem: Use right triangle trig equations.
Mental and written→	**(2) Sort by vector**
	Draw vector diagram(s), and sort information *by vector.*
Mental and written→	**(3) Equations & unknowns**
	For each vector: Make a right triangle (mentally or on paper) and write trig equations. Then circle the unknowns.
Mental→	**(4) Outline solution**

Don't forget: These steps are here to help you, not restrict you. We give a more detailed summary later in the chapter.

EXERCISE 3.2

The displacement vector $\vec{B}$ goes from airport 2 (in the previous exercise) to airport 3. The y-component of the vector is 41 km south, and the direction of the vector is 55 degrees south of east. Determine the x-component and the magnitude of the vector.

VECTOR = TWO PIECES OF INFORMATION!

Problem statements often give either *magnitude* and *direction,* or *x-* and *y-components.* However, if we know *any two* of these four pieces of information, we know enough to figure out the other two using the trig equations. This exercise illustrates the idea.

Solution

(1) Type of problem

Vector problem: Use right triangle trig equations.

(2) Sort by vector

We make a diagram of the vector. Just as with motion quantities, we use the sign to indicate direction. However, since this vector is 2D (i.e., not along just one axis, but having both x- and y-components), we do the sign *for each component separately.* The y-component is *south,* and so we make it *negative.*

You try the gray boxes, and check your answers at the end:

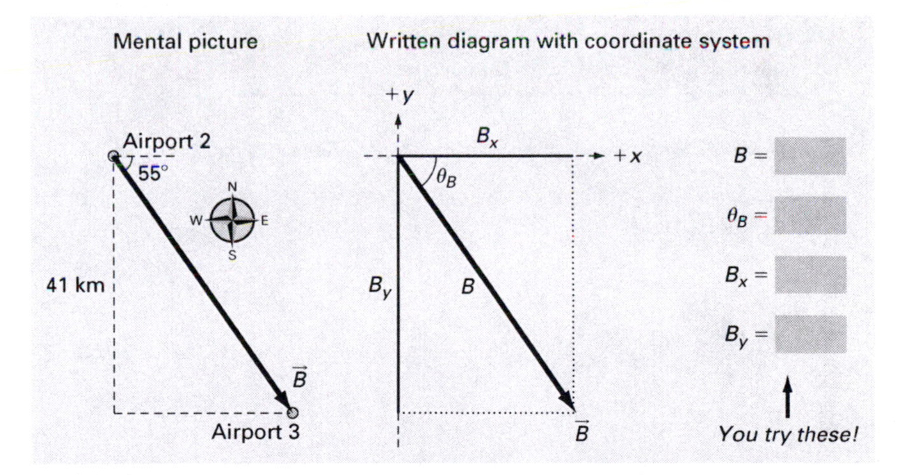

ALWAYS MEASURE THE ANGLE AT THE BEGINNING OF THE VECTOR!

For $\vec{B}$, we measure the angle at the upper left end, not at the lower right. Think of each vector as having its own "private" origin at its beginning, and measure the angle at that origin.

(3) Equations & unknowns and (4) Outline solution

We redraw the right triangle (either in our minds or on paper) and write the trig equations:

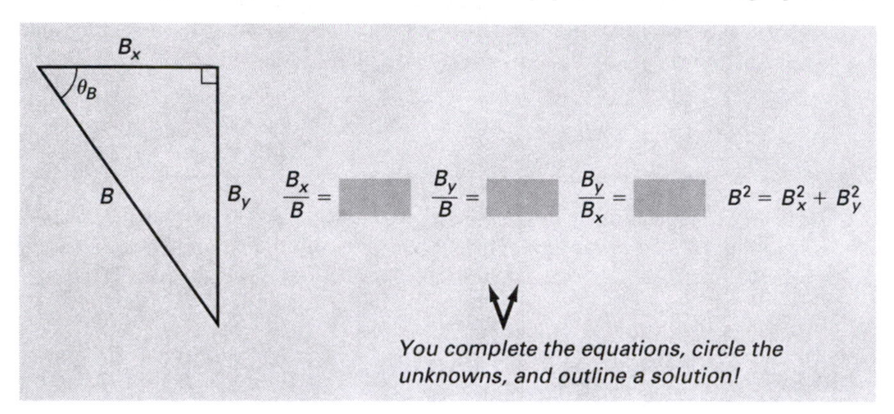

Answers for gray boxes

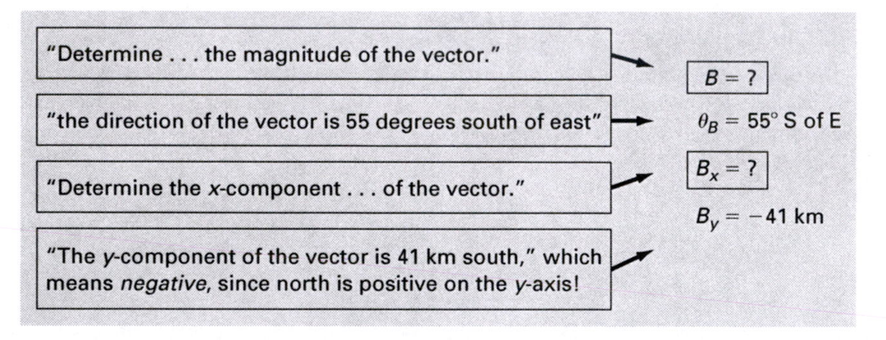

Answers for (3) Equations & unknowns and (4) Outline solution

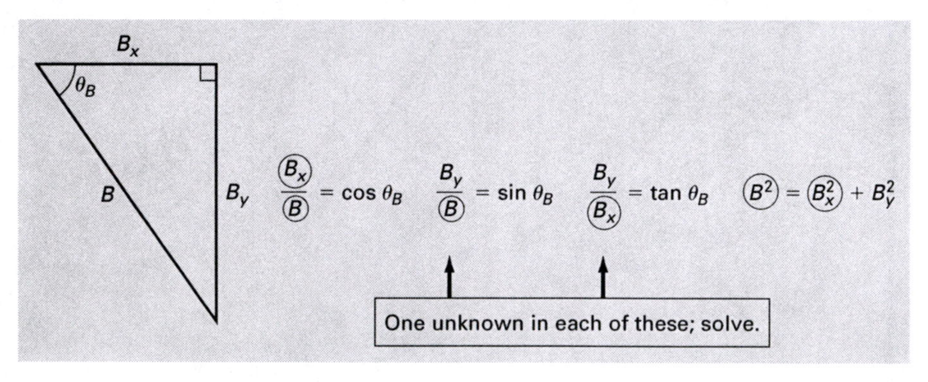

Answers

Again we show the math. But first, an important note: In this book we *ignore* positives and negatives in the calculations with *right triangle trig equations* (sine, cosine, tangent) and then just *make the answer have the correct sign.* Why? We can usually determine the correct sign from our vector diagram, which is much easier than being precise with signs in trig equations. Key idea: EASIER, but still correct!

We use this approach to calculate B_x and B:

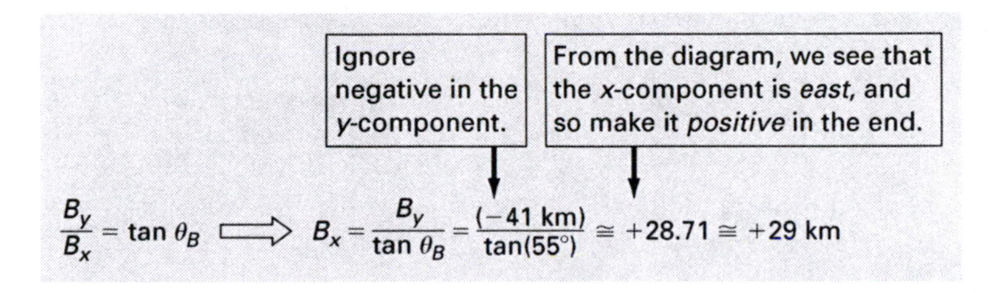

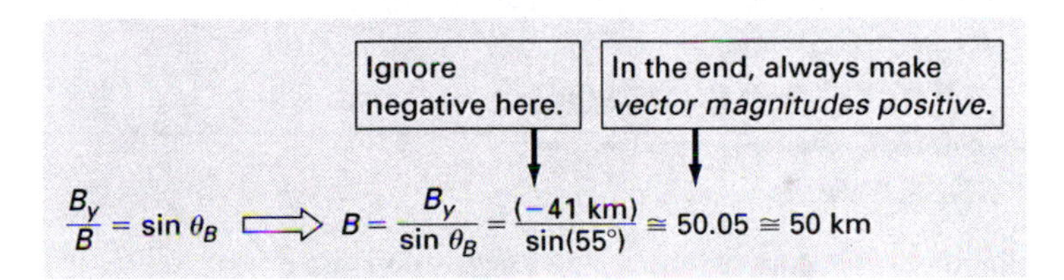

THE MAGNITUDE OF A VECTOR IS ALWAYS POSITIVE!

We don't even bother writing the + sign with the value of B because the *magnitude* of a vector is ALWAYS *positive*. The x- and y-components can be either positive or negative, but the magnitude is the "size" of the vector, which is only positive.

■

3.2 VECTORS ALONG ONE AXIS

If a vector is pointing directly along either the x- or y-axis, then the components are very easy to deal with. For example, consider these two vectors:

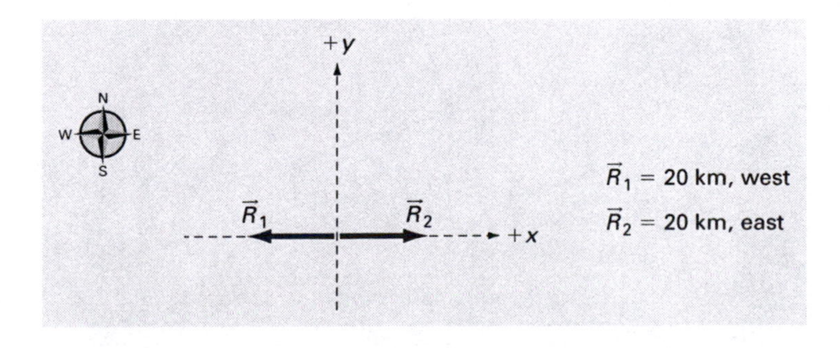

Now compare their magnitudes and x- and y-components:

Magnitudes:	$R_1 = 20$ km $R_2 = 20$ km	Magnitude is ALWAYS positive.
x-components:	$R_{1x} = -20$ km $R_{2x} = +20$ km	*Sign* of x-component tells us if it is east (positive) or west (negative).
y-components:	$R_{1y} = 0$ $R_{2y} = 0$	Nothing north or south for either vector.

We take a similar approach for vectors that are along the positive or negative y-axis. For a vector along an axis, we determine the components *just by looking*. No calculations!

3.3 VECTOR ADDITION

EXERCISE 3.3

The pilot (in the previous exercises) wants to go from airport 1 to airport 2 along vector $\vec{A}$ ($A = 25$ km, $\theta_A = 70°$ N of E), and then from airport 2 to airport 3 along vector $\vec{B}$ (say we are only given that $B = 50$ km and $\theta_B = 55°$ S of E). Calculate the magnitude and direction of the total displacement vector, $\vec{C}$, from airport 1 to airport 3.

Solution

(1) Type of problem

Vector problem: Use right triangle trig equations *for each vector.* We also have vector addition ($\vec{C} = \vec{A} + \vec{B}$), so we use equations *relating the vectors.*

(2) Sort by vector

We make a diagram of all three vectors together, including information about the two known vectors ($\vec{A}$ and $\vec{B}$), and write the unknowns for the other vector ($\vec{C}$).

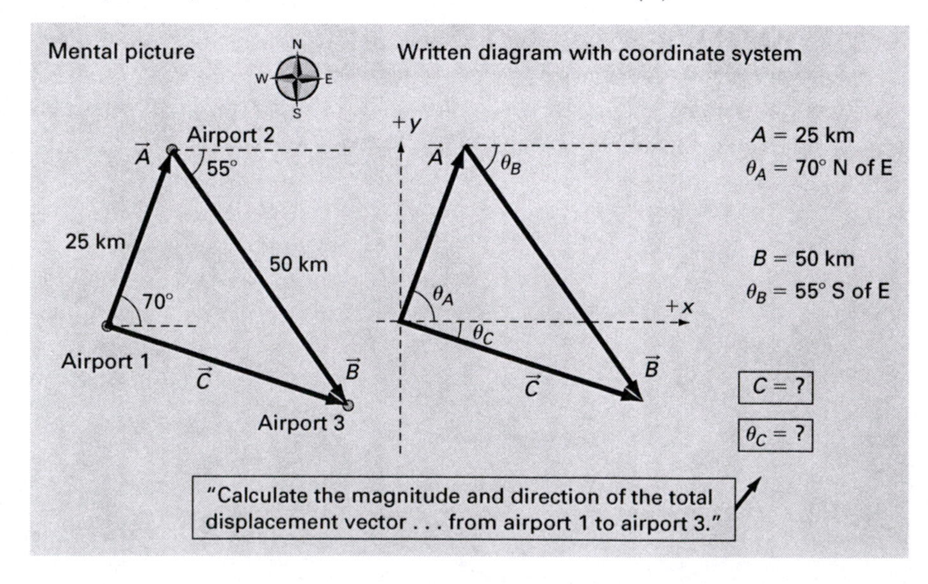

We draw all vectors in a problem approximately *to the same scale* when possible. Here, $\vec{B}$ is very roughly twice as long as $\vec{A}$ since $B = 50$ km and $A = 25$ km. This helps us to solve the problem and evaluate the answers.

DON'T ADD THE MAGNITUDES!

We can see from the previous figure that the magnitude C is NOT the sum of the magnitudes A and B ($C \neq A + B = 25 + 50 = 75$ km). That would be the total *distance covered* for this trip, but not the displacement.

The previous figure shows a "head-to-tail" or "tip-to-tail" vector addition diagram: At the end (head, or tip) of $\vec{A}$, we draw the beginning (tail) of $\vec{B}$, and then $\vec{C}$ goes *from the beginning of* $\vec{A}$ *to the end of* $\vec{B}$.

An alternate diagram shows the "parallelogram" vector addition method:

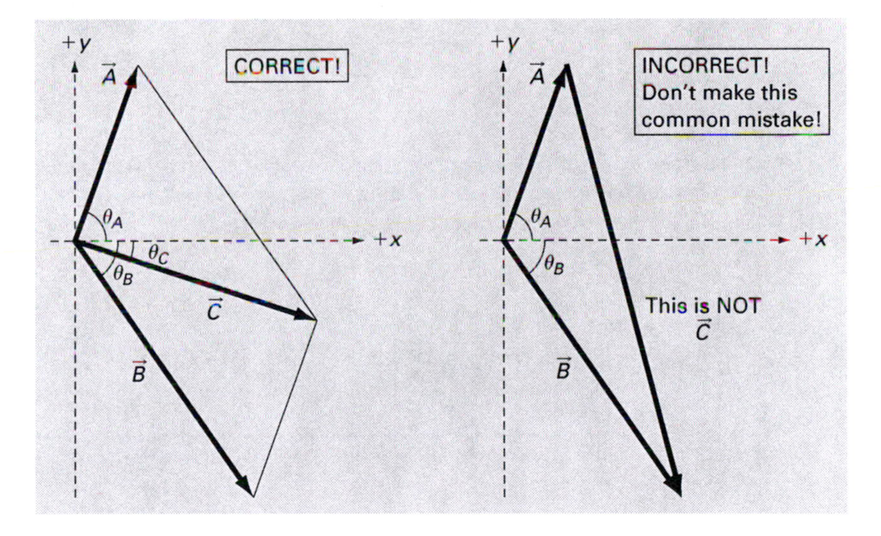

We *move* $\vec{B}$ so it begins at the same origin as $\vec{A}$, and then $\vec{A}$ and $\vec{B}$ are two sides of a *parallelogram.* In the correct figure on the left, we draw $\vec{C}$ *from the origin to the opposite corner of the parallelogram.*

VECTORS ARE MOVABLE!

When adding or subtracting vectors, we can *move* a vector to anywhere! We just keep its *magnitude* (length of arrow) and *direction* (which way the arrow points) the same. Each vector is "fixed" in length and direction, but not "stuck" to a certain point. We can "slide" it around to wherever we want it to be. This is very useful when drawing vector addition or subtraction diagrams.

Finally, either in our minds or on paper, we picture each vector with its own "private" origin at its beginning point. We already did this for $\vec{A}$ and $\vec{B}$ in previous exercises. Here are vector $\vec{C}$ and its components:

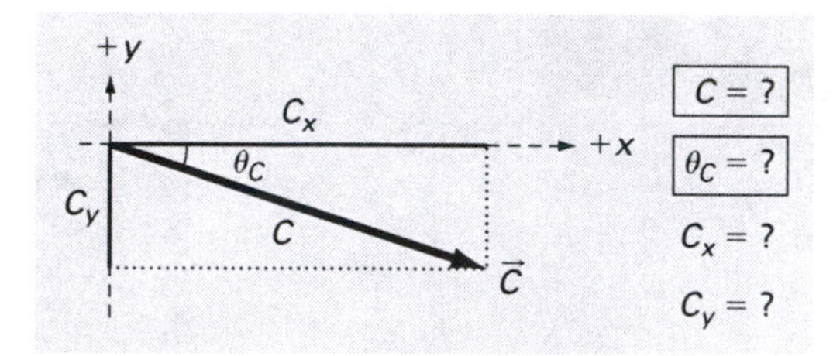

(3) Equations & unknowns and (4) Outline solution

In vector addition problems, we combine these steps. There are equations *relating the vectors* and equations *for each vector.* We write down and solve the equations in this order:

- For each *known* vector
- Relating the vectors
- For the *unknown* vector

We begin with the equations.

For each* known *vector

Vectors $\vec{A}$ and $\vec{B}$ are "known" because we are given two pieces of information (magnitude and direction) about each and can calculate everything else (their x- and y-components). For each of these, we make a right triangle and write down the trig equations:

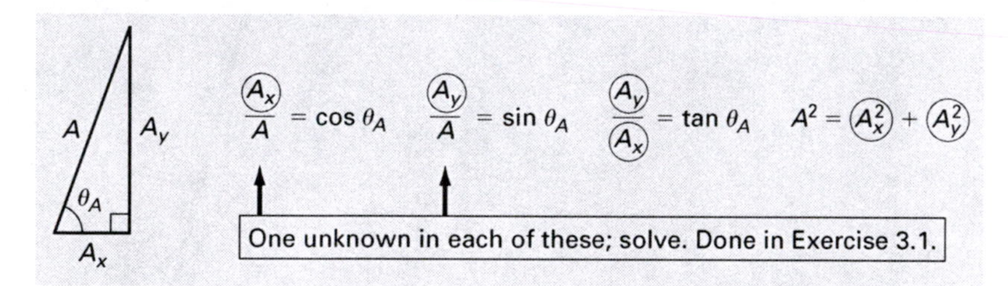

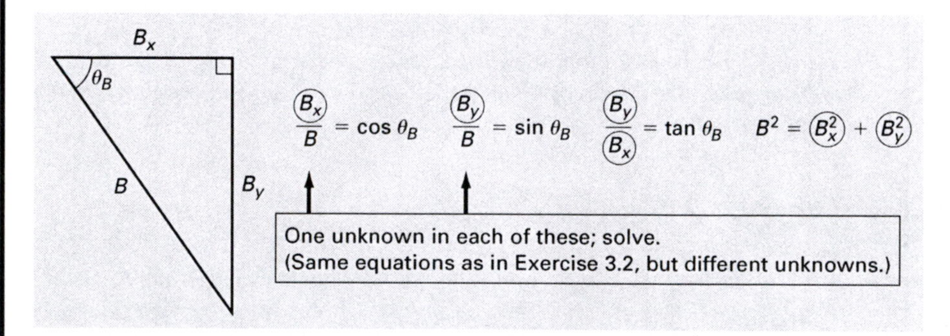

Don't forget that with trig equations, we ignore signs during the calculations, and choose *positive* or *negative* in the end.

Relating the vectors

Vector $\vec{C}$ is the *vector sum* of the other two, or the *resultant vector:* $\vec{C} = \vec{A} + \vec{B}$. We don't actually use this vector equation for calculations, but rather the two *component equations* that come from it:

$\vec{C} = \vec{A} + \vec{B}$ "contains" both component equations

$C_x = A_x + B_x$ — The total east–west displacement is the sum of the east–west displacements of the parts.

and

$C_y = A_y + B_y$ — The total north–south displacement is the sum of the north–south displacements of the parts.

Now that we have solved for A_x, A_y, B_x, and B_y, these equations each have only one unknown:

$$\textcircled{C_x} = A_x + B_x$$

$$\textcircled{C_y} = A_y + B_y$$

One unknown in each; solve. Use correct *signs.*

If we have correct signs for A_x, A_y, B_x, and B_y, then the equations give us the correct signs for C_x and C_y.

USE COMPONENT EQUATIONS TO ADD VECTORS!

When adding (or subtracting) vectors, we ALWAYS do *calculations* with the *component equations*—NEVER with the vector equation.

For the* unknown *vector

We make a right triangle and write down the trig equations for $\vec{C}$, the "unknown" vector in the problem. We have calculated C_x and C_y, so they are now known:

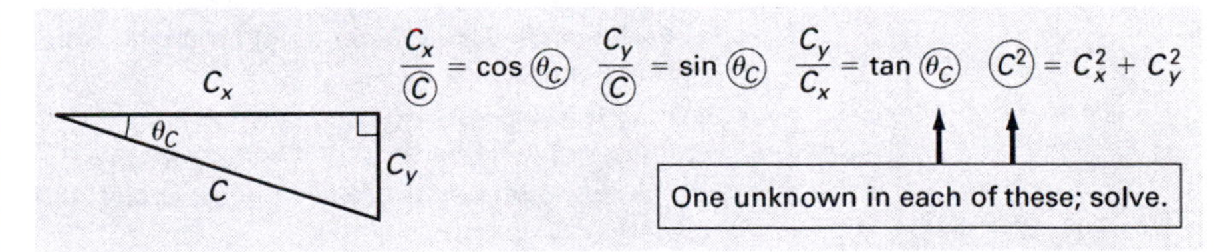

Answers

We show the math not already done in previous exercises. The components for $\vec{C}$:

$C_x = A_x + B_x \cong (+8.55) + (+28.71) \cong +37.3 \cong +37$ km (*positive* means *east*)

$C_y = A_y + B_y \cong (+23.49) + (-41) \cong -17.51 \cong -18$ km (*negative* means *south*)

Use correct *signs.*

Remember it is only with *trig* equations (sine, cosine, tangent) that we ignore signs in the calculations. These numbers and signs make sense with our vector diagram.

Now the magnitude and direction of $\vec{C}$:

$$C^2 = C_x^2 + C_y^2 \Longrightarrow C = \sqrt{C_x^2 + C_y^2} \cong \sqrt{(+37.3)^2 + (-17.51)^2} \cong 41 \text{ km}$$

$$\tan \theta_C = \frac{C_y}{C_x} \Longrightarrow \theta_C = \tan^{-1}\left(\frac{C_y}{C_x}\right) = \tan^{-1}\left(\frac{(-17.51)}{(+37.3)}\right) \cong 25°$$

Ignore signs in trig calculations.

The right triangle shows us that this is 25° S of E. ■

3.4 HOW TO SET UP VECTOR PROBLEMS

Here is a more detailed summary of the steps for solving vector problems, tailored to vector addition or subtraction problems:

Vector Addition or Subtraction Problems—Mental and Written Steps

Mental →

(1) Type of problem

Adding or subtracting vectors:

- Use right triangle trig *for each vector.*
- Use vector component addition or subtraction equations to *relate the vectors.*

Mental and written →

(2) Sort by vector

Draw vector diagram(s) showing all vectors together, and sort the information *by vector.* For components, think of each vector as having its own "private" origin.

Mental and written →

(3) Equations & unknowns and (4) Outline solution

Combine these steps in three parts:

- **For each *known* vector**: Make a right triangle; write and solve trig equations.
- **Relating the vectors**: Write and solve *x*- and *y*-component addition or subtraction equations.
- **For the *unknown* vector**: Make a right triangle; write and solve trig equations.

For each part, simplify, circle the unknowns, and outline the solution.

3.5 "BACK WHERE YOU STARTED"— WHEN VECTORS ADD TO ZERO

EXERCISE 3.4

Now the pilot (in the previous exercises) wants to go from airport 1 to airport 2 along vector $\vec{A}$, then from airport 2 to airport 3 along vector $\vec{B}$, and finally from airport 3 back to airport 1 along vector $\vec{D}$. Calculate the magnitude and direction of $\vec{D}$.

Solution

(1) Type of problem

Vector problem: Use right triangle trig equations *for each vector*. Because the trip ends up back where it started, the total displacement is zero, which gives us the vector equation: $\vec{A} + \vec{B} + \vec{D} = 0$. This is vector addition and gives us equations *relating the vectors*.

(2) Sort by vector

Vector $\vec{D}$ is the same size but in the exact opposite direction of $\vec{C}$ from the previous exercise:

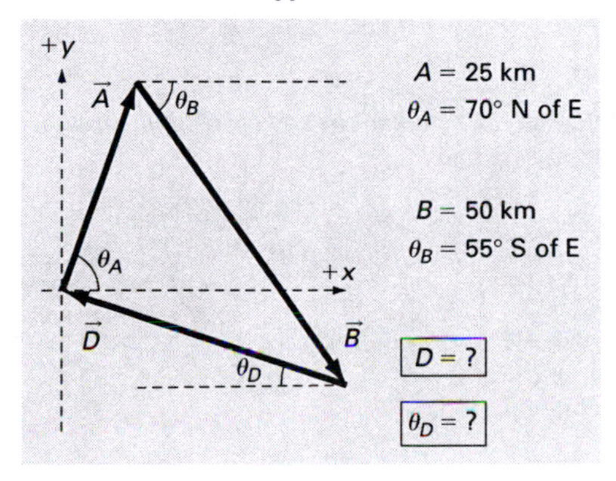

As we did for $\vec{A}$ and $\vec{B}$ in previous exercises, we picture vector $\vec{D}$ with its own "private" origin at its beginning point, and its components:

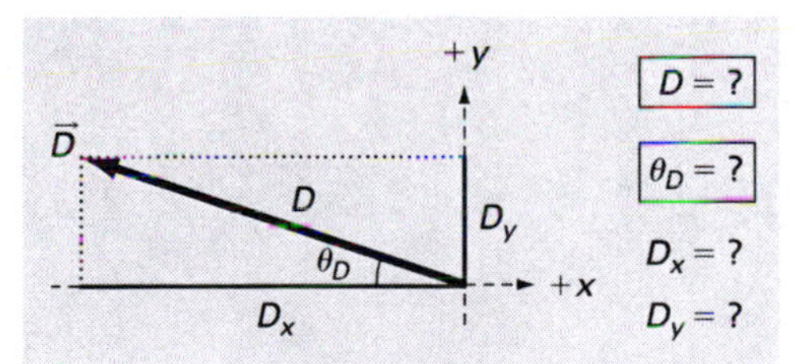

(3) Equations & unknowns and (4) Outline solution

This is very similar to the previous exercise. The numbers are the same, but the *signs* are often different.

For each* known *vector

We calculate A_x, A_y, B_x, and B_y, just as in the previous exercise.

Relating the vectors

The vector equation contains *two component equations* that we use *for calculations:*

$\vec{A} + \vec{B} + \vec{D} = 0$ "contains" both component equations:

$$\begin{cases} A_x + B_x + D_x = 0 \\ \text{and} \\ A_y + B_y + D_y = 0 \end{cases}$$

$A_x + B_x + D_x = 0$ — Total east–west displacement is zero.

$A_y + B_y + D_y = 0$ — Total north–south displacement is zero.

↑ *You circle the unknowns and outline a solution for this part!*

For the* unknown *vector

Vector $\vec{D}$ is the "unknown" vector:

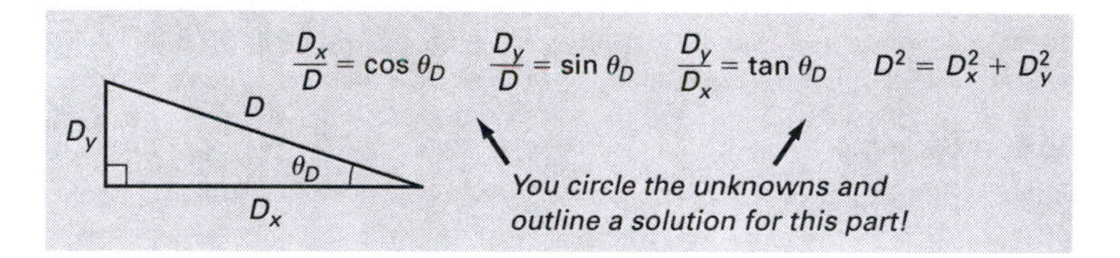

Answers for (3) Equations & unknowns and (4) Outline solution

Once we know A_x, A_y, B_x, and B_y:

$\vec{A} + \vec{B} + \vec{D} = 0$ "contains" both component equations

$$A_x + B_x + D_x = 0$$

and

$$A_y + B_y + D_y = 0$$

One unknown in each; solve. Use correct *signs*.

Then, once we know D_x and D_y:

$$\frac{D_x}{D} = \cos\theta_D \qquad \frac{D_y}{D} = \sin\theta_D \qquad \frac{D_y}{D_x} = \tan\theta_D \qquad D^2 = D_x^2 + D_y^2$$

D, D_y, θ_D, D_x

One unknown in each of these; solve.

Answers

- $D_x \cong -37.3 \cong -37$ km (*negative* means *west*).
- $D_y \cong +17.51 \cong +18$ km (*positive* means *north*).
- $D \cong 41$ km (magnitude is *always positive,* so we don't bother writing + sign).
- $\theta_D \cong 25°$ N of W.

THE "NEGATIVE," OR "OPPOSITE," OF A VECTOR

The vectors $\vec{C}$ and $\vec{D}$ are *opposite,* or *negative,* of each other: $\vec{C} = -\vec{D}$ or $\vec{D} = -\vec{C}$.

- The magnitudes have the same positive value: $C = D$.
- The directions are opposite: $\theta_C = 25°$ S of E, but $\theta_D = 25°$ N of W.
- The components are opposite: $C_x = -D_x$ and $C_y = -D_y$.

■

3.6 SUBTRACTING VECTORS, OR, WHEN ONE OF THE ADDED VECTORS IS UNKNOWN

Some problems ask us to determine the unknown vector for one of the following situations:

- Vector 1 (known) is subtracted from vector 2 (known) to get vector 3 (unknown).
- Vector 1 (known) is added to vector 2 (unknown) to get vector 3 (known).

These are very similar to one another, and the next exercise shows how to set up and solve either of them.

EXERCISE 3.5

Starting at a certain tree, you walk through an initial displacement of 75 m at an angle of 60 degrees west of south. From this new position, you hike through an unknown second displacement. You end up at a position 120 m from the tree, at an angle of 40 degrees north of west. Determine the unknown displacement (both magnitude and direction).

Solution

(1) Type of problem

Vector problem: Use right triangle trig equations *for each vector* and component equations *relating the vectors.*

Say $\vec{R}_1$ is the (known) initial displacement, $\vec{R}_2$ is the (unknown) second displacement, and $\vec{R}$ is the (known) total displacement. The vector addition equation: $\vec{R}_1 + \vec{R}_2 = \vec{R}$. If we solve for the unknown vector, we can think of it as *vector subtraction:* $\vec{R}_2 = \vec{R} - \vec{R}_1$.

(2) Sort by vector

This diagram illustrates either equation, $\vec{R}_1 + \vec{R}_2 = \vec{R}$ or $\vec{R}_2 = \vec{R} - \vec{R}_1$:

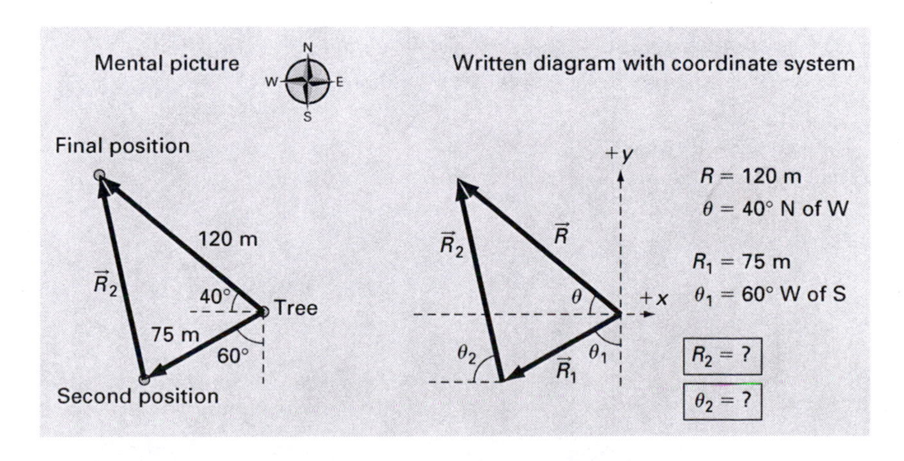

For determining components, we think of each vector as beginning at its own origin:

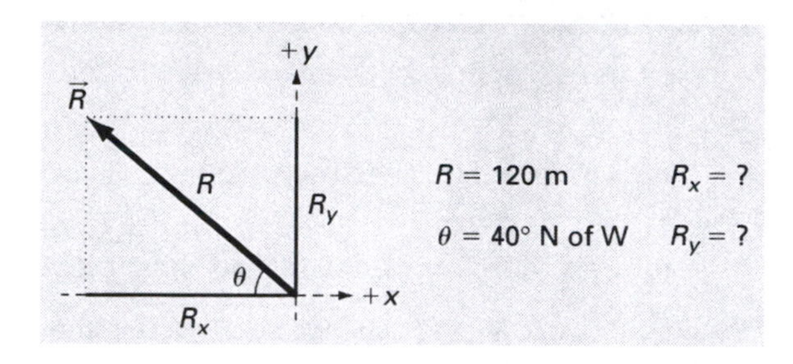

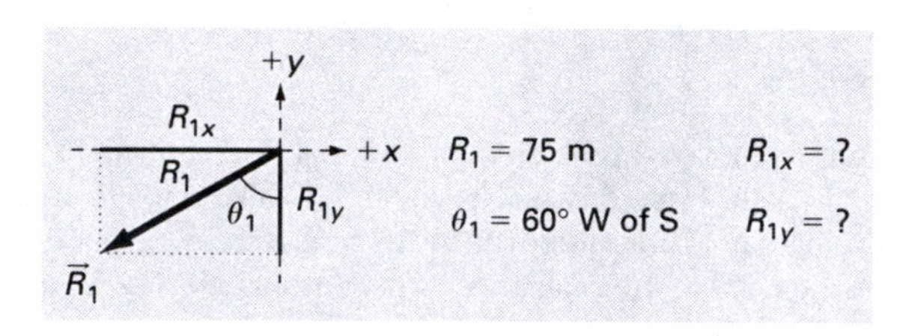

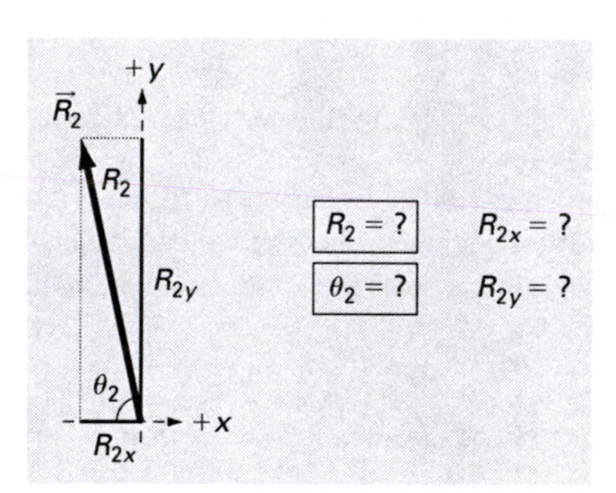

(3) Equations & unknowns and (4) Outline solution

We follow the same outline here as for vector addition.

For each* known *vector

The two vectors for which we are given information are $\vec{R}$ and $\vec{R}_1$:

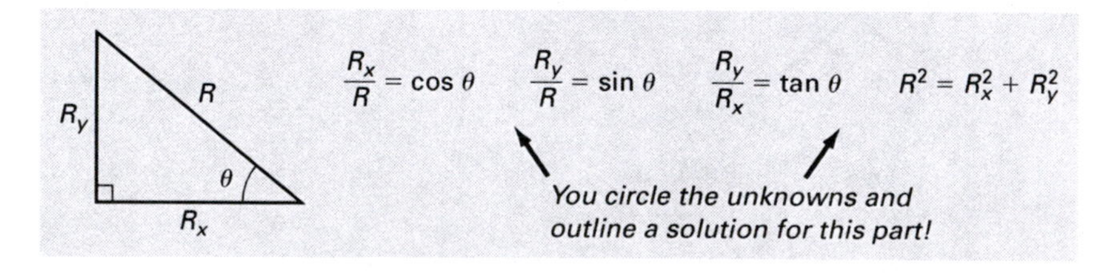

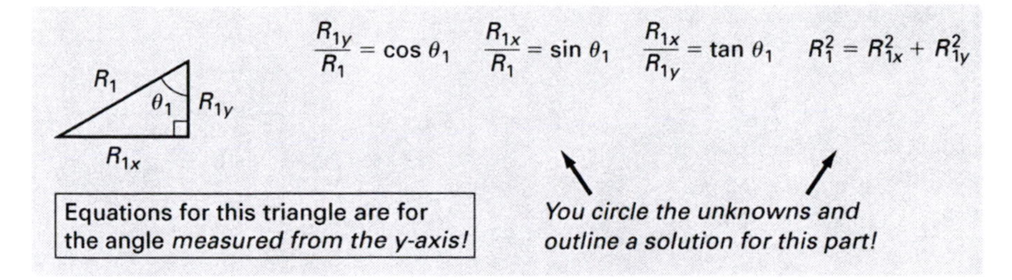

Relating the vectors

We use the vector subtraction equation to get the component equations that relate the vectors:

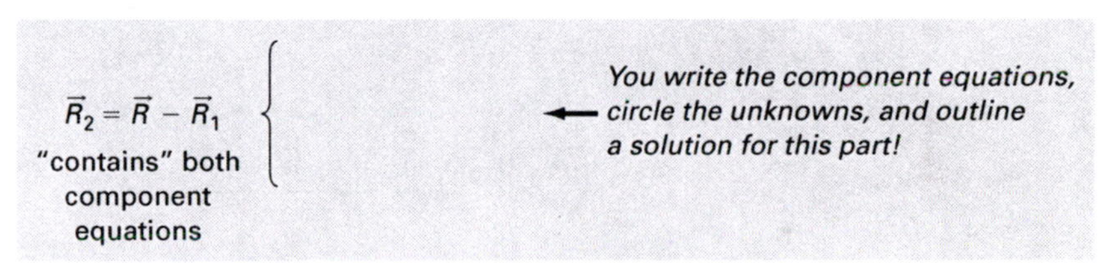

For the* unknown *vector

Vector $\vec{R}_2$:

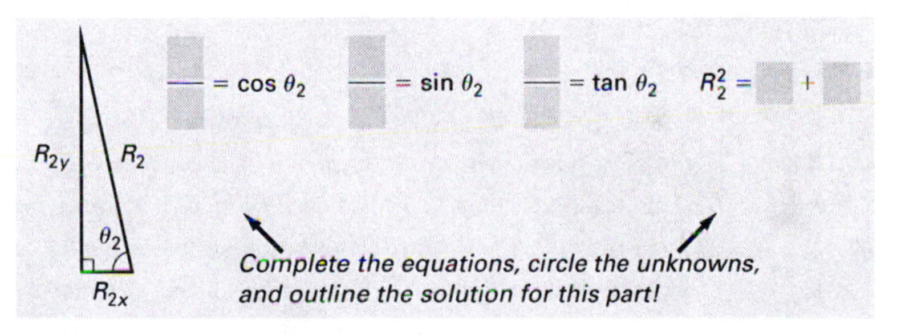

Answers for (3) Equations & unknowns and (4) Outline solution

To solve for R_x and R_y:

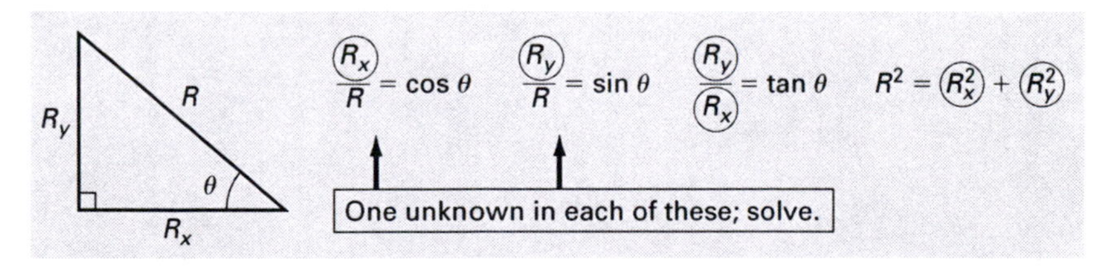

To solve for R_{1x} and R_{1y}:

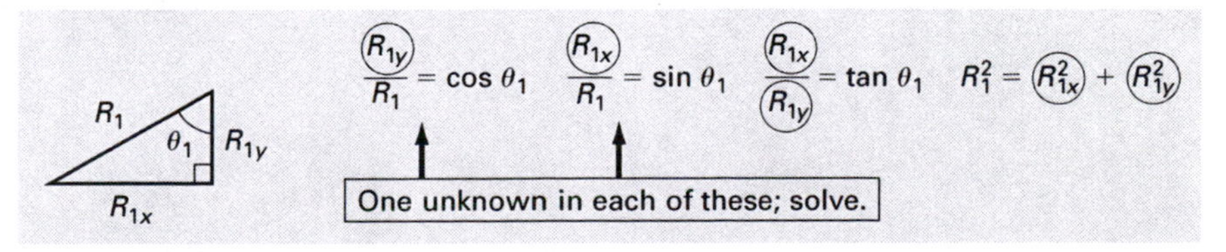

Once we know R_x, R_y, R_{1x}, and R_{1y}:

$\vec{R}_2 = \vec{R} - \vec{R}_1$ "contains" both component equations

$$\begin{cases} R_{2x} = R_x - R_{1x} \\ R_{2y} = R_y - R_{1y} \end{cases}$$

One unknown in each; solve. Be careful with *signs.*

Once we know R_{2x} and R_{2y}:

$$\frac{R_{2x}}{R_2} = \cos(\theta_2) \quad \frac{R_{2y}}{R_2} = \sin(\theta_2) \quad \frac{R_{2y}}{R_{2x}} = \tan(\theta_2) \quad R_2^2 = R_{2x}^2 + R_{2y}^2$$

One unknown in each of these; solve.

Answers

- $R_2 \cong 118 \cong 120$ m (magnitude *always positive*).
- $\theta_2 \cong 76.8 \cong 77°$ N of W.

Intermediate answers:

- $R_x \cong -91.9$ m (Trig calculation: Make this *negative* since it is *west.*)
- $R_y \cong +77.1$ m (Trig calculation: Make this *positive* since it is *north.*)
- $R_{1x} \cong -65.0$ m (Trig calculation: Make this *negative* since it is *west.*)
- $R_{1y} \cong -37.5$ m (Trig calculation: Make this *negative* since it is *south.*)
- $R_{2x} \cong -26.9$ m (*Negative* means *west,* which agrees with our diagrams.)
- $R_{2y} \cong +115$ m (*Positive* means *north,* which agrees with our diagrams.) ■

CHAPTER 4

PROJECTILE MOTION

TOSS A BALL straight up or down and it has 1D (one-dimensional) free-fall motion. It moves only vertically. Toss a ball horizontally, or upward at an angle, or downward at an angle (i.e., in *any direction other than vertical*), and it has 2D (two-dimensional) *projectile motion.* It is still free-fall, but it now has horizontal motion along with the vertical.

Say you toss a ball horizontally so that it eventually hits the ground. Gravity pulls it into a parabolic path that looks something like this:

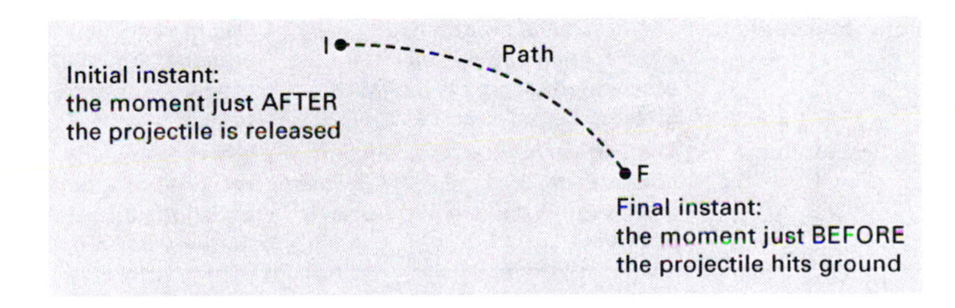

The interval for the flight path does NOT include the toss (or push, or launch, etc.) that gets the ball going or the hit at the ground that stops it. We consider only the motion *in between.*

4.1 PROJECTILE MOTION: COMBINING THREE BASIC CONCEPTS

There are no new equations in this chapter. We just combine three concepts, which we already know from Chapters 2 and 3, into one motion.

Here is the idea: We mentally *separate* the 2D motion interval into two 1D motion intervals, like so:

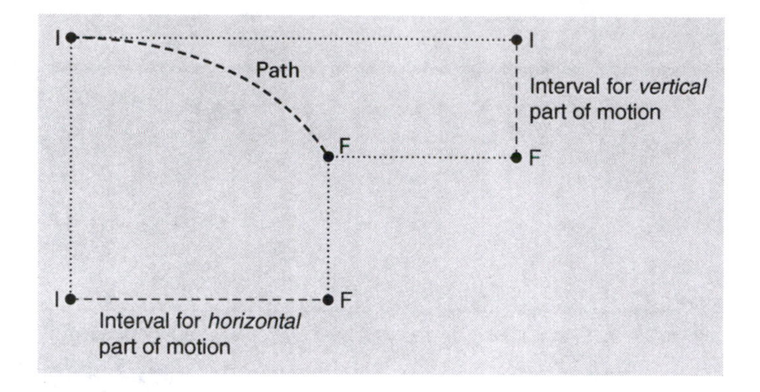

Keeping this diagram in mind, we see from the following table how to apply three basic concepts to solve projectile motion problems:

Concept	How it applies to projectile motion	What we DO with it
Vectors	For *2D* motion, we separate *x*- and *y*-components for the *initial* and *final velocity vectors.*	Use right triangle trig equations for initial and final velocity vectors.
Horizontal motion	The *horizontal* motion has *constant velocity.* It is NOT accelerated by gravity, because gravity acts only *vertically.*	Use the constant velocity equation, Equation 2.2, modified for the *x*-direction only.
Vertical motion	The *vertical* motion has *constant acceleration,* $a_y = -9.8\frac{m}{s^2}$ (*negative* means *down* if we define *up* as *positive*).	Use the constant acceleration equations, Equations 2.3a–d, modified for the *y*-direction only.

The details become much easier to handle when we mentally separate this motion into these three bite-sized parts. For the rest of the chapter, we will illustrate how to do this.

4.2 WHEN INITIAL VELOCITY IS HORIZONTAL

As usual, the exercises in this chapter are longer than what you would ever actually write out on paper because we show all the mental steps as well as the much shorter written steps.

ISN'T THERE A "MAGIC" EQUATION WE CAN PLUG INTO?

We won't use a "magic" equation meant to solve *one specific kind* of projectile motion problem, such as the horizontal range equation given in many textbooks. This equation is great *if that is the only kind of projectile problem you ever see.* But what if the problem is slightly different? Learn the basic concepts in this chapter, and you can solve any type of projectile motion problem.

EXERCISE 4.1

A helicopter traveling horizontally at 3.0 m/s releases a ball from a height of 4.0 m above the ground. (a) How far horizontally has the ball traveled by the time it strikes the ground? (b) What is its velocity when it hits the ground?

Solution

(1) Type of problem

The ball is released with an initial horizontal velocity equal to that of the helicopter; it is *just as if* someone tossed the ball horizontally at 3.0 m/s. This is projectile motion. Ignore air resistance, as we always will do in projectile problems. We use the following:

- Right triangle trig for velocity vectors
- Equation 2.2 for the constant velocity horizontal motion
- Equations 2.3a–d for the constant acceleration vertical motion

(2) Sort by . . . and (3) Equations & unknowns

We sort the information and write equations for each of the three parts described earlier:

- Velocity vectors
- Horizontal motion
- Vertical motion

> *FOCUS ON THE BIG PICTURE!*
>
> Skim ahead through this solution quickly to see the overall outline. Then come back and study the details, which are all based on ideas we already know.

Velocity vectors

We focus on the initial and final velocity vectors, $\vec{v}_0$ and $\vec{v}$, mentally picturing each of these as tangent to the curved path:

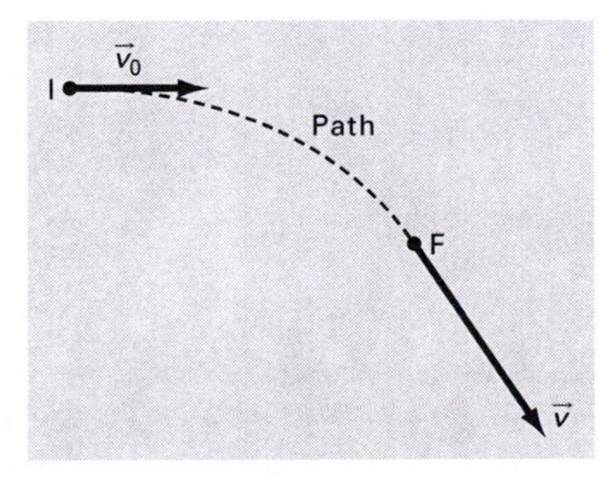

Putting each velocity vector at its own origin, we list the magnitude, direction (angle), and x- and y-components:

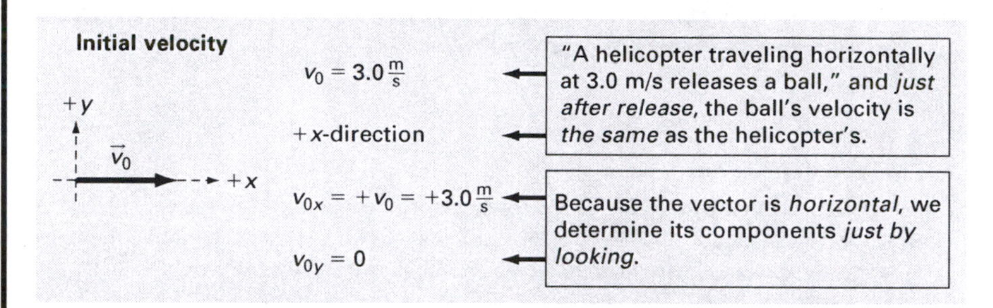

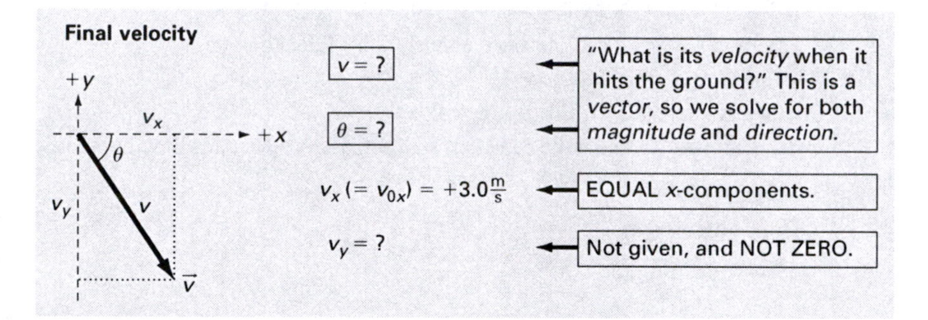

THE FINAL VELOCITY IS NOT ZERO!

A common mistake is to assume that the final velocity of the ball is zero. We are looking at the moment just before the ball hits the ground, so it is *still moving*.

Since $\vec{v}$ is at an angle, we use a right triangle and trig equations to determine its components:

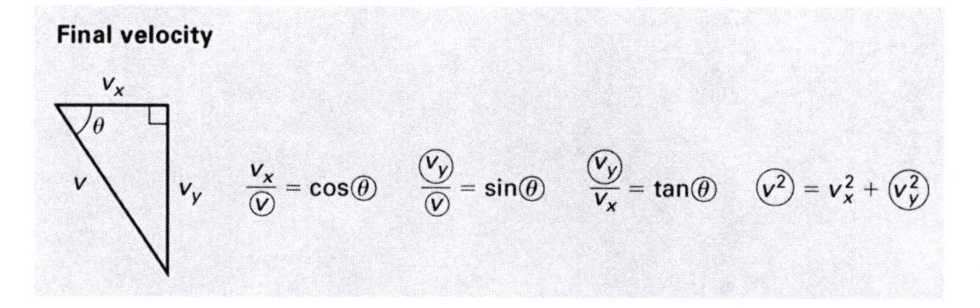

Horizontal motion

Now we mentally separate the *horizontal* and *vertical* parts of the motion by setting up *separate* motion intervals. Watch the x and y subscripts on the *velocities* and *accelerations* in the next several figures.

TREAT PROJECTILE MOTION AS TWO SEPARATE MOTIONS!

Even though the ball is only one object, we treat it as having *two separate motions*. These *simultaneous* motions have the *same* elapsed time, *t*.

First, we set up the *horizontal* motion interval with *constant velocity* (in the *x*-direction only) for Equation 2.2:

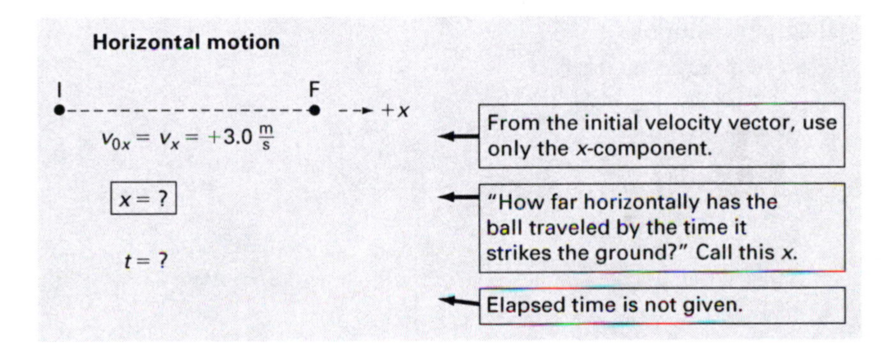

The equation and unknowns:

Horizontal motion
Equation 2.2, modified to have only the *x*-component of the velocity

$$v_x = \frac{ⓧ}{ⓣ}$$

Vertical motion

The *vertical* motion interval with *constant acceleration* (in the *y*-direction only) is set up for Equations 2.3a–d:

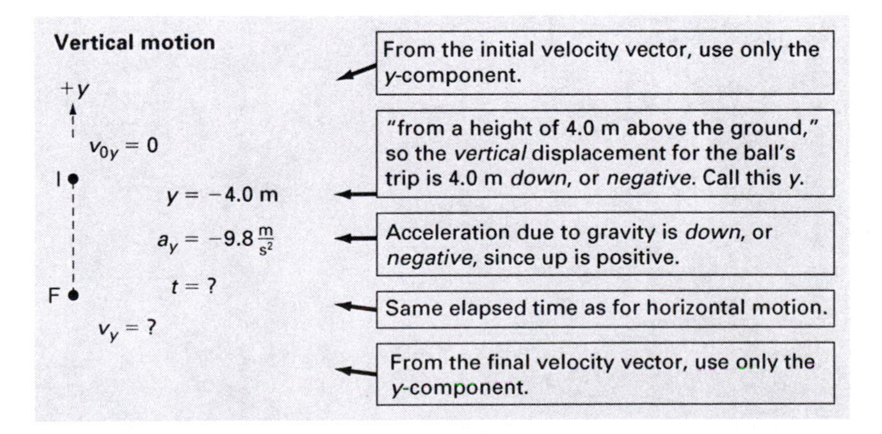

Equations 2.3a–d and unknowns:

Vertical motion
Equations 2.3a–d, modified to have only the *y*-components

$$y = v_{0y}(t) + \tfrac{1}{2}a_y(t^2)$$

$$y = \left(\frac{v_{0y} + (v_y)}{2}\right)(t)$$

$$(v_y) = v_{0y} + a_y(t)$$

$$(v_y^2) = v_{0y}^2 + 2a_y y$$

(4) Outline solution

We put all the equations together:

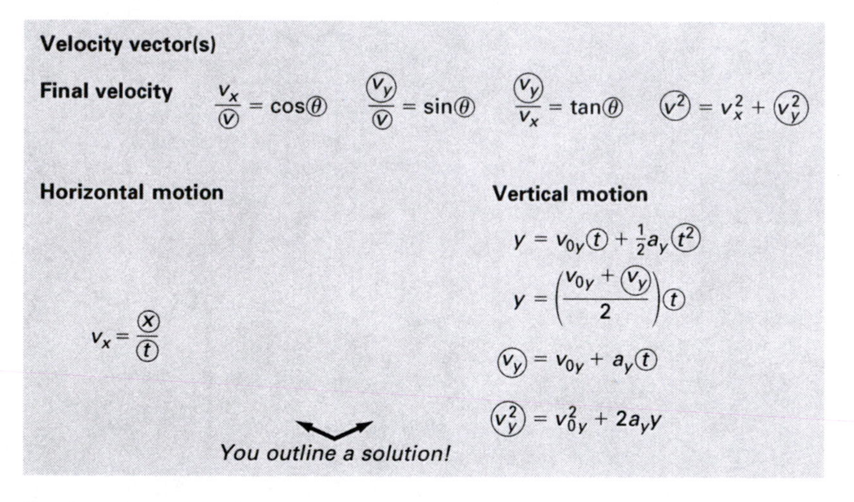

We also show the math in the answers section.

Answers for (4) Outline solution
Here is one possible outline:

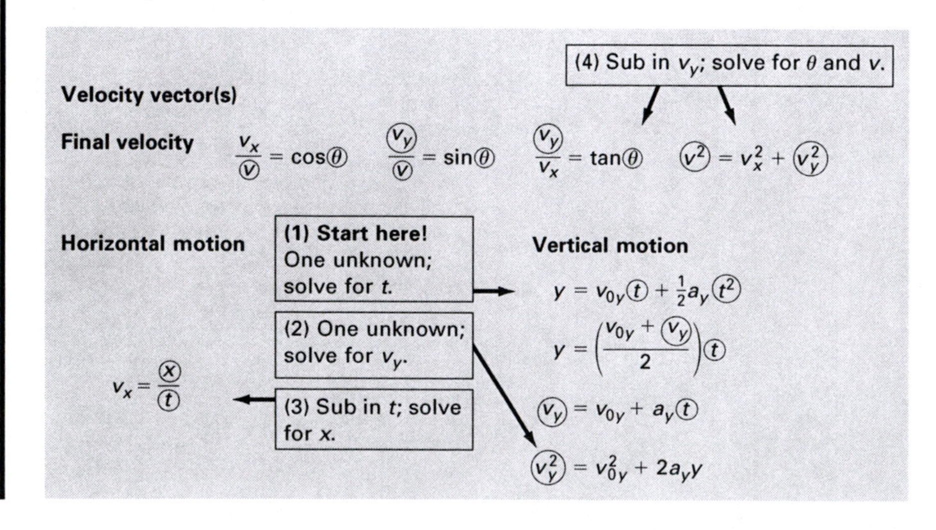

Answers for the math
Watch the *signs!*

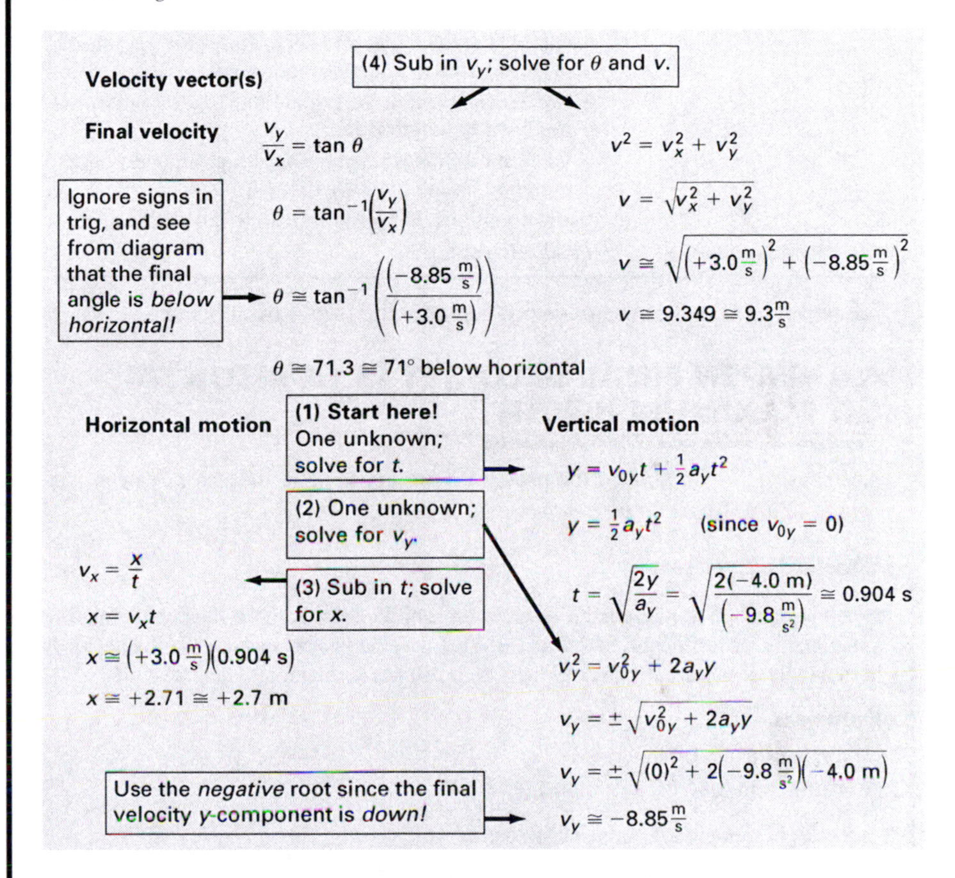

Answers

- $x \cong +2.7$ m (*positive* means to the *right*).
- Magnitude, v, and direction, θ, are both included when *velocity* (a vector) is asked for:
  - $v \cong 9.3\frac{m}{s}$ (vector magnitude should always be positive).
  - $\theta \cong 71°$ below horizontal. ■

4.3 HOW TO SET UP PROJECTILE MOTION PROBLEMS

Here we summarize the setup steps for projectile motion.

Projectile Motion Problems—Mental and Written Steps

Mental→	**(1) Type of problem** For projectile motion, use: • Right triangle trig for velocity vectors • Equation 2.2 for constant velocity horizontal motion • Equations 2.3a–d for constant acceleration vertical motion

Mental and written → **(2) Sort by . . . and (3) Equations & unknowns**

We combine these steps in three parts:

- **Velocity vectors:** For $\vec{v}_0$ and $\vec{v}$, make a right triangle if the vector is at an angle, and separately sort magnitude, direction, and *x*- and *y*-components.
- **Horizontal motion:** Sort quantities in *x*-direction only for modified Equation 2.2.
- **Vertical motion:** Sort quantities in *y*-direction only for modified Equations 2.3a–d.

Write equations, simplify, and circle the unknowns.

Mental and written → **(4) Outline solution**

We will use this pattern to solve problems for the rest of this chapter.

4.4 WHEN FINAL VELOCITY IS HORIZONTAL (AT MAXIMUM HEIGHT)

At the *maximum height,* or the peak, of a projectile path, the *velocity vector is horizontal,* and so the *velocity y-component is zero.*

EXERCISE 4.2

A ball is launched from ground level at an angle of 55 degrees above the horizontal. It reaches a maximum vertical height of 2.5 m. (a) What was the launch speed of the ball? (b) How far horizontally does the ball travel by the time it reaches its maximum height?

Solution

(1) Type of problem

Projectile motion, so use the following:

- Right triangle trig for velocity vectors
- Equation 2.2 for the constant velocity horizontal motion
- Equations 2.3a–d for the constant acceleration vertical motion

We separate the horizontal and vertical motion into two separate but simultaneous intervals:

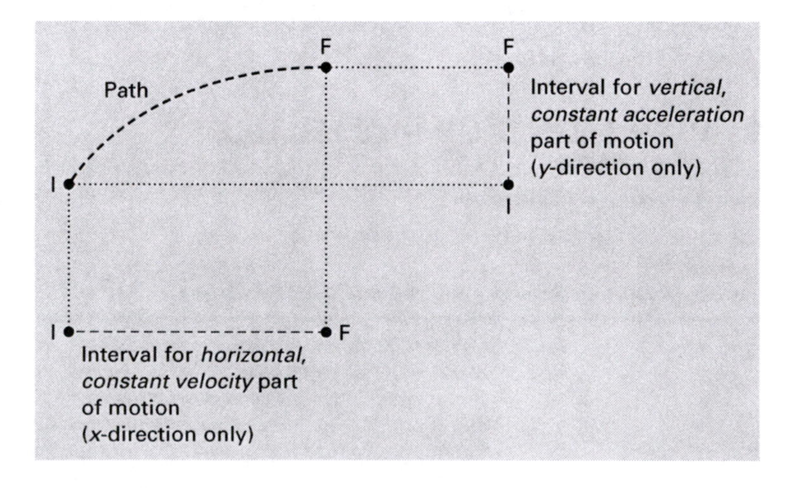

(2) Sort by . . . and (3) Equations & unknowns

Velocity vectors

The "maximum vertical height" at the end of the path means that the vertical component of the final velocity vector is zero ($v_y = 0$) and the final velocity vector $\vec{v}$ *is horizontal.* We can picture the initial and final velocity vectors like this:

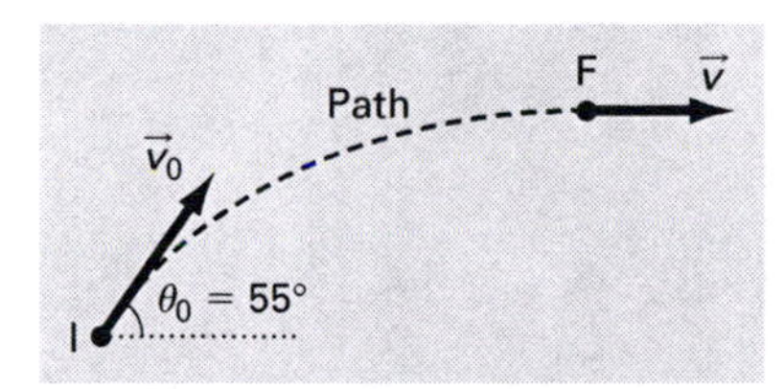

THE LAUNCH ANGLE IS NOT THE LINE-OF-SIGHT ANGLE!

The launch angle, $\theta_0 = 55°$, is shown in the previous figure as the angle for $\vec{v}_0$. The line-of-sight, drawn *straight* from I to F, is at an unknown angle:

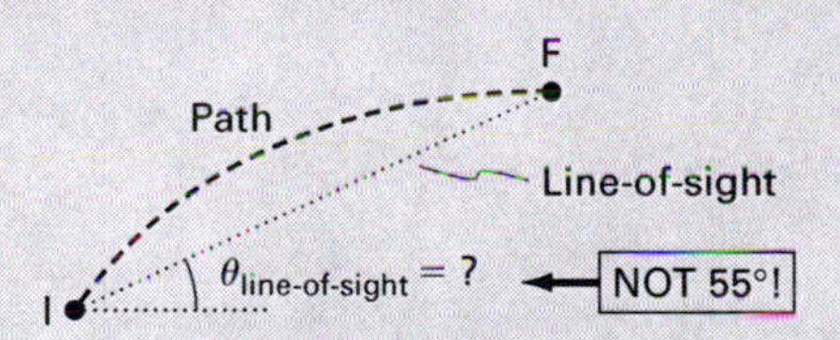

The two angles are *not even close* to being equal!

Now we put each velocity vector at its own origin, with its known and unknown quantities:

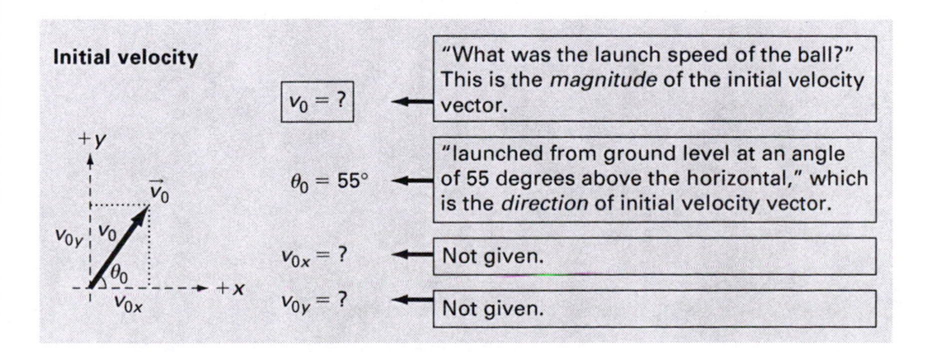

Since $\vec{v}_0$ is at an angle, we use a right triangle and trig equations:

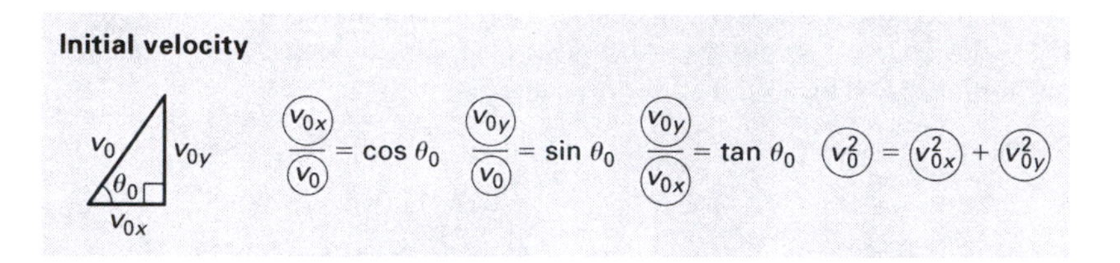

Next:

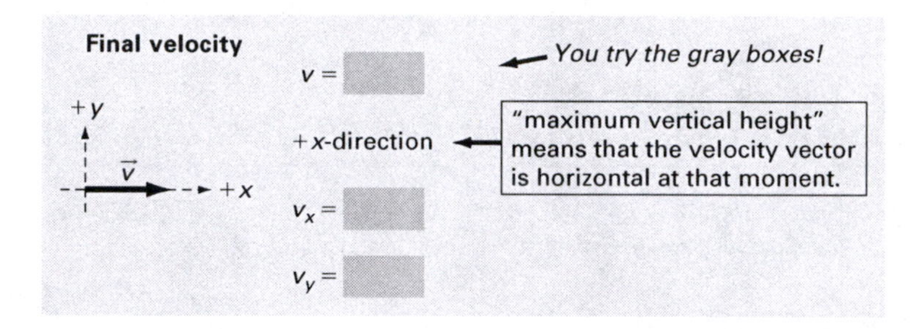

Horizontal motion

The interval for the constant velocity part of the motion:

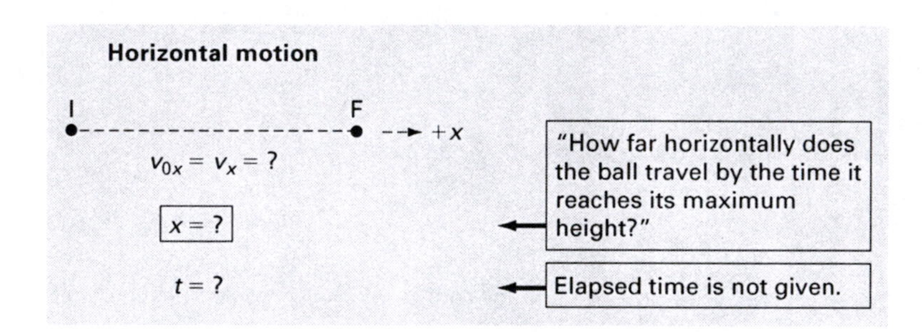

The constant velocity equation:

Horizontal motion
Modified Equation 2.2

$$v_x = \frac{x}{t}$$

Vertical motion

The interval for the constant acceleration part of the motion:

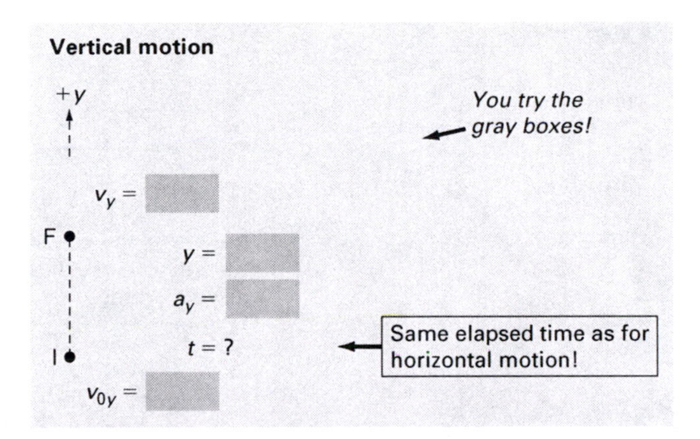

Check your answers at the end. The constant acceleration equations:

Vertical motion
Modified Equations 2.3a–d

$$y = v_{0y} t + \frac{1}{2} a_y t^2$$

$$y = \left(\frac{v_{0y} + v_y}{2}\right) t$$

$$v_y = v_{0y} + a_y t$$

$$v_y^2 = v_{0y}^2 + 2 a_y y$$

(4) Outline solution

Now we put together the three sets of equations:

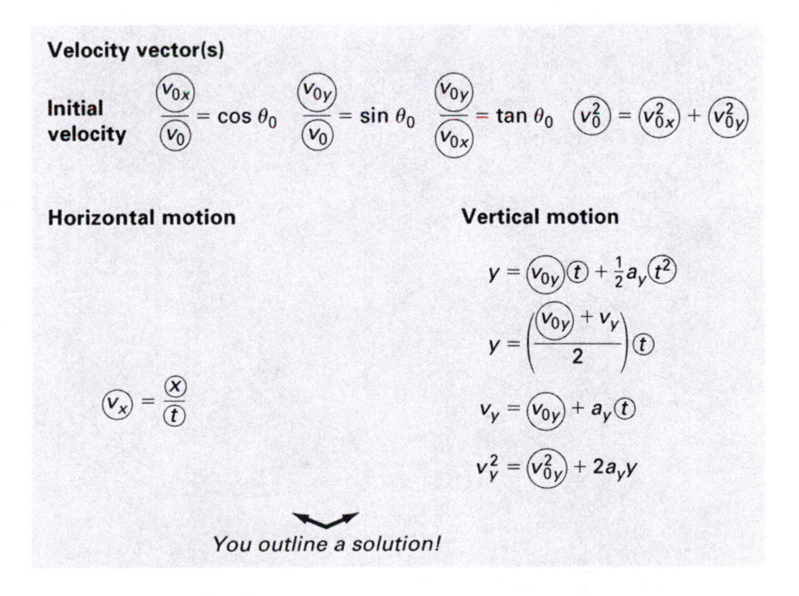

Be careful with negative signs!

Answers for gray boxes

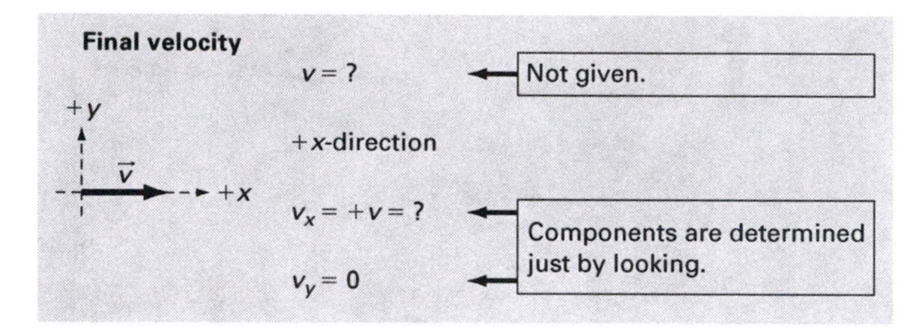

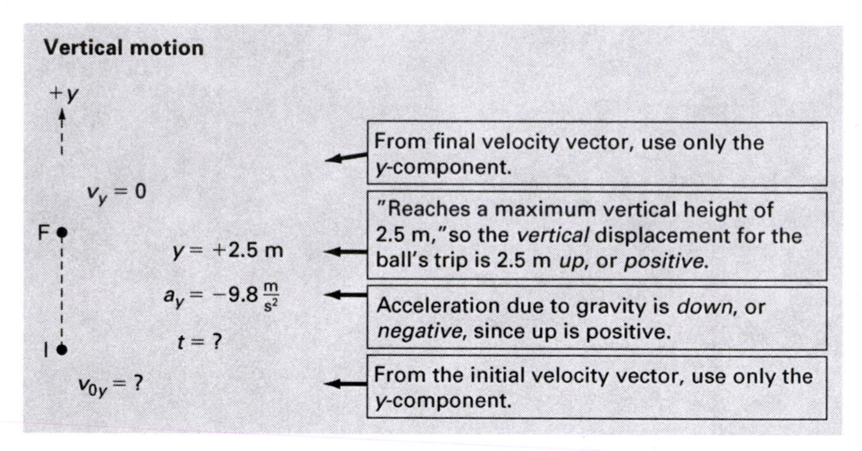

Answers for (4) Outline solution

Here is one possible outline (start at the bottom):

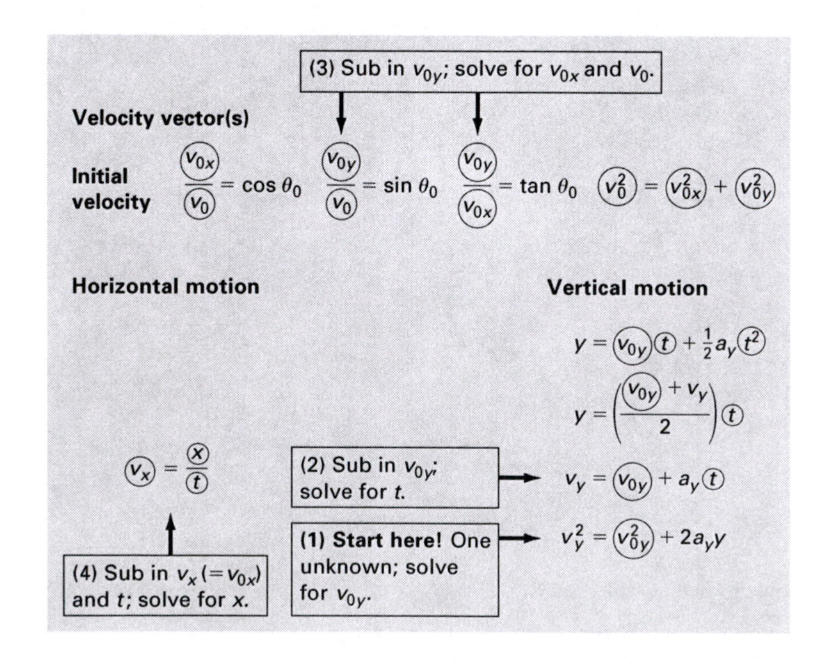

Answers $v_0 \cong 8.545 \cong 8.5\frac{\text{m}}{\text{s}}$, $x \cong +3.50 \cong +3.5$ m
Intermediate answers: $v_{0y} = +7.0\frac{\text{m}}{\text{s}}$, $v_{0x} \cong +4.90 \cong +4.9\frac{\text{m}}{\text{s}}$, $t \cong 0.714 \cong 0.71$ s ■

SOLVE FOR t WHETHER IT IS ASKED FOR OR NOT!

Even if it is not asked for, we need to solve for t in almost every projectile motion problem. Follow one of these two patterns (the previous exercises both followed the first):

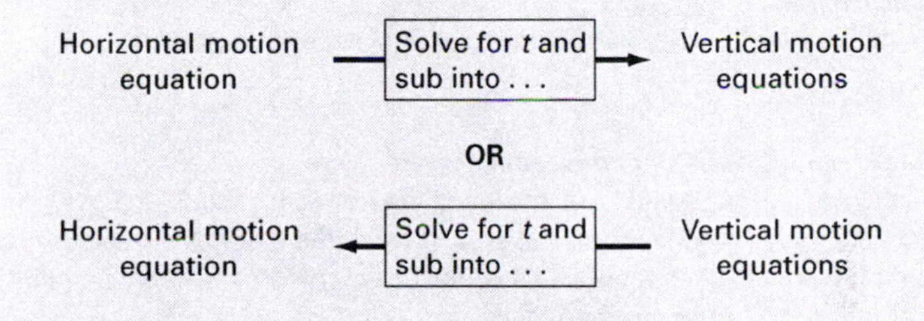

4.5 WHEN INITIAL AND FINAL HEIGHTS ARE EQUAL

EXERCISE 4.3

The ball from the previous exercise (launched from ground level at 55 degrees above the horizontal, reaching a maximum height of 2.5 m) continues to travel until it returns to ground level. (a) What is its total horizontal displacement? (b) What is the total vertical displacement? (c) How much time elapses for the entire trip? (d) What is the velocity of the ball just before it hits the ground?

Solution

(1) Type of problem

Projectile motion, so use the following:

- Right triangle trig for velocity vectors
- Equation 2.2 for the constant velocity horizontal motion
- Equations 2.3a–d for the constant acceleration vertical motion

Easier solution: Use symmetry if it applies

The questions here might have shown up as parts (c) and (d) in the previous exercise. The second half of the motion is *identical* to the first half, except it goes *down* instead of *up*. So we will use the results from the previous exercise to answer the questions here. Quantities for the first half get the subscript *1*, everything for the second half gets the subscript *2*, and everything for the total trip gets the subscript *total*.

(a) Total horizontal displacement

On the way up, $x_1 \cong +3.50$ m, so the total must be twice this amount:

$$x_{\text{total}} = 2(x_1) \cong +7.00 \cong +7.0 \text{ m}.$$

Or we could think of it as the sum of the horizontal displacements of the two equal halves. Both are *positive* since the *horizontal* motion is always to the *right*:

$$x_{\text{total}} = x_1 + x_2 \cong (+3.50 \text{ m}) + (+3.50 \text{ m}) \cong +7.0 \text{ m}$$

(b) Total vertical displacement
This must be *zero* because it starts and ends at the *same height*:

$$y_{\text{total}} = 0$$

Or we could add the vertical displacements of the two halves. Vertical displacement is *positive* on the way *up* and *negative* on the way *down*:

$$y_{\text{total}} = y_1 + y_2 = (+2.50\text{ m}) + (-2.50\text{ m}) = 0$$

(c) Time elapsed
On the way up, $t_1 \cong 0.714$ s, so the total time elapsed must be twice this:

$$t_{\text{total}} = 2(t_1) \cong 1.428 \cong 1.4\text{ s}$$

(d) Final velocity (both magnitude and direction)
On the way up, at ground level, the initial velocity is $v_0 \cong 8.5\frac{\text{m}}{\text{s}}$ at $\theta_0 = 55°$ *above* the horizontal. On the way down, also at ground level, the ball has the same speed, but now the angle is *below* the horizontal. So the final velocity is $v \cong 8.5\frac{\text{m}}{\text{s}}$ (*magnitude* is always *positive*) at $\theta = 55°$ *below* the horizontal.

The End!

Alternate solution: Detailed setup
If we had not already done the previous exercise, we would need to use a detailed setup here. You may want to try it for practice: Use the entire trip interval (*from* just after launch at ground, *to* just before impact at ground) or the interval on the way down (*from* peak, *to* just before impact at ground). ■

4.6 WHEN BOTH INITIAL AND FINAL VELOCITIES ARE AT ANGLES

If both velocity vectors are at angles, then we need right triangles for both. In this kind of problem, it is especially easy to make a mistake with the *signs* of v_{0y}, v_y, and y. Here are the four possible kinds of path, along with the signs of these quantities for each path:

Path	Sign of . . . (Assume *up* is *positive.*)		
	v_{0y}	v_y	y
(i) Moves *upward,* then *downward;* ends *higher* than it starts.	+ (Moves *upward*)	− (Moves *downward*)	+ (Ends *higher*)

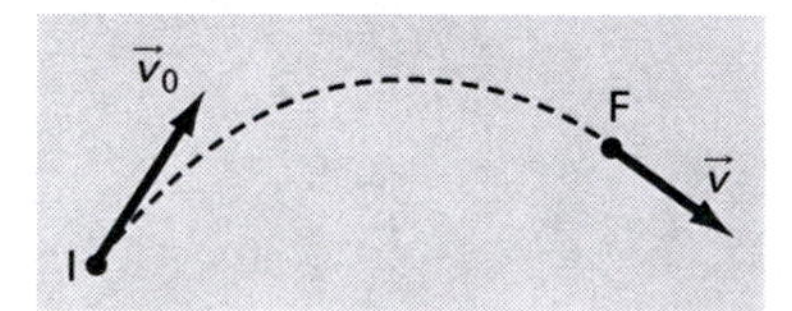

Path	**Sign of . . .** You fill in the blanks. Answers at end.		
	v_{0y}	v_y	y
(ii) Moves *upward,* then *downward;* ends *lower* than it starts. 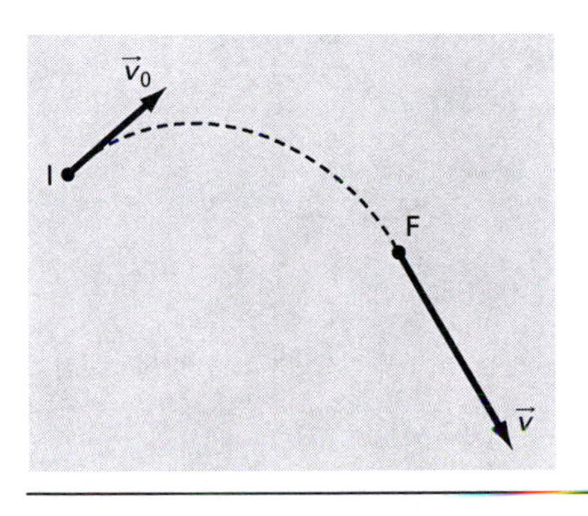	*You try!*	*You try!*	*You try!*
(iii) Always moves *upward;* ends *higher* than it starts. 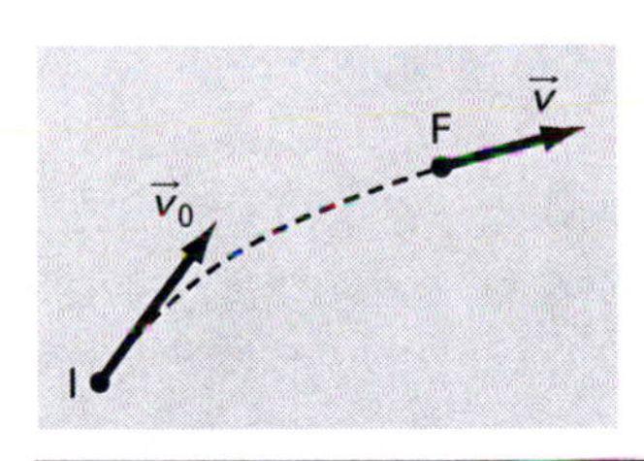	*You try!*	*You try!*	*You try!*
(iv) Always moves *downward;* ends *lower* than it starts. 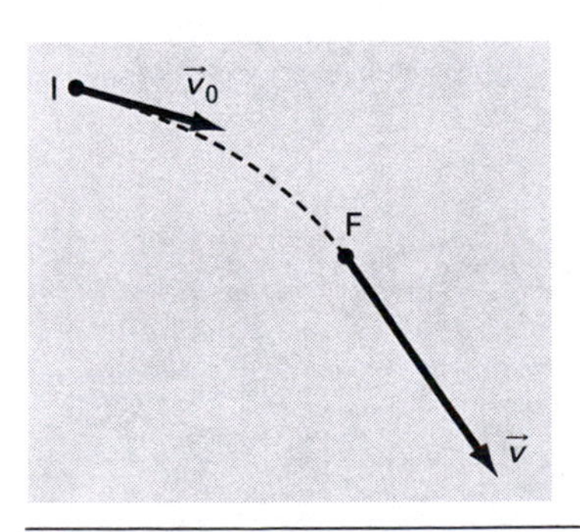	*You try!*	*You try!*	*You try!*

Answers

	Sign of . . .		
Path	v_{0y}	v_y	y
(ii)	+ (Moves *upward*)	– (Moves *downward*)	– (Ends *lower*)
(iii)	+ (Moves *upward*)	+ (Moves *upward*)	+ (Ends *higher*)
(iv)	– (Moves *downward*)	– (Moves *downward*)	– (Ends *lower*)

EXERCISE 4.4

A ball is launched from ground level toward a tall vertical wall that is a horizontal distance of 7.0 m away. The initial velocity is 11 m/s at an angle of 30 degrees above the horizontal. (a) How high on the wall does the ball hit? (b) When it hits the wall, is the ball still on its way up, or is it coming back down?

Solution

(1) Type of problem

Projectile motion, so use the following:

- Right triangle trig for velocity vectors
- Equation 2.2 for the constant velocity horizontal motion
- Equations 2.3a–d for the constant acceleration vertical motion

(2) Sort by . . . and (3) Equations & unknowns

Velocity vectors

Look back at the four possible paths in the table. The motion here is either (i) or (iii) because it *initially moves upward,* ending on the wall *higher* than it started. When it finally hits the wall, we don't know whether it is moving *downward* (i) or *upward* (iii)—that is one of the questions. The answer comes from the *sign* of v_y, the final velocity y-component.

We start with the initial velocity vector at its own private origin, with its known and unknown quantities:

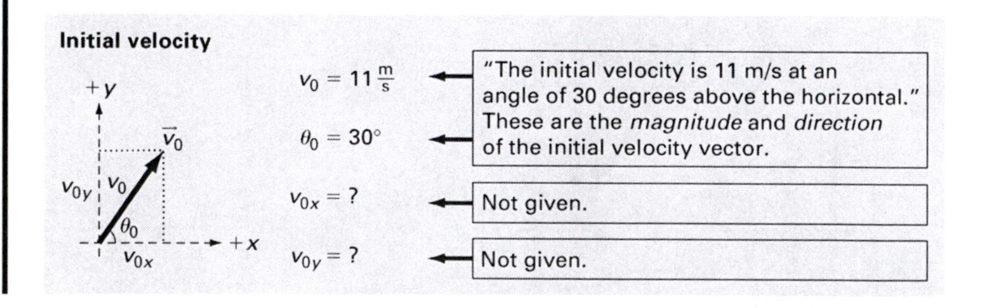

The right triangle and trig equations:

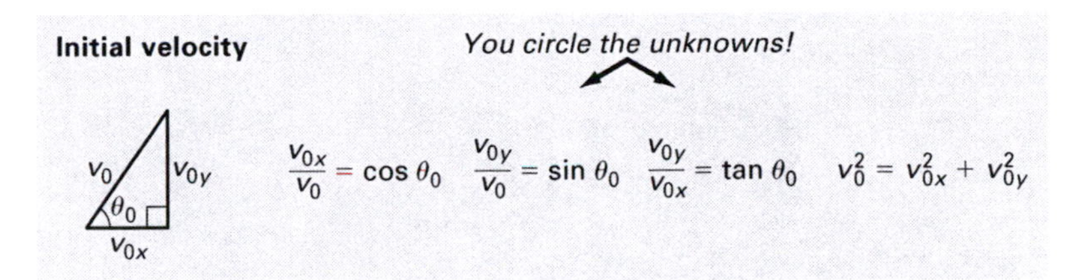

We show both possibilities for the final velocity vector, with positive OR negative v_y, since we don't know which one:

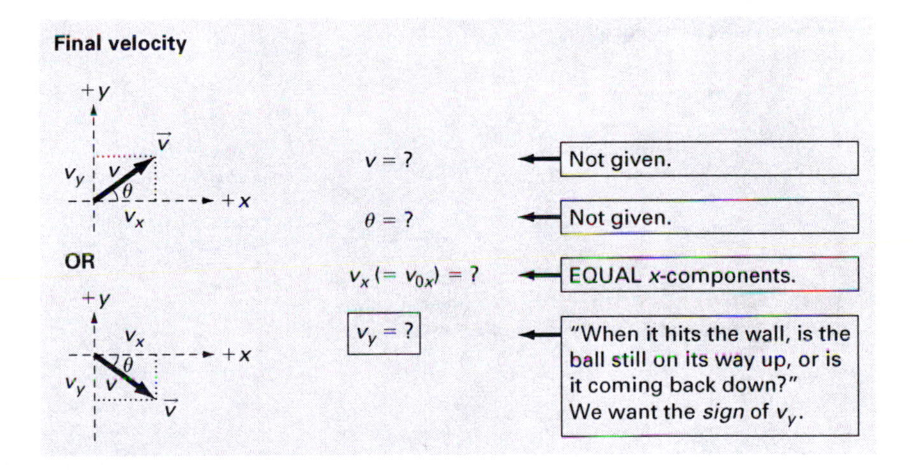

The right triangle and equations:

Final velocity

OR

$$\frac{v_x}{v} = \cos\theta \quad \frac{v_y}{v} = \sin\theta \quad \frac{v_y}{v_x} = \tan\theta \quad v^2 = v_x^2 + v_y^2$$

Horizontal motion

After mentally separating the horizontal and vertical motions, we draw the interval for the horizontal, constant velocity motion:

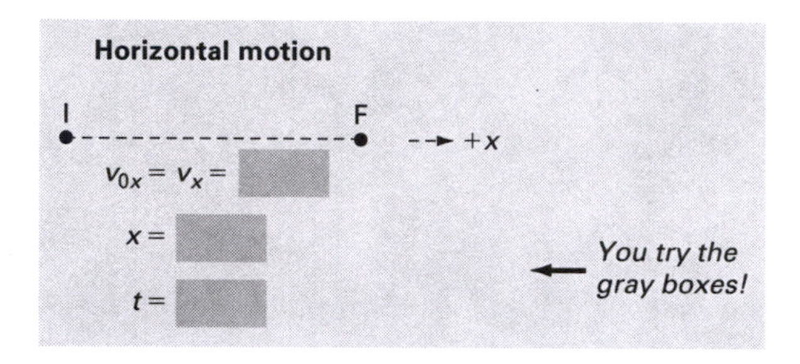

The constant velocity equation:

Horizontal motion
Modified Equation 2.2

$$v_x = \frac{x}{t}$$

You circle the unknowns!

Vertical motion

We show both possibilities for the vertical motion interval:

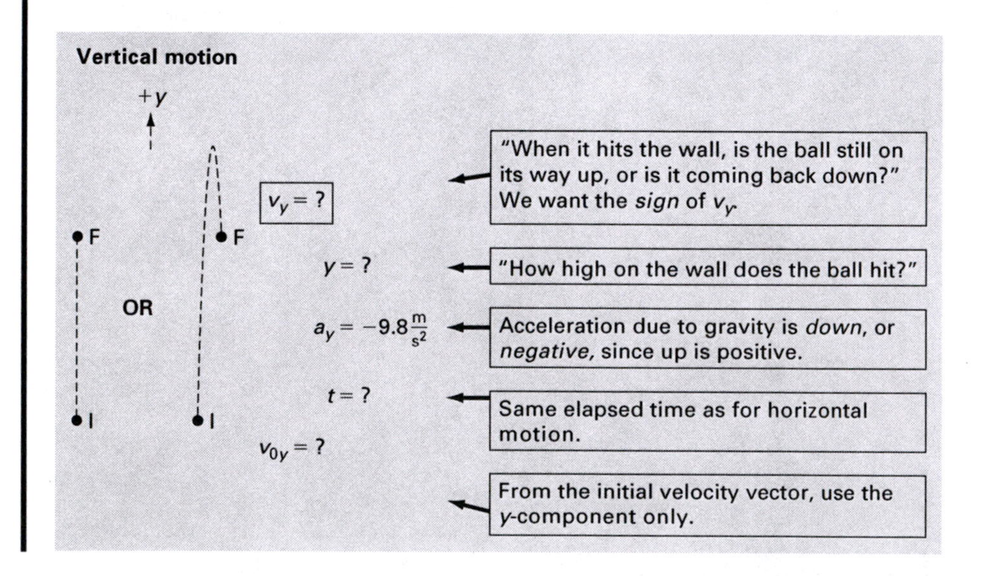

The constant acceleration equations:

Vertical motion
Modified Equations 2.3a–d

$$y = v_{0y}t + \frac{1}{2}a_y t^2$$

$$y = \left(\frac{v_{0y} + v_y}{2}\right)t$$

$$v_y = v_{0y} + a_y t$$

$$v_y^2 = v_{0y}^2 + 2a_y y$$

(4) Outline solution

The three sets of equations:

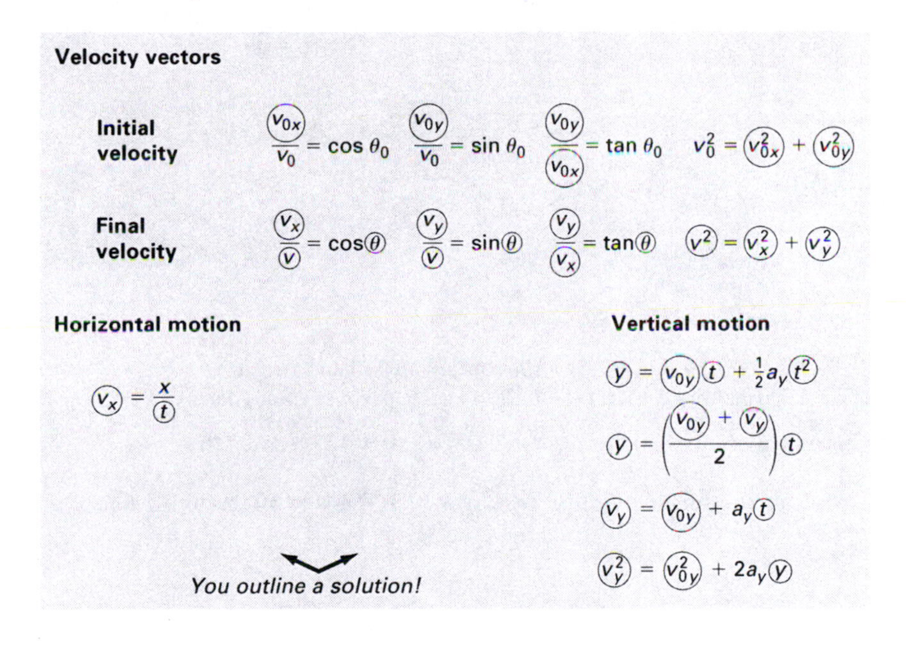

Answers for gray boxes

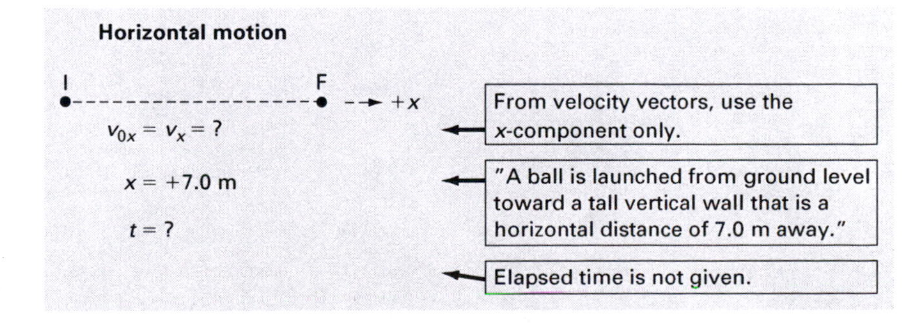

Answers for circle unknowns & (4) Outline solution
Here is one possible outline:

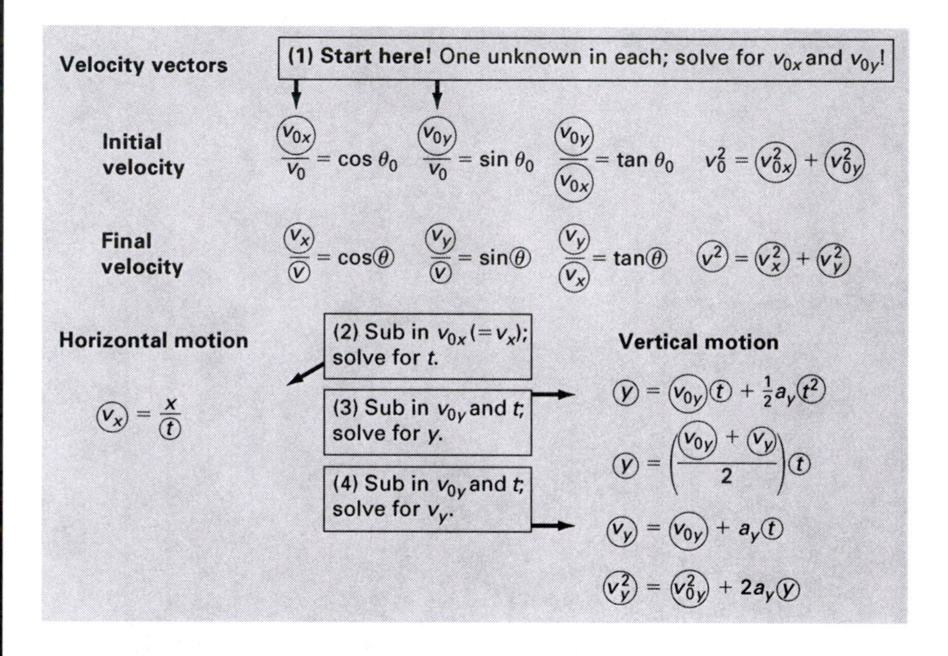

Answers

- $y \cong +1.39 \cong +1.4$ m (*positive* means *above* launch height).
- On the way *down* ($v_y \cong -1.70 \cong -1.7\frac{\text{m}}{\text{s}}$, *negative* means *down*).

Intermediate answers: $v_x = v_{0x} \cong +9.53\frac{\text{m}}{\text{s}}$, $v_{0y} = +5.5\frac{\text{m}}{\text{s}}$, $t \cong 0.735$ s

We could also solve for the final velocity: $v \cong 9.7\frac{\text{m}}{\text{s}}$ at $\theta \cong 10°$ below the horizontal. ■

CHAPTER 5

FORCE AND NEWTON'S LAWS OF MOTION

NEWTON'S THREE LAWS of motion are used in problems that involve force and acceleration. We will occasionally discuss Newton's first and third laws, but our focus will be on how to solve problems using Newton's second law:

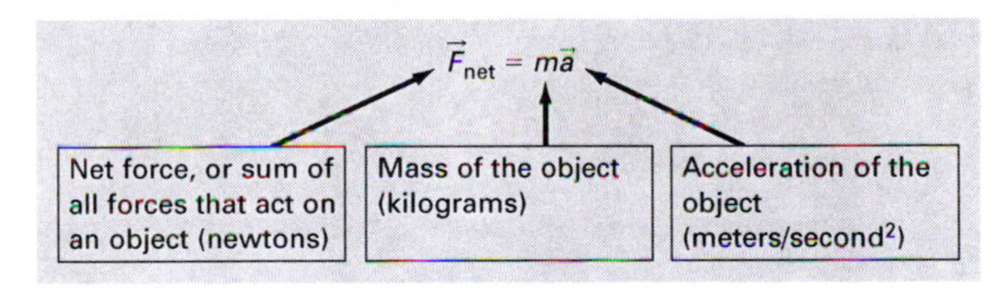

For doing calculations, we do not use this *vector* equation, but rather the two *component equations* that come from it:

$\vec{F}_{net} = m\vec{a}$ contains both component equations

$$F_{net,x} = ma_x \qquad (5.1a)$$

and

$$F_{net,y} = ma_y \qquad (5.1b)$$

Newton's second law is often MISUNDERSTOOD as equating two different forces, $\vec{F}_{net}$ and $m\vec{a}$. The right side, $m\vec{a}$, is NOT a force. The equation is one of *cause and effect:*

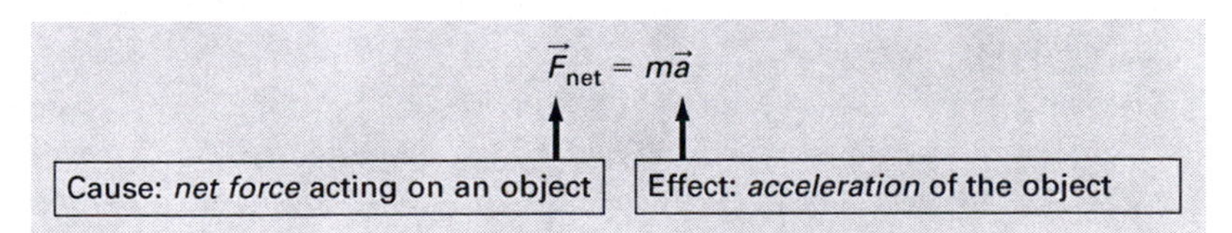

How do we identify the *actual forces* acting on an object? Before solving any problems, we will answer this question in the next section.

5.1 HOW TO DRAW A FREE-BODY DIAGRAM (FBD)

The first and most important step in solving any force problem is to mentally identify all the force vectors that act *on an object,* and then draw these vectors together in a free-body diagram, or FBD. There are a few main types of force that show up in physics problems, all of which act *on the block* here:

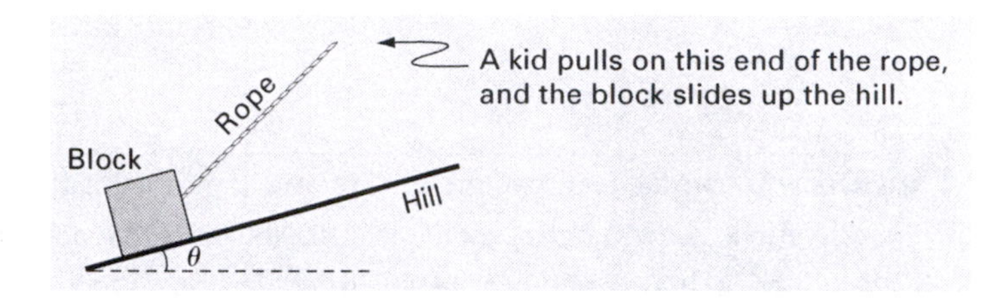

We look for two types of force: *long-range* and *contact.* And for each force that acts on the block, we can identify some *other specific object* that causes it.

Long-Range Forces

There is one *other specific object* that causes a long-range force to act on the block:

- The *entire earth* causes the gravitational force, or weight, $\overrightarrow{Wt}$, to act on the block. Direction: *vertically downward.*

Weight is a long-range force because the block does NOT need to be in direct contact with the earth in order for this force to act.

For now, gravitational force is the only long-range force. Later in physics, there are others, like electrostatic force and magnetic force.

Contact Forces

Here we look for objects that *touch* the block: the *rope* and the *hill,* but NOT the kid or the entire earth. The rope and hill cause a total of three contact forces to act *on the block:*

- The *rope* causes a "tension" force, $\vec{T}$, on the block. Direction: *along the rope away from the block.*
- The hill *surface* causes a "normal" force, $\vec{n}$, on the block. Direction: *perpendicular to the contact surface, and toward the block. Normal* means "perpendicular."
- The hill *surface* also exerts a "friction" force on the block. Direction: parallel to the surface. More specifically, the block is sliding, so it is a "kinetic friction" force, $\vec{f}_k$. Direction: *opposite the block's sliding motion* or *parallel to the surface, down the hill.*

  (We could instead have "static friction," if the block were at rest. More on that later.)

The *kid* has *no direct contact* with the block and so exerts *no force on the block*. The kid pulls on the rope, which in turn pulls on the block, and so we indirectly take the kid into account when we include the tension force caused by the rope.

EACH FORCE ACTING ON AN OBJECT IS CAUSED BY SOME OTHER SPECIFIC OBJECT!

If it is a real force acting on our object (the block), then we *must* be able to identify some *other specific object* (the earth, the surface, or the rope) as *the cause* of the force. With this rule, we can be sure about which forces act on an object. Think concretely in identifying forces.

Now we picture all of the force vectors that we have identified as acting on the block:

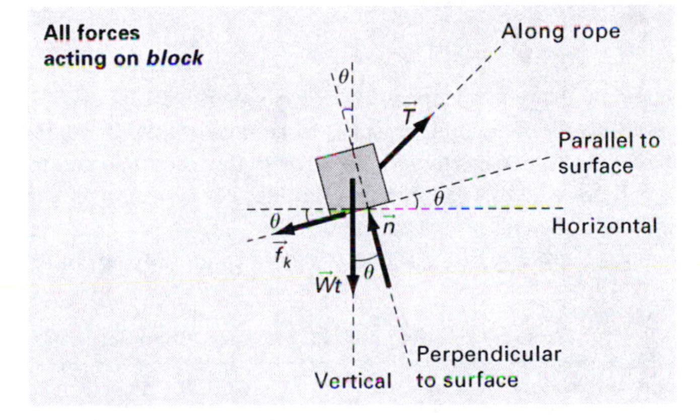

Drawing the Free-Body Diagram (FBD)

To solve physics problems, we draw a modified version of the previous figure. We call it a free-body diagram or, FBD:

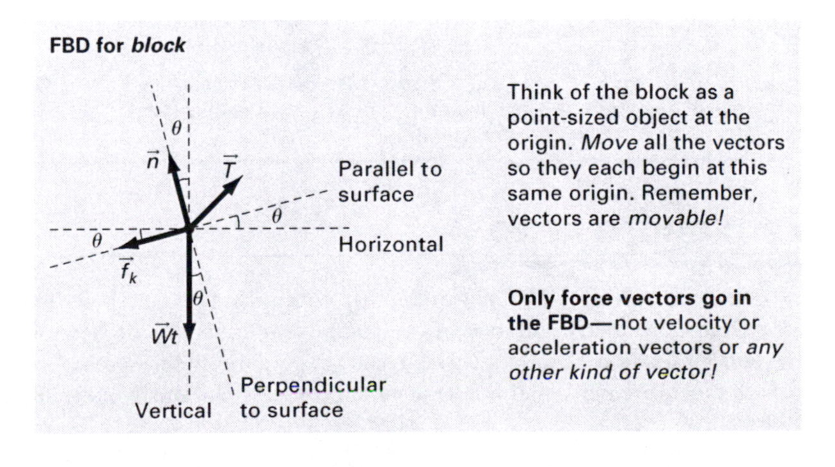

The dotted lines are just for reference.

HOW DOES NEWTON'S THIRD LAW FIT IN?

Look at the tension force: The *rope* pulls *on the block.* Direction: along the rope, away from the block. Newton's third law says that there is also another equally sized tension force: The *block* pulls *on the rope,* in the opposite direction, along the rope, toward the block. That is absolutely true!

However, we care only about the first force and completely ignore the second. Why? We are focusing on the *block,* and the only forces that affect the *block* are the forces that act *on the block.*

The second force acts *on the rope* and it affects the *rope.* We might consider this force in a problem in which there are questions about forces that act *on the rope* or on something attached to the other end of the rope, like the kid.

Make sure that you understand how we constructed the FBD. If you do, you are well on your way to being able to solve force problems with confidence! If there is anything that is still unclear to you, go through this section again to clear up the details. We will not take this example any further, but an exercise with a similar situation is done near the end of this chapter.

Here is a summary of the forces we most commonly put in FBDs:

Type	Name	Magnitude	Direction	Cause
Long-range	Weight	$Wt = mg$	Vertically down	Entire earth
Contact	Tension	T	Parallel to string, away from object	String, cord, rope, chain, spring, etc.
	Normal	n	Perpendicular to surface, toward object	Surface
	Friction	$f_k = \mu_k n$ (kinetic friction)	Parallel to surface, opposite motion	Surface
		$f_{s,\max} = \mu_s n$ (maximum static friction)	Parallel to surface (more details later)	Surface

5.2 FORCES IN 1D

For one-dimensional (1D) force problems, with *only vertical* or *only horizontal* force vectors, we need only one of the component equations of Newton's second law: either Equation 5.1a *or* 5.1b. As usual, in solving these exercises, we will go through much greater detail and take much longer than you would on your own paper. Focus on the thought process which will really help when working more complicated problems.

EXERCISE 5.1

A block is suspended from a vertical cable. Determine the force with which the cable pulls on the block of mass 2.0 kg when the block has an upward acceleration of 3.0 m/s^2.

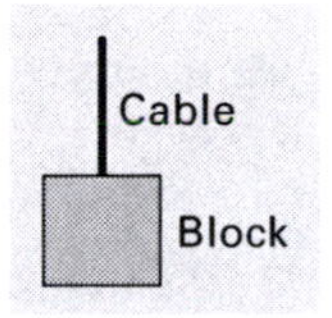

Solution

(1) Type of problem

Force and acceleration, vertical vectors only: Use Equation 5.1b.

(2) Sort by object

The question is about force acting *on the block,* so we focus on the *block* for which we sort the quantities in Equation 5.1b: *force, mass,* and *acceleration.*

We first mentally identify all the force vectors acting *on the block:*

Long-range forces

- Weight: Magnitude $Wt = mg$, direction *down* [caused by *the earth*].

Contact forces: The block *touches* only the cable:

- Tension: Magnitude $T = ?$, direction *up* (*along cable, away from block*) [caused by *the cable*].

Here is a summary of this *force* list, along with known and unknown details about the *mass* and *acceleration* of the object:

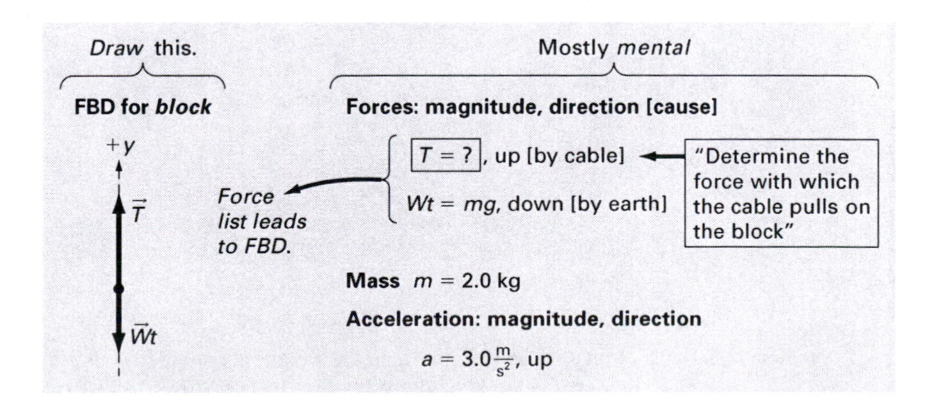

TENSION FORCE IS NOT ALWAYS EQUAL TO WEIGHT IN MAGNITUDE!

Do NOT *assume* that $T = Wt$ for an object supported by a string. Here, the tension magnitude must be larger than the weight magnitude in order to produce a net *upward* force that causes *upward* acceleration.

As much as possible, draw the vectors roughly to the same scale. This will help you solve the problem and understand the result.

(3) Equations & unknowns and (4) Outline solution

We first get the components of each force or acceleration vector, *using the axes on the FBD*. In this exercise, it is easy because all the vectors are along the y-axis:

- For any vector pointing in the positive y-direction, its y-component is equal to its positive magnitude (so $T_y = +T$ and $a_y = +a$). The acceleration vector is NOT part of the FBD, but we use the *same axis directions.*
- For any vector pointing in the negative y-direction, its y-component is equal to the negative of its magnitude (so $Wt_y = -Wt = -mg$).

We substitute these components into Equation 5.1b, simplify, circle the unknowns, and outline the solution:

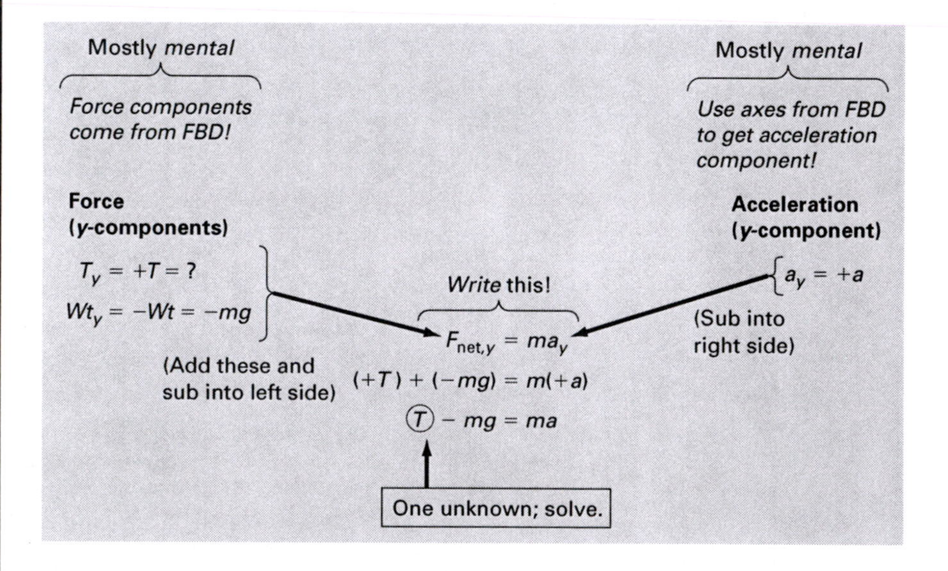

Answer $T \cong 25.6 \cong 26$ N

As predicted, the tension magnitude is *larger* than the weight magnitude (=19.6 N)! If we are consistent with signs, the algebra should always give us a *positive* result for a vector *magnitude.* ■

IS g POSITIVE OR NEGATIVE IN $Wt = mg$?

POSITIVE: $g = +9.8\frac{m}{s^2}$. The magnitude, $Wt = mg$, of the weight vector, $\overrightarrow{Wt}$, is *always positive.* We took care of the *downward* direction of $\overrightarrow{Wt}$ with the *negative* sign in the y-component: $Wt_y = -mg$. The math for this exercise gives us:

$$T = ma + mg = (2.0\text{ kg})\left(3.0\tfrac{\text{m}}{\text{s}^2}\right) + (2.0\text{ kg})\left(+9.8\tfrac{\text{m}}{\text{s}^2}\right) \cong 26\text{ N}$$

Sign written for emphasis

The negative sign went away in the algebra, but the downward direction was already accounted for. In the end, we substitute in $+9.8\frac{m}{s^2}$ for g.

5.3 HOW TO SET UP FORCE PROBLEMS

Here is a summary of the steps we followed in the previous exercise. As always, don't be restricted by them, but let them guide your thinking.

Force Problems—Mental and Written Steps

Mental→	**(1) Type of problem** If a problem involves force: Use right triangle trig for vectors at angles, and Equations 5.1a and/or 5.1b.
Mental and written→	**(2) Sort *by object*** For *each object* for which we know (or want to know) about forces, we sort out: • **Forces vectors (long-range and contact):** magnitude and direction of each; draw FBD with *x*- and/or *y*-axes • **Mass** • **Acceleration vector**: magnitude and direction
Mental and written→	**(3) Equations & unknowns** For each object: • **Force components:** Determine from FBD, then add and sub into *left side* of Equation 5.1a and/or 5.1b. • **Mass:** Sub into *right side* of Equation 5.1a and/or 5.1b. • **Acceleration components:** Use axes from FBD, then sub into *right side* of Equation 5.1a and/or 5.1b. Simplify and circle the unknowns.
Mental→	**(4) Outline solution**

Remember that most of this is done in your mind without ever making it onto paper. Write enough to be clear, for yourself and your instructor, but be efficient in what you actually write down. We will use these steps throughout the rest of the chapter.

5.4 MOTION INTERVALS IN FORCE PROBLEMS

If a force problem also includes questions related to a motion interval, then we use Chapter 2 methods to get motion equations to help solve the problem.

EXERCISE 5.2

A girl of mass 25 kg stands on a scale on the floor of an elevator that is initially at rest. The elevator starts accelerating, and during the acceleration the scale reads 210 N. (a) Is the acceleration upward or downward? (b) What is the magnitude of the acceleration? (c) What is the elevator's velocity after it has traveled 1.0 m?

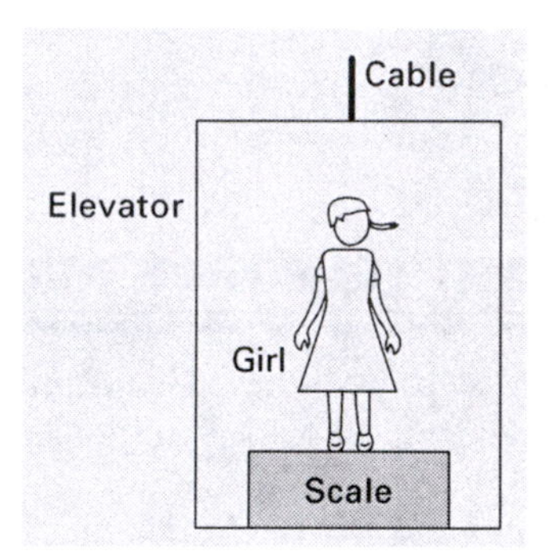

Solution

(1) Type of problem

This problem has two parts to set up separately, and later combine in the algebra at the end:

- **Force and acceleration:** Vertical only; use Equation 5.1b.
- **Motion interval:** Use modified Equations 2.3a–d.

(2) Sort by object and (3) Equations & unknowns

The question is about motion of the elevator. However, the problem gives the mass *of the girl*, $m = 25$ kg, which also tells us the gravitational force *on the girl* ($Wt = mg$). So we will actually focus *on the girl* whose motion is identical to that of the elevator.

WHAT DOES THE SCALE READING MEAN?

The 210-N scale reading means there is a downward force of 210 N acting *on the scale*, caused by the girl. However, by Newton's third law, there is also an upward force of 210 N acting *on the girl*, caused by the scale.

The force acting *on the girl* goes in our FBD *for the girl* as the *normal force*. The force acting *on the scale* would go in a FBD *for the scale* if we made one (we won't).

Force and acceleration

The force vectors acting *on the girl:*

Long-range forces

- Weight: Magnitude $Wt = mg$, direction *down* [caused by *the earth*].

Contact forces: The girl *touches* only the scale:

- Normal: Magnitude $n = 210$ N, direction *up* (*perpendicular to top surface of scale, toward girl*) [caused by *the scale*].

This leads to:

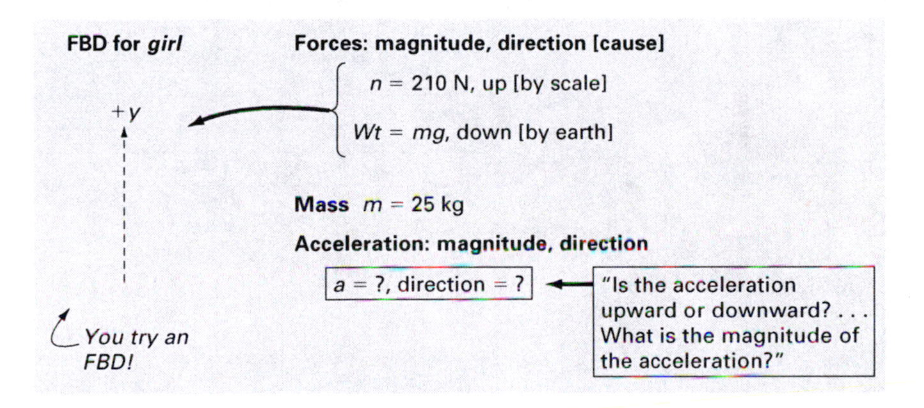

We can go ahead and answer part (a). The true weight of the girl is $Wt = mg = (25\text{ kg})(9.8\frac{\text{m}}{\text{s}^2}) \cong 245$ N. But she *feels* only 210 N, the *scale reading,* or her *apparent weight.* Your own experience in an elevator might tell you the acceleration must be downward. The numbers also tell us this: Since there is a 245-N downward force and a 210-N upward force, then the net force is 35 N downward. A *downward* net force means a *downward* acceleration.

Putting the components into Equation 5.1b:

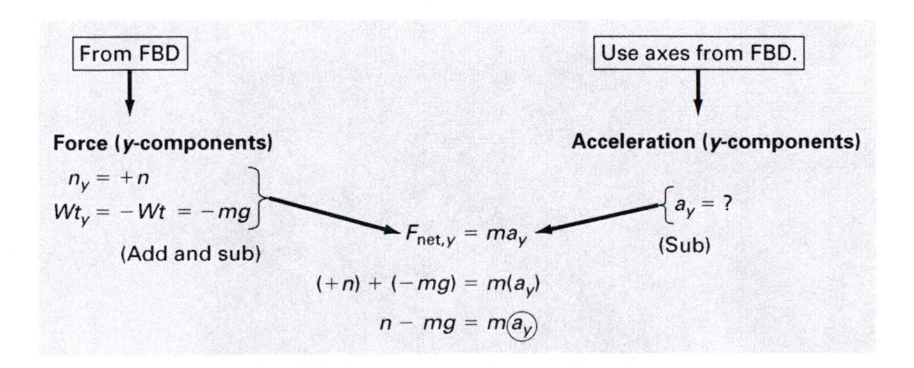

Motion interval

Now we sort the motion information:

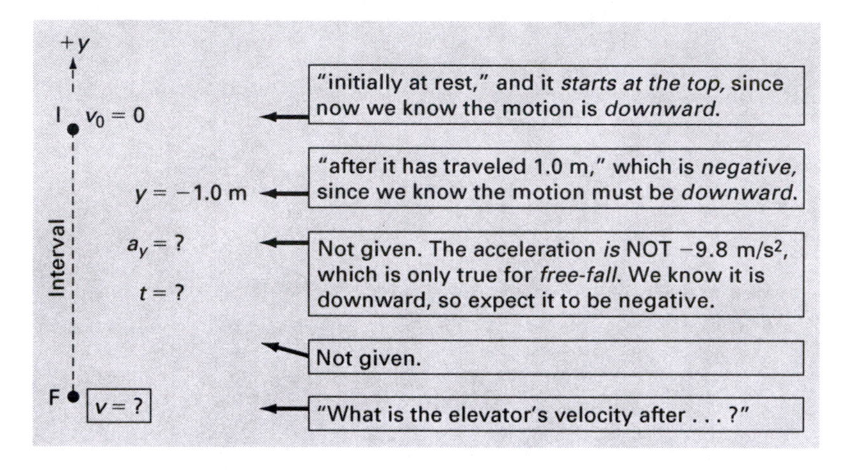

Modified Equations 2.3a–d:

$$y = v_0 t + \frac{1}{2} a_y t^2$$

$$y = \left(\frac{v_0 + v}{2}\right) t$$

$$v = v_0 + a_y t$$

$$v^2 = v_0^2 + 2 a_y y$$

(4) Outline solution

We put all the equations together:

Force and acceleration	Motion interval
$n - mg = m a_y$	$y = v_0 t + \frac{1}{2} a_y t^2$ $y = \left(\frac{v_0 + v}{2}\right) t$ $v = v_0 + a_y t$ $v^2 = v_0^2 + 2 a_y y$

You outline a solution!

Answers for . . .

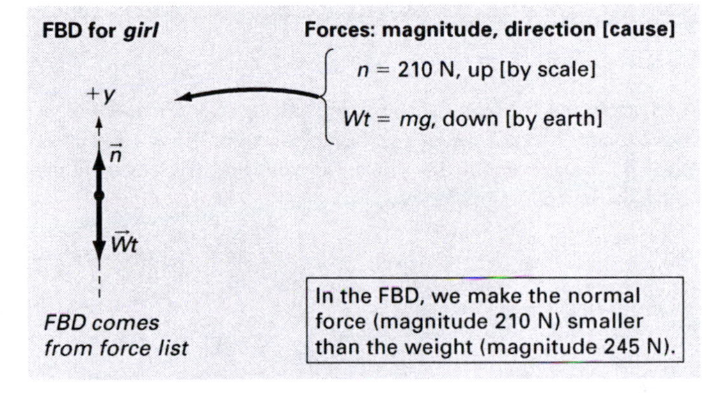

Answers for (4) Outline solution

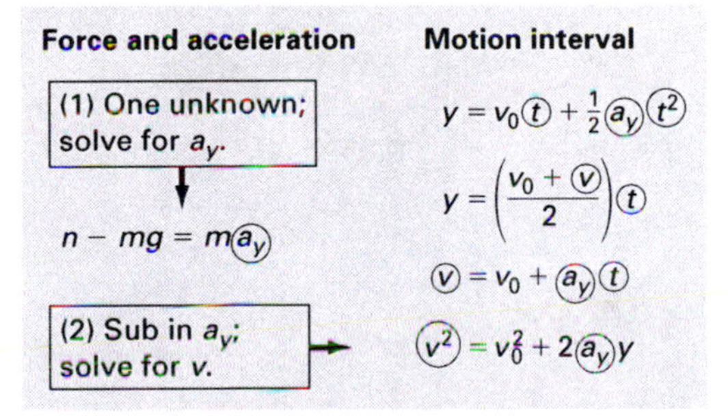

Answers

- Direction of acceleration: down. (We have already discussed other reasons, but now we also have $a_y = -1.4\frac{\text{m}}{\text{s}^2}$, where *negative* means *down.*)
- Magnitude of acceleration: $a = 1.4\frac{\text{m}}{\text{s}^2}$. (A vector magnitude is always positive.)
- $v \cong -1.67 \cong -1.7\frac{\text{m}}{\text{s}}$ (Be careful here! When solving for velocity using Equation 2.3d, taking the square root gives two possibilities: positive and negative. Take the *negative* root because the velocity is *down.* Check with the other motion equations: They give consistent results *only* if the velocity is *negative.*) ■

In any problem involving *force and acceleration* and a *motion interval,* we can relate the two parts of the problem through the *acceleration.* There are two possibilities (the previous exercise used the first):

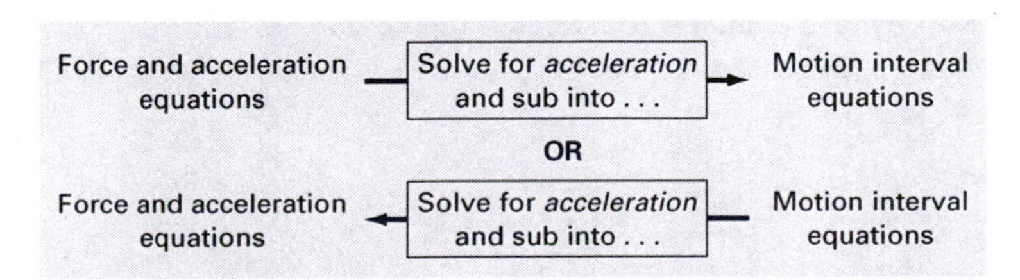

5.5 OBJECTS CONNECTED BY STRINGS, ROPES, AND SO ON

EXERCISE 5.3

Two blocks are connected by a light (massless) string over a massless, frictionless pulley as shown in the figure. Block 2 sits on a horizontal tabletop. Block 1 has mass 2.0 kg, and block 2 has mass 5.0 kg, and everything is at rest. Determine (a) the tension in the string and (b) the normal force on the bottom of block 2.

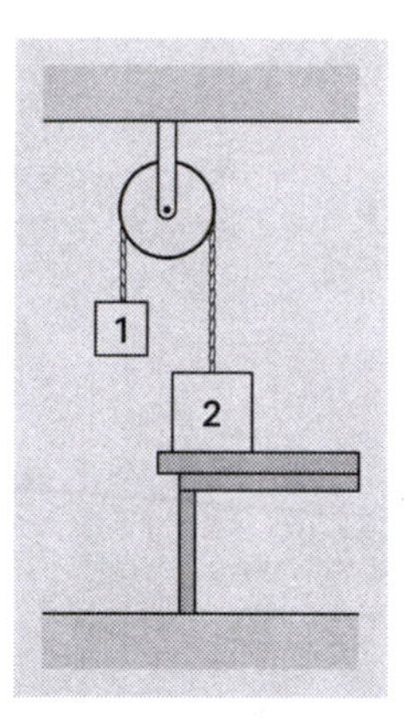

EQUILIBRIUM

Any object with *constant velocity,* which could be *constant zero velocity,* has *zero acceleration.* Remember: Acceleration is the rate of change of velocity. This means there is *zero net force* acting on the object ($\vec{F}_{\text{net}} = m\vec{a} = 0$).

$$\text{Equilibrium} \begin{cases} \vec{v} = \text{constant (may be zero)} \\ \vec{a} = 0 \\ \vec{F}_{\text{net}} = 0 \end{cases}$$

These three always go together. *Any one* of these means *the other two* are also true.

Solution

(1) Type of problem

Force, vertical only: Use Equation 5.1b.

(2) Sort by object and (3) Equations & unknowns

We have two objects for which we know, or want to know, about the forces: block 1 and block 2. We start with block 1.

Block 1

The forces:

Long-range forces acting *on block 1*

- Weight: Magnitude $Wt_1 = m_1 g$, direction *down* [caused by *earth*].

Contact forces acting *on block 1:* Block 1 *touches* only the string:

- Tension: Magnitude $T = ?$, direction *up* (*along string, away from block 1*) [caused by *string*].

This leads to:

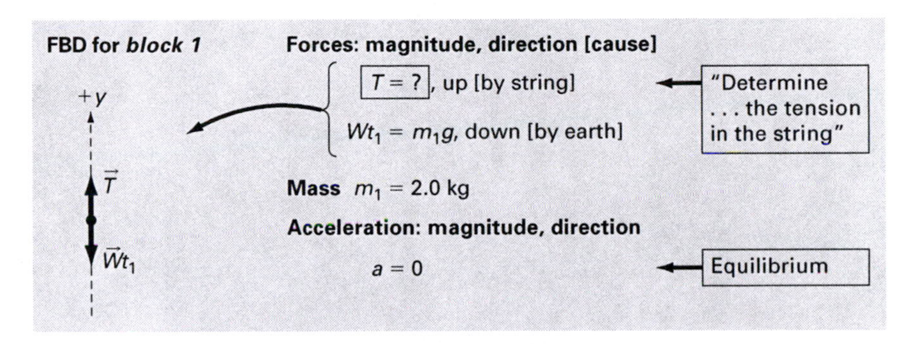

The tension pulling on an object (block 1) at one end of a string is the same magnitude as the tension pulling on another object (block 2) at the other end of the string, so we just call them both T (rather than T_1 and T_2). The accelerations are both zero.

Equation 5.1b for block 1:

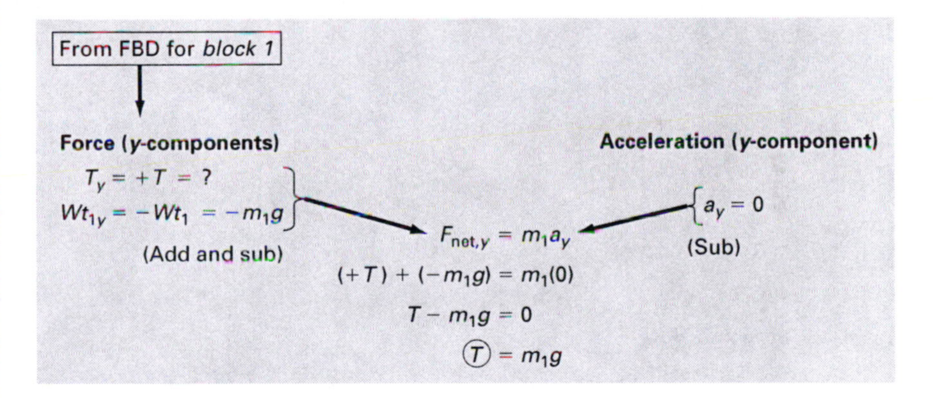

The simplified equation, $T = m_1g$, just says what we already know from the FBD: The magnitudes of the tension and the weight of block 1 are equal. If you see shortcuts like this, use them, but only if you are sure!

Block 2

The forces:

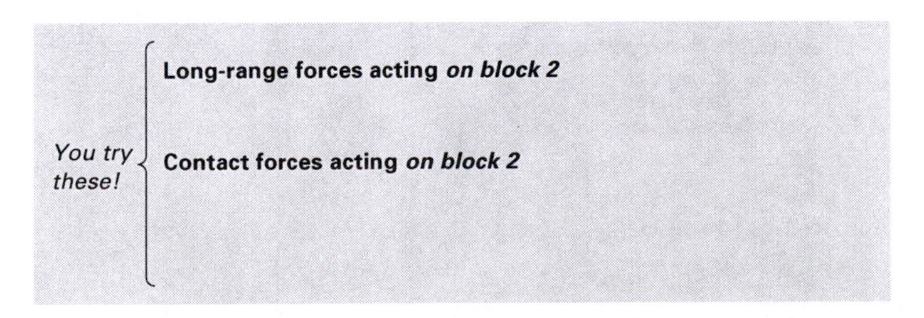

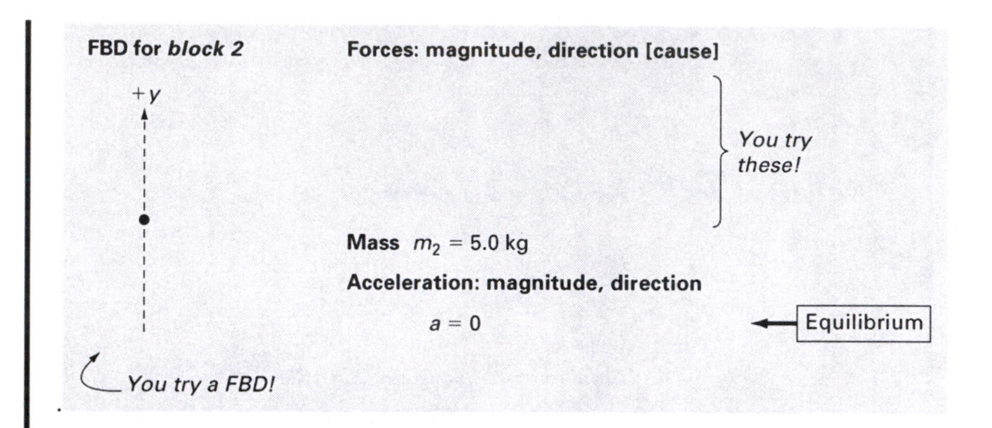

The y-components and Equation 5.1b for block 2:

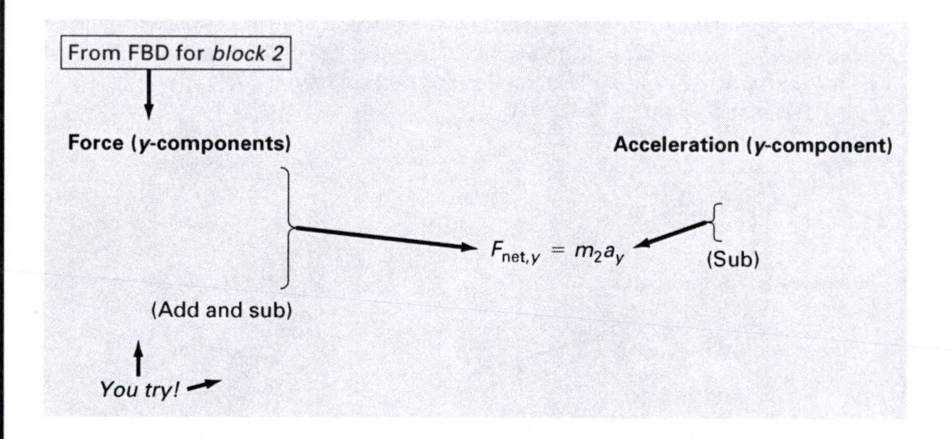

(4) Outline solution

Once *set up separately* for each block, the simplified equations are *solved together:*

Block 1 $\textcircled{T} = m_1 g$ ← (1) One unknown; solve for T.

Block 2 $\textcircled{T} + \textcircled{n} - m_2 g = 0$ ← (2) Sub in T; solve for n.

Answers for . . .

Long-range forces acting *on block 2*

- Weight: Magnitude $Wt_2 = m_2 g$, direction *down* [caused by *earth*].

Contact forces acting *on block 2:* Block 2 *touches* both the string and the tabletop surface:

- Tension: Magnitude $T = ?$, direction *up* (*along string, away from block 2*) [caused by *string*].
- Normal: Magnitude $n = ?$, direction *up* (*perpendicular to surface, toward block 2*) [caused by *tabletop*].

Answers for . . .

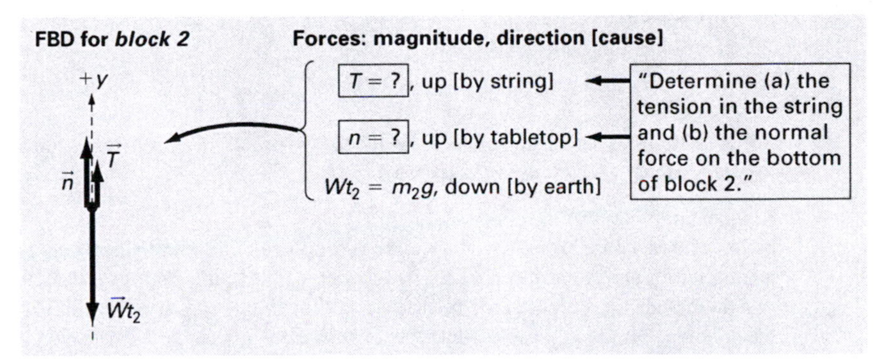

Answers for block 2

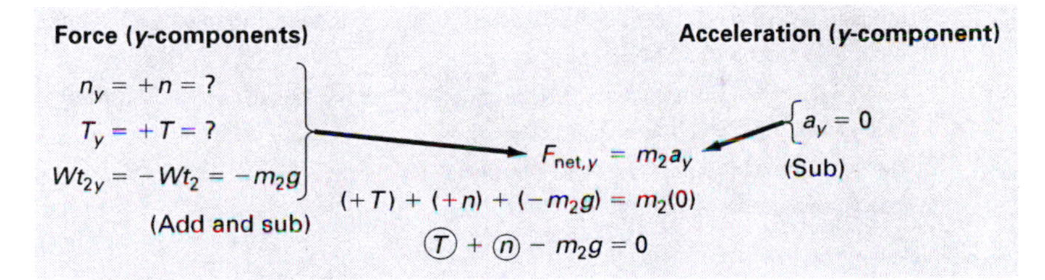

Answers $T \cong 19.6 \cong 20$ N, $n \cong 29.4 \cong 29$ N ■

NORMAL FORCE IS NOT ALWAYS EQUAL TO WEIGHT!

Do NOT assume that $n = Wt$ for an object supported by a surface. In this exercise, for block 2, they are clearly *not equal* in magnitude: $n \cong 29$ N, $Wt_2 = m_2g \cong 49$ N.

5.6 FORCES IN 2D

When we have forces at angles, or both vertical and horizontal forces, the problem is two-dimensional (2D), and we need both Equations 5.1a *and* 5.1b.

EXERCISE 5.4

A block with a weight of 25 N hangs from vertical cable 1, which is attached to cables 2 and 3, which are attached to the ceiling, as shown in the figure. Determine the tensions in cables 2 and 3.

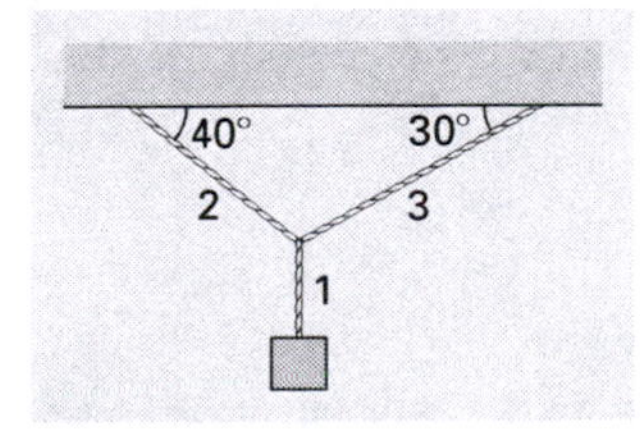

Solution

(1) Type of problem

Force, 2D: Use Equations 5.1a and 5.1b.

(2) Sort by object

We are looking at forces that all act *on the point where the cables meet.* It helps to think of that point as a *knot,* so that we look at forces *on the knot:*

Long-Range Forces
None! This is a little bit subtle: The weight is the force of the earth pulling *on the block,* which in turn pulls on cable 1. Then, *by contact,* cable 1 pulls *on the knot.* However, if you do include the weight of the block here, although not *exactly* correct, it is okay for the solution. Just do not *also* include the tension caused by cable 1 in the next list.

Contact Forces: The knot *touches* all three cables, so there are three tension forces:

- Tension: Magnitude $T_1 = 25$ N, direction *down* (*along cable 1, away from the knot*) [caused by *cable 1*]. Because we have *equilibrium,* the tension in cable 1 is equal to the weight of the block, 25 N.
- Tension: Magnitude $T_2 = ?$, direction 40° above negative x-axis (*along cable 2, away from the knot*) [caused by *cable 2*].
- Tension: Magnitude $T_3 = ?$, direction 30° above positive x-axis (*along cable 3, away from the knot*) [caused by *cable 3*].

Before we draw the FBD:

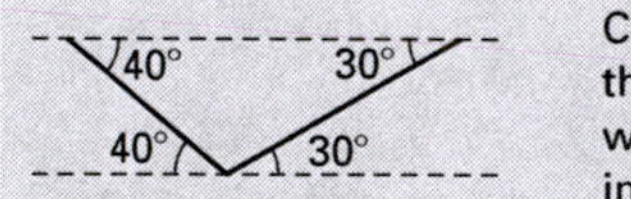

Compare these angles to those in the figure given with the problem and those in the FBD that follows.

The FBD and information:

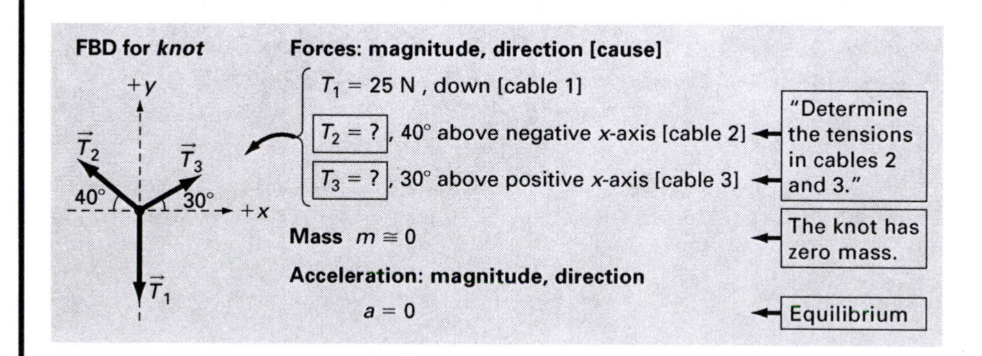

(3) Equations & unknowns

In 2D, we need to get the x- AND y-components of each vector so we can substitute into Equations 5.1a AND 5.1b. The components of $\vec{T}_1$ are simple. However, we need right triangles to get the components of $\vec{T}_2$ and $\vec{T}_3$ in terms of the unknown magnitudes, T_2 and T_3. Enlarged views of the triangles:

T_{2y}, T_2, $40°$, T_{2x}

$$\frac{T_{2x}}{T_2} = \cos 40° \Rightarrow T_{2x} = -T_2 \cos 40°$$ ← Make this *negative,* since it is *left.*

$$\frac{T_{2y}}{T_2} = \sin 40° \Rightarrow T_{2y} = +T_2 \sin 40°$$ ← Make this *positive,* since it is *up.*

T_3, $30°$, T_{3x}, T_{3y}

$$\frac{T_{3x}}{T_3} = \cos 30° \Rightarrow T_{3x} = +T_3 \cos 30°$$ ← Make this *positive,* since it is *right.*

$$\frac{T_{3y}}{T_3} = \sin 30° \Rightarrow T_{3y} = +T_3 \sin 30°$$ ← Make this *positive,* since it is *up.*

As usual, we ignore signs during *trig calculations,* and then put the correct sign on each component in the end. All x-components go into Equation 5.1a:

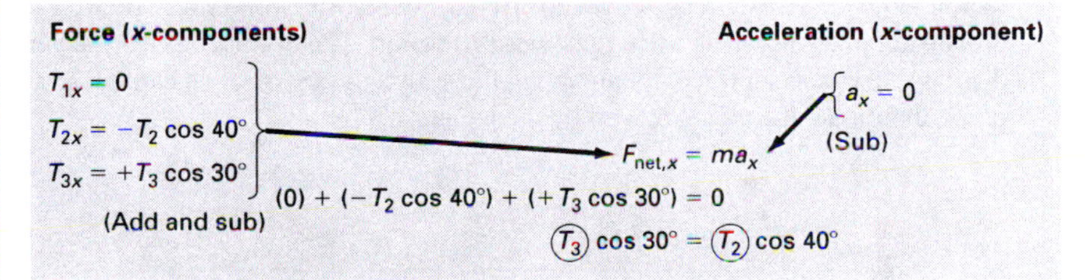

And y-components into Equation 5.1b:

Force (y-components) **Acceleration (y-component)**

$T_{1y} = -T_1$

$T_{2y} = +T_2 \sin 40°$

$T_{3y} = +T_3 \sin 30°$

(Add and sub) → $F_{net,y} = ma_y$ ← (Sub) $a_y = 0$

$$(-T_1) + (+T_2 \sin 40°) + (+T_3 \sin 30°) = 0$$

$$T_2 \sin 40° + T_3 \sin 30° = T_1$$

(4) Outline solution

We *set up* each component *separately,* and then *solve together* the simplified equations:

From x-component equation $T_3 \cos 30° = T_2 \cos 40°$

You try an outline!

From y-component equation $T_2 \sin 40° + T_3 \sin 30° = T_1$

Answers for (4) Outline solution

One possible outline:

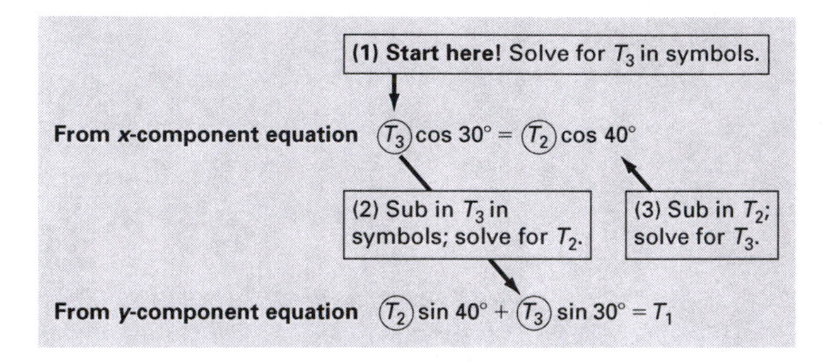

Answers $T_2 \cong 23.04 \cong 23$ N, $T_3 \cong 20.4 \cong 20$ N ■

5.7 SLIDING—KINETIC FRICTION

If a block slides along a surface with friction, the *sliding,* or *kinetic* friction force, $\vec{f}_k$, is parallel to the surface and opposes the motion. The normal force, $\vec{n}$, is perpendicular to the same surface. Even though these two forces are at right angles to each other, their magnitudes are related by the equation:

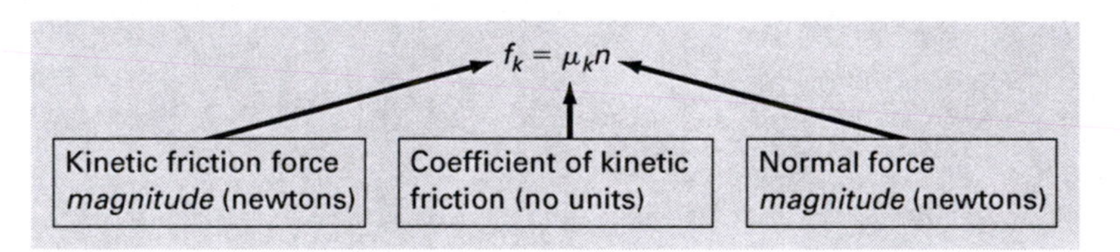

Because these two forces are at right angles, we need to deal with both x- and y-directions to get our equations.

EXERCISE 5.5

Two blocks are connected by a light string over a massless, frictionless pulley as shown in the figure. Block 1 sits on a horizontal table and slides toward the pulley as the blocks each move with the same constant speed. The coefficient of kinetic friction between table and block 1 is $\mu_k = 0.15$. Block 1 has mass 2.0 kg. Determine the mass of block 2.

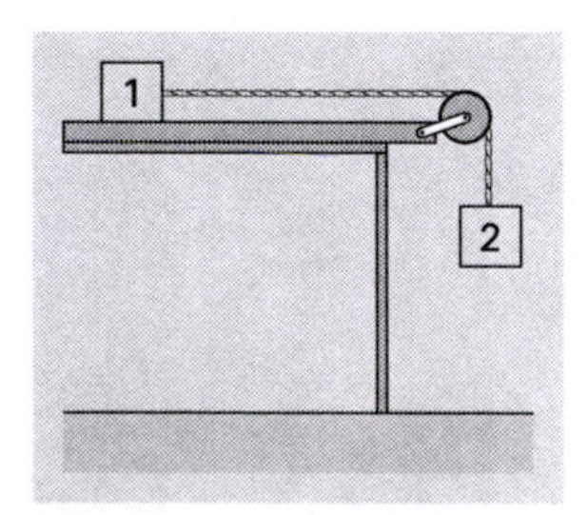

Solution

(1) Type of problem

Force, including friction: Use Equations 5.1a *and* 5.1b.

(2) Sort by object and (3) Equations & unknowns

We have two objects for which we know, or want to know, about the forces: blocks 1 and 2.

Block 1

We start with force vectors acting on block 1, which moves to the right.

Long-range forces acting *on block 1*

- Weight: Magnitude $Wt_1 = m_1 g$, direction *down* [caused by *earth*].

Contact forces acting *on block 1:* Block 1 *touches* both the string (one force) and the table surface (two forces):

- Tension: Magnitude T = ?, direction *right* (*along string, away from block 1*) [caused by *string*].
- Normal: Magnitude n = ?, direction *up* (*perpendicular to surface, toward block 1*) [caused by *table surface*].
- Kinetic friction: Magnitude $f_k = \mu_k n$ = ?, direction *left* (*parallel to surface, opposite the motion of block 1*) [caused by *table surface*].

THE TABLE SURFACE CAUSES TWO FORCES ON BLOCK 1!

When friction is present, the table surface causes both the normal force and the friction force.

The FBD and other information for block 1:

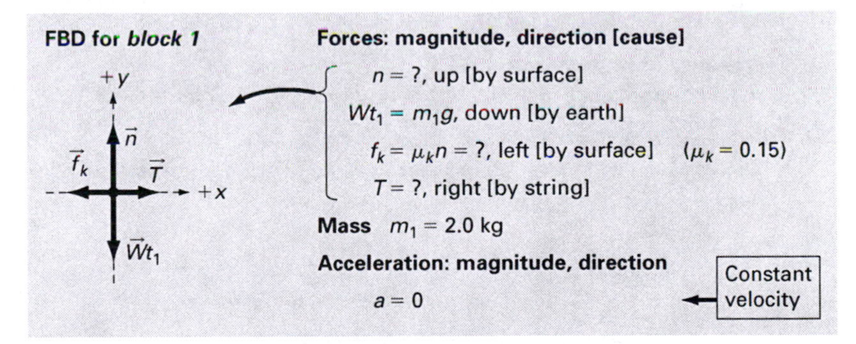

Because the acceleration is zero, we know the net force is zero, and so draw $\vec{T}$ and $\vec{f}_k$ the same size, and $\vec{n}$ and $\vec{Wt}_1$ the same size.

CONSTANT VELOCITY MEANS ZERO ACCELERATION AND ZERO NET FORCE!

Once the block is moving it does NOT need any net force to *keep it moving* with constant velocity! This is due to the block's inertia (Newton's first law).

We need both Equations 5.1a (for x-components) and 5.1b (for y-components) for block 1:

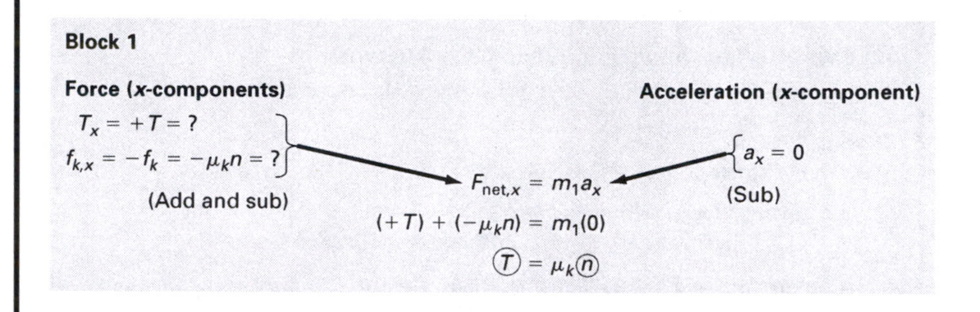

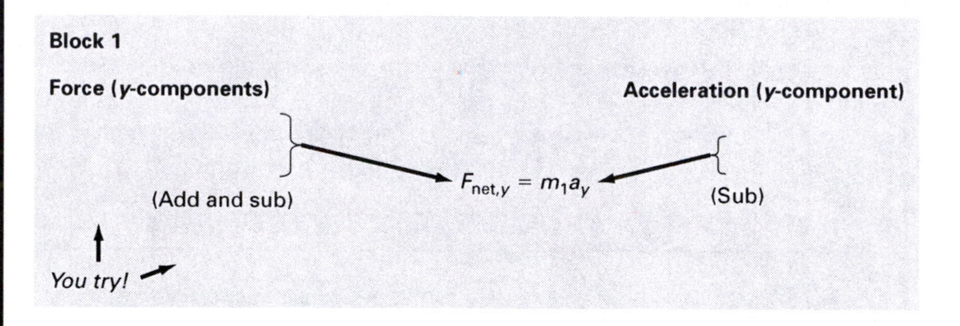

Block 2

The forces:

Long-range forces acting *on block 2*

- Weight: Magnitude $Wt_2 = m_2 g$, direction *down* [caused by *earth*].

Contact forces acting *on block 2*: Block 2 *touches* only the string:

- Tension: Magnitude $T = ?$, direction *up* (*along string, away from block 2*) [caused by *string*].

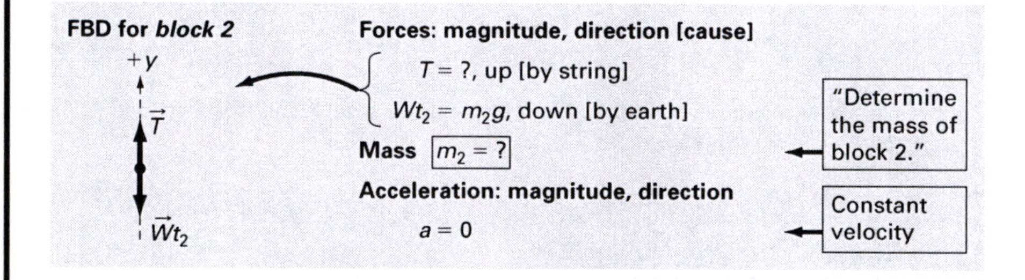

Everything is along the y-axis, so we need only Equation 5.1b:

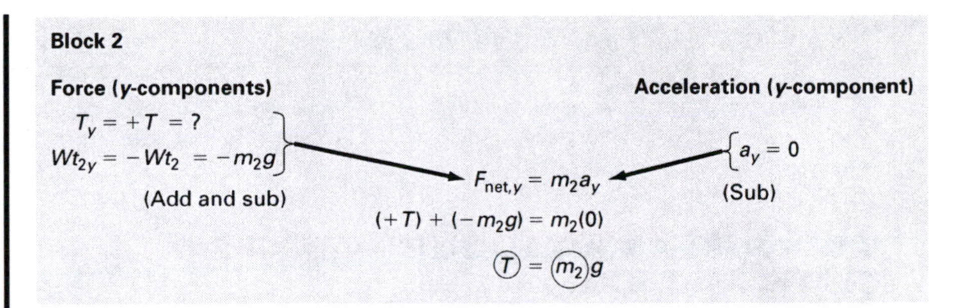

(4) Outline solution

Once the simplified equations are *set up separately* for each block, we solve them *together:*

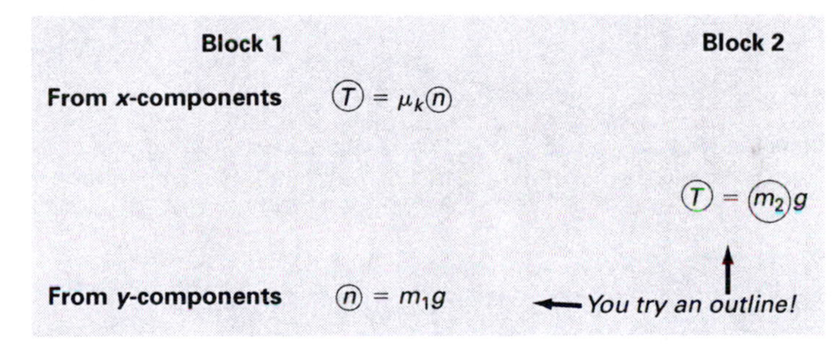

Answers for . . .

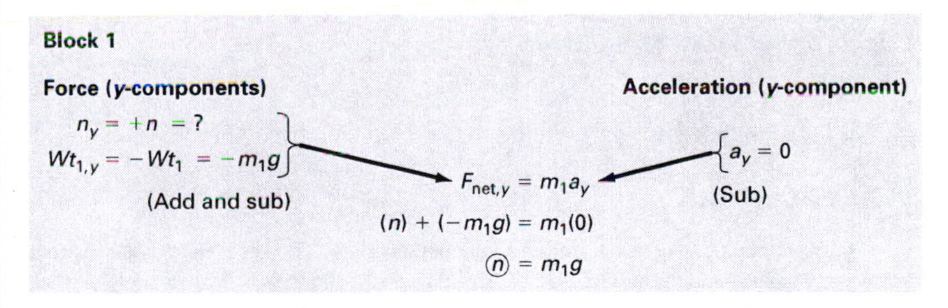

Answers for (4) Outline solution

	Block 1		Block 2
From x-components	$T = \mu_k n$		
	(1) Sub in for n in symbols; solve for T.	(2) Sub in T; solve for m_2. →	$T = m_2g$
From y-components	$n = m_1g$		

Answer $m_2 = 0.30$ kg ■

HOW TO SOLVE FRICTION PROBLEMS

Use the y-component equation and simplify to get an equation for n. Then substitute n into the friction term ($\mu_k n$) in the x-component equation. This gets rid of n as an unknown.

5.8 "JUST ABOUT TO SLIP"—MAXIMUM STATIC FRICTION

When an object is "just about to slip" or "on the verge of slipping," there is a *static* friction force that is at its *maximum possible value:*

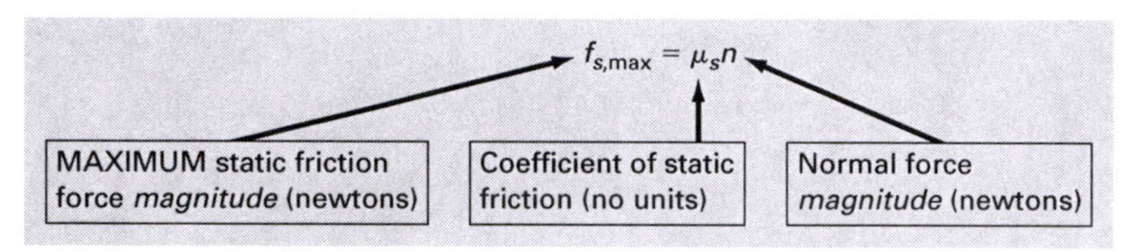

THIS EQUATION ONLY WORKS FOR MAXIMUM STATIC FRICTION!

For static friction that is *not* at its maximum value, there is no special equation: It is just $\vec{f}_s$ and we deal with it like any other known or unknown force.

The solution process is very similar to that in kinetic friction problems.

EXERCISE 5.6

A block of mass 2.0 kg sits at rest on a horizontal surface. The coefficient of static friction between the block and the surface is 0.35. A worker pushes on the block at an angle of 30 degrees below the horizontal. With what force must the worker push so that the block just begins to slide?

Solution

(1) Type of problem

Force, including friction: Use Equations 5.1a *and* 5.1b.

(2) Sort by object

It helps to picture the situation:

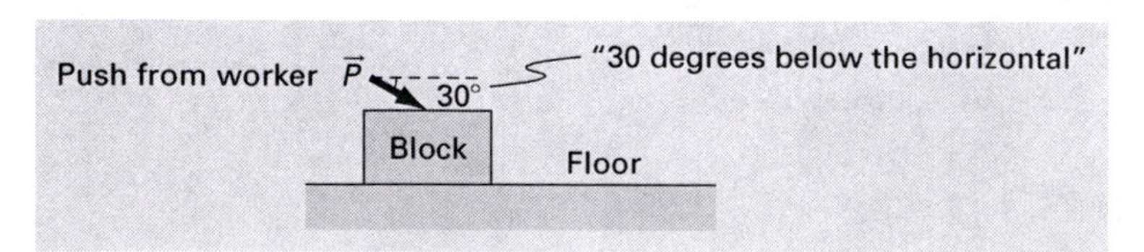

Because the block "just begins to slide," it is on the verge of slipping but not actually moving yet, so the *velocity and acceleration are both zero!*

If the block *does* slip, it will move to the *right.* The static friction force *opposes* any slipping, and so must be to the *left.*

The forces acting on the block:

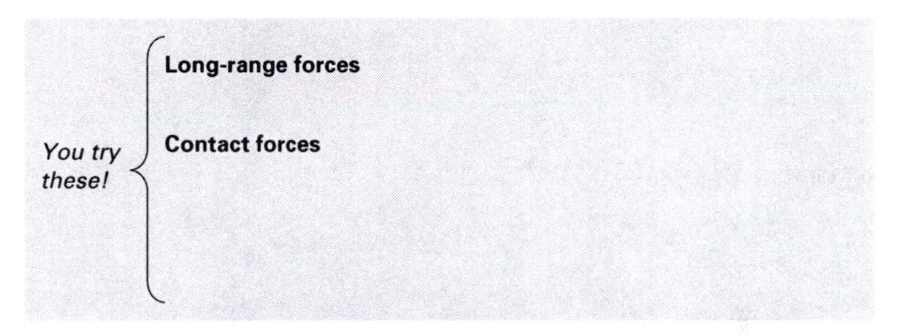

Long-range forces

Contact forces

You try these!

This leads to:

FBD for *block*

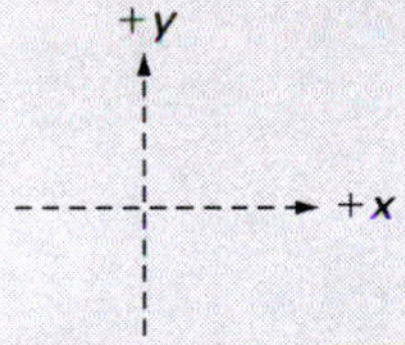

You try a FBD!

Forces: magnitude, direction [cause]

You try these!

Mass $m = 2.0$ kg

Acceleration: magnitude, direction

$a = 0$ ← Still at rest

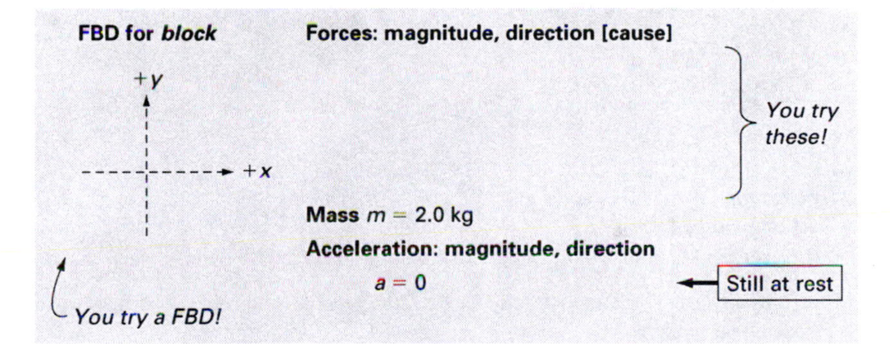

(3) Equations & unknowns

We need a right triangle to get components for the push, $\vec{P}$, in terms of its unknown magnitude, P:

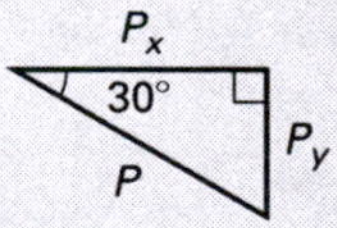

$$\frac{P_x}{P} = \cos 30° \quad \Rightarrow \quad P_x = +P\cos 30°$$ ← Make this *positive,* since it is *right.*

$$\frac{P_y}{P} = \sin 30° \quad \Rightarrow \quad P_y = -P\sin 30°$$ ← Make this *negative,* since it is *down.*

Now we use both Equations 5.1a (for x-components) and 5.1b (for y-components):

Force (x-components)

$$P_x = +P\cos 30°$$

$$f_{s,\max,x} = -f_{s,\max} = -\mu_s n = ?$$

(Add and sub)

Acceleration (x-component)

$$a_x = 0$$

(Sub)

$$F_{\text{net},x} = ma_x$$

$$(+P\cos 30°) + (-\mu_s n) = m(0)$$

$$\textcircled{P}\cos 30° = \mu_s \textcircled{n}$$

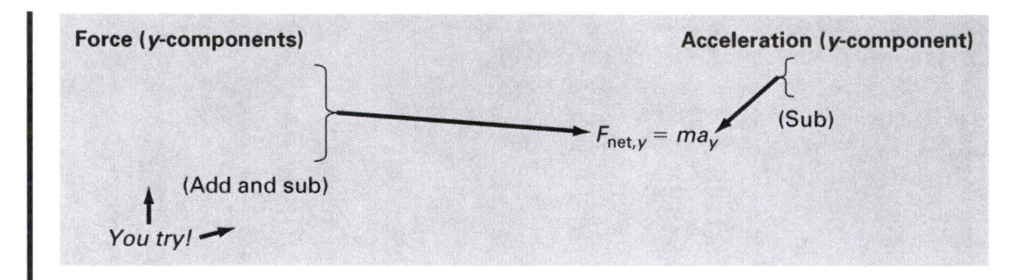

(4) Outline solution

Putting the equations together:

From x-components $(P) \cos 30^\circ = \mu_s (n)$

You try an outline!

From y-components $(n) = mg + (P) \sin 30^\circ$

Answers for . . .

Long-range forces

- Weight: Magnitude $Wt = mg$, direction *down* [caused by *earth*].

Contact forces: The block touches both the worker (one force) and the horizontal surface (two forces):

- Pushing force: Magnitude $P = ?$, direction 30° below positive x-axis [caused by *worker*].
- Normal: Magnitude $n = ?$, direction *up* (*perpendicular to surface, toward block*) [caused by *surface*].
- Maximum static friction: Magnitude $f_{s,\max} = \mu_s n = ?$, direction *left* (*parallel to surface*) [caused by *surface*].

Answers for . . .

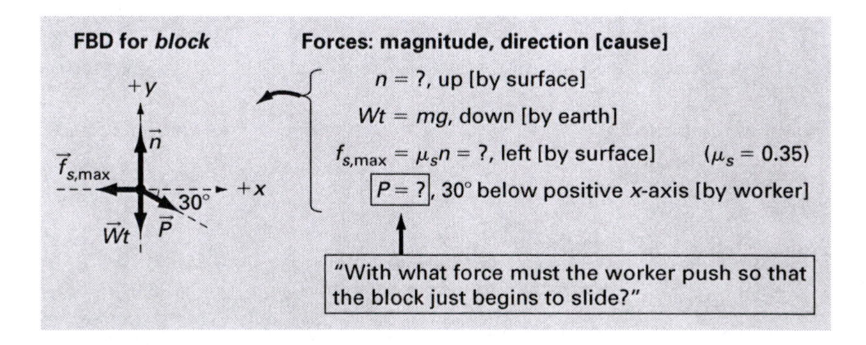

Answers for . . .

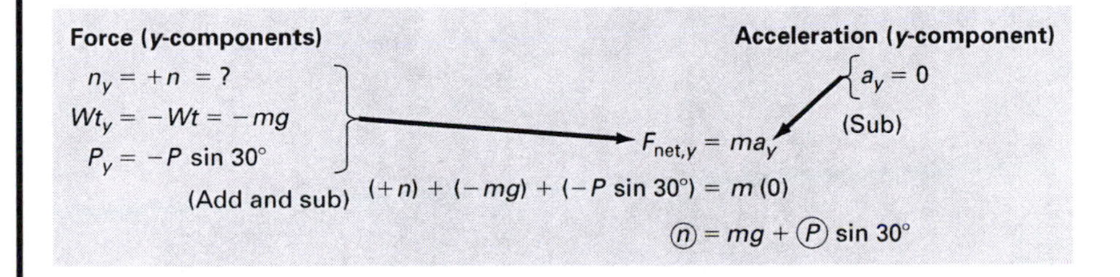

Answers for (4) Outline solution

The two simplified equations have the same two unknowns: P and n. Again, we solve the y-component equation for n, and then substitute this into the friction term ($\mu_s n$) in the x-component equation:

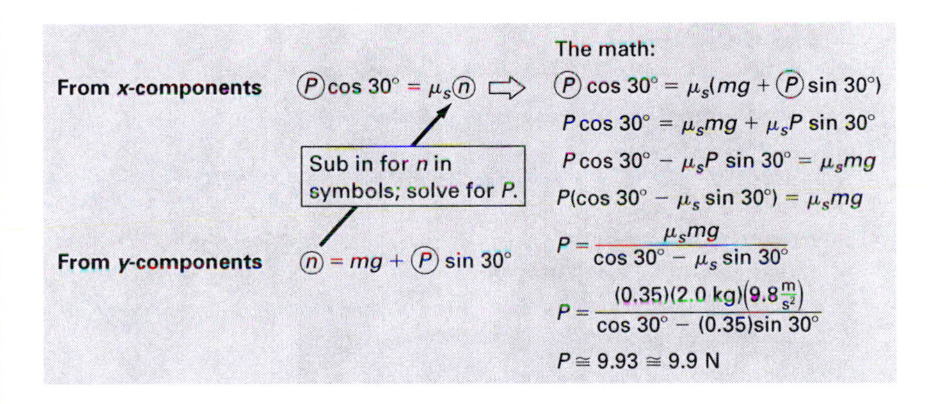

Answer $P \cong 9.93 \cong 9.9$ N ■

5.9 INCLINES OR RAMPS

With objects on inclines, ramps, and so on, *the normal force is no longer vertical,* because it is perpendicular to the surface. In this case, the math is much easier if we use *axes parallel and perpendicular to the ramp,* rather than the usual horizontal and vertical axes.

EXERCISE 5.7

A block of mass 3.0 kg is being pulled up a ramp by a rope that is parallel to the ramp surface, which is inclined at 25° from the horizontal. The block accelerates at 1.0 m/s^2 up the ramp, and the coefficient of kinetic friction between the block and the surface is 0.20. Determine the tension in the rope.

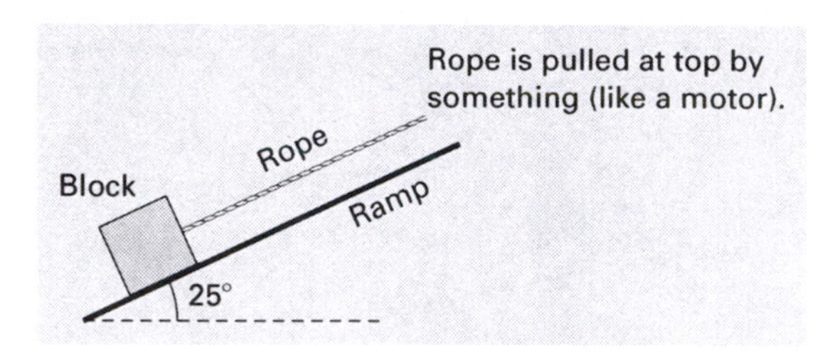

Solution

(1) Type of problem

Force, including friction: Use Equations 5.1a *and* 5.1b.

(2) Sort by object

Forces acting *on the block:*

Long-range forces

- Weight: Magnitude $Wt = mg$, direction *down* [caused by *earth*].

Contact forces: The block *touches* both the rope (one force) and the ramp surface (two forces):

- Tension: Magnitude $T = ?$, direction *up ramp* (*along rope, away from block*) [caused by *rope*].
- Normal: Magnitude $n = ?$, direction *perpendicular to ramp surface, toward block 1* (NOT vertical!) [caused by *surface*].
- Kinetic friction: Magnitude $f_k = \mu_k n = ?$, direction *down ramp* (*parallel to surface, opposite the motion of block*) [caused by *surface*].

Notice all the 25° angles in this first version of the FBD:

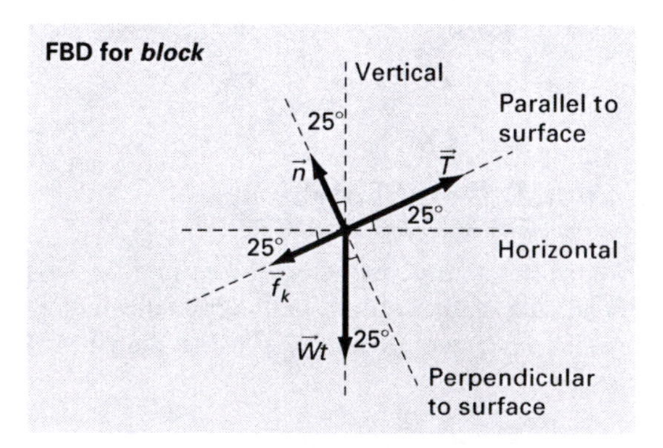

In other exercises up to now, we have made the x-axis *horizontal* and the y-axis *vertical.* But here, the acceleration is parallel to the inclined surface. In addition, three of the four force vectors are either parallel or perpendicular to the surface. The math is usually simpler if we make the x- and y-axes *parallel and perpendicular to the acceleration.* In this exercise, that means parallel and perpendicular to the surface.

MAKE AXES PARALLEL AND PERPENDICULAR TO THE ACCELERATION!

This almost always makes the math much easier. Key word: EASIER! There are fewer vector components to calculate than with horizontal/vertical axes, and one of the acceleration components is zero. This means simpler equations!

Here is the FBD with the "easier math" axes and the other information:

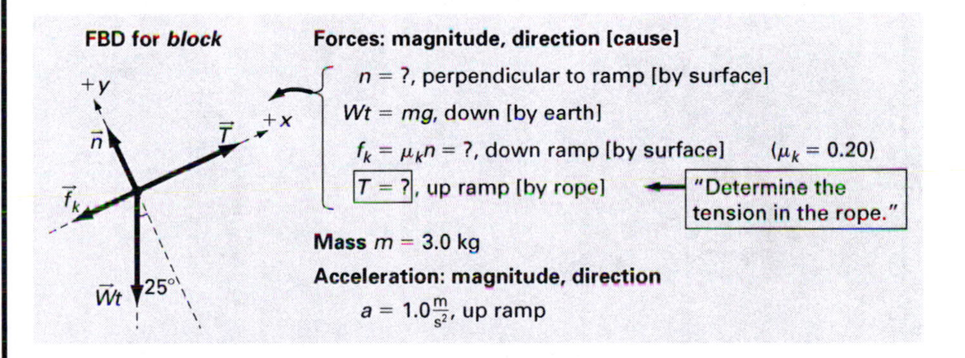

If you don't like the looks of these axes, try rotating your head by 25°. But don't hurt your neck! If it helps, just rotate the whole diagram by 25° on your paper, but be very careful to get all the vectors right:

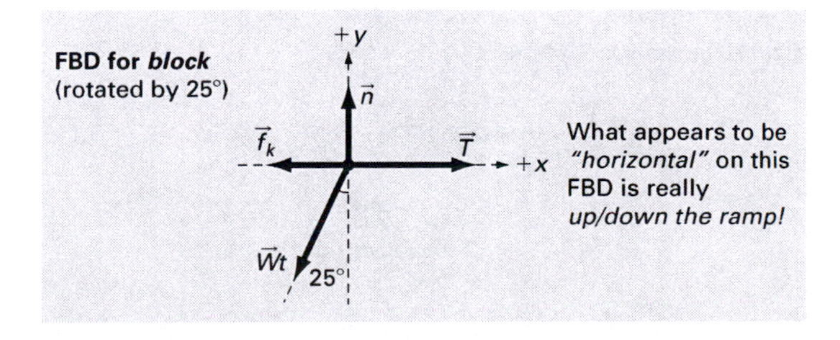

(3) Equations & unknowns

The right triangle for $\vec{Wt}$ (enlarged view):

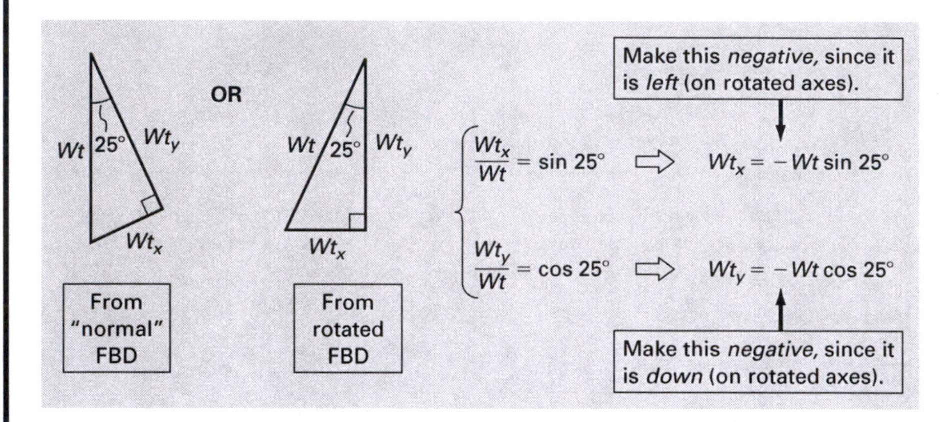

Be careful here: The angle is measured from the y-axis, and so the y-component goes with the cosine of the angle (not the sine). Now we put the components into Equations 5.1a and 5.1b:

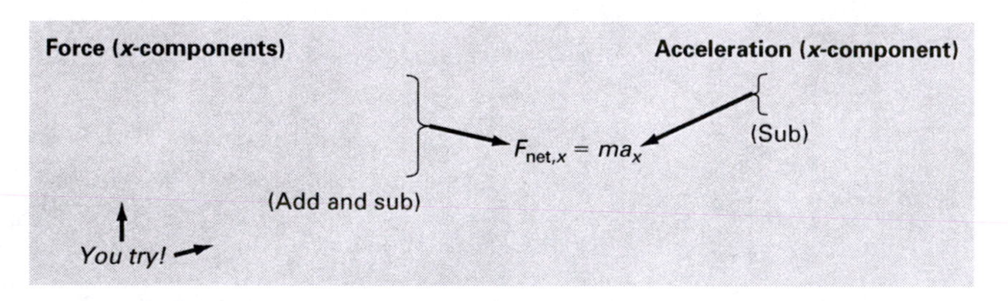

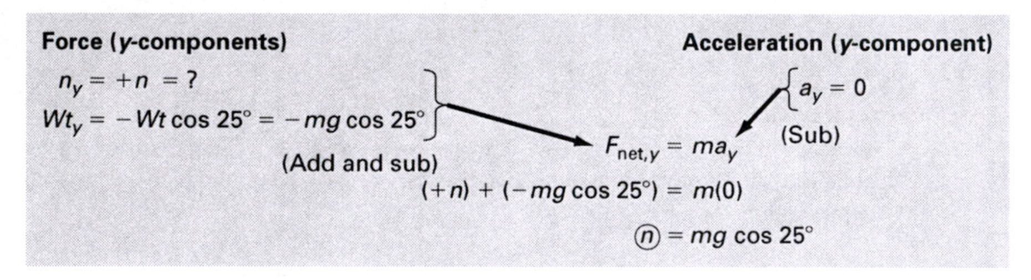

(4) Outline solution

The simplified equations and an outline:

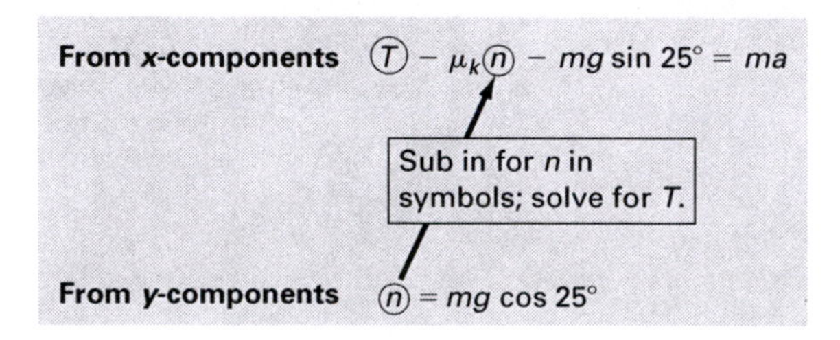

Answers for . . .

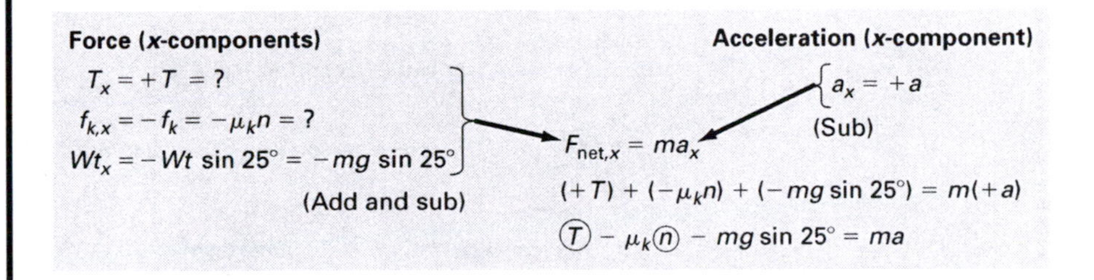

Answer $T \cong 20.8 \cong 21$ N ■

5.10 OBJECTS PUSHING ON EACH OTHER

EXERCISE 5.8

Five blocks, 0.50 kg each, are placed end to end on a frictionless horizontal table as shown in the figure. A person pushes the blocks to the right so that they accelerate at a rate of 1.0 m/s^2. Determine the magnitudes of (a) the pushing force and (b) the force exerted on block 3 by block 2.

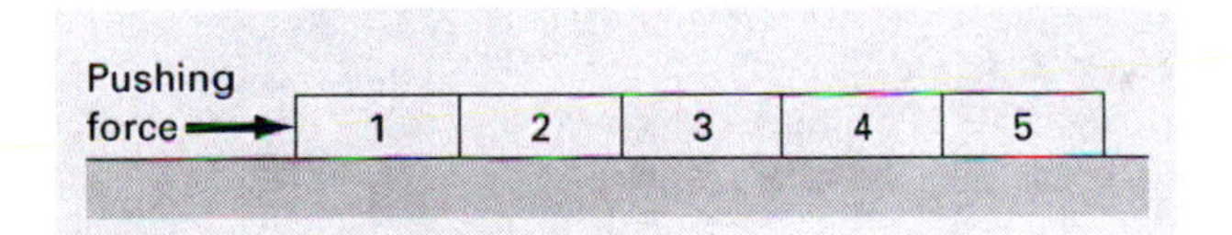

Solution

(1) Type of problem

Question about horizontal force: Use Equation 5.1a.

(2) Sort by object, (3) Equations & unknowns, and (4) Outline solution

It helps to think of the blocks in groups.

Group of all five blocks

First think of all five blocks as one big block:

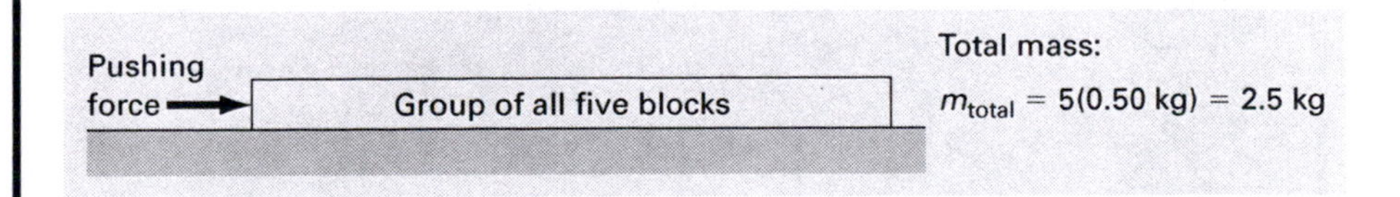

We look only at horizontal forces because (i) the question is about horizontal forces, (ii) there are no inclines at angles and so on, and (iii) there is no friction, so we don't need to determine the normal force from the vertical component equation.

So, *ignoring vertical forces* (the weight caused by earth and the normal force caused by the table), we see that the group of five blocks is only in contact with the person:

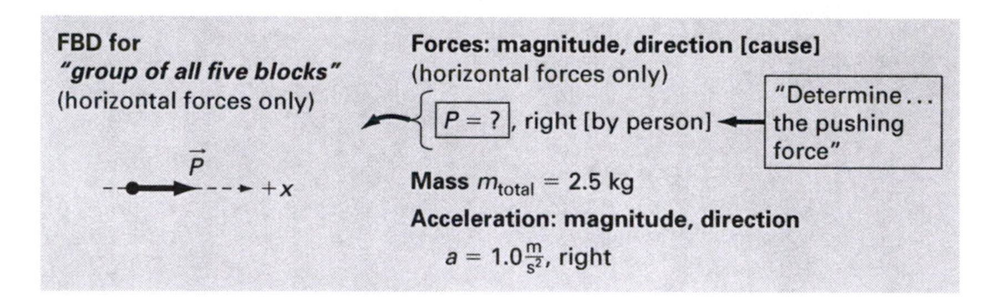

We can use this to easily find P with Equation 5.1a:

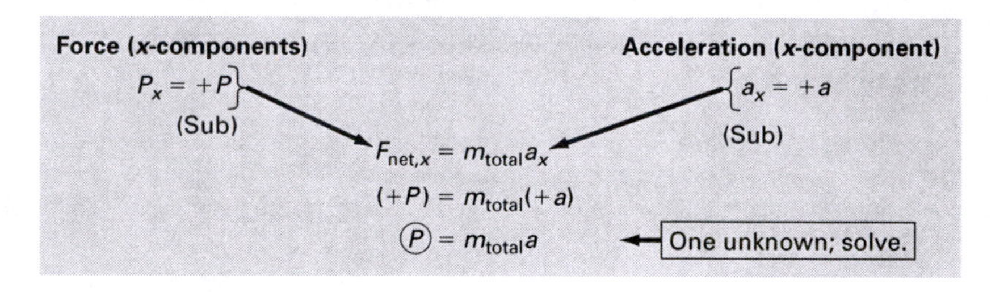

Group of blocks 3, 4, and 5

To find the magnitude of the force (which we will call $\vec{F}$) exerted on block 3 by block 2, we can think of the blocks in these two groups:

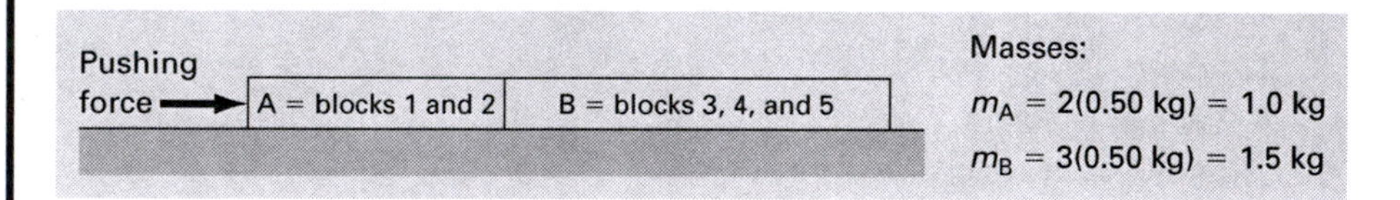

We use $\vec{F}$ for the force exerted *on block 3* by block 2, which is the *same* as the force exerted *on group B* by group A. Again ignoring vertical forces, we see that $\vec{F}$ is the only force acting *on group B:*

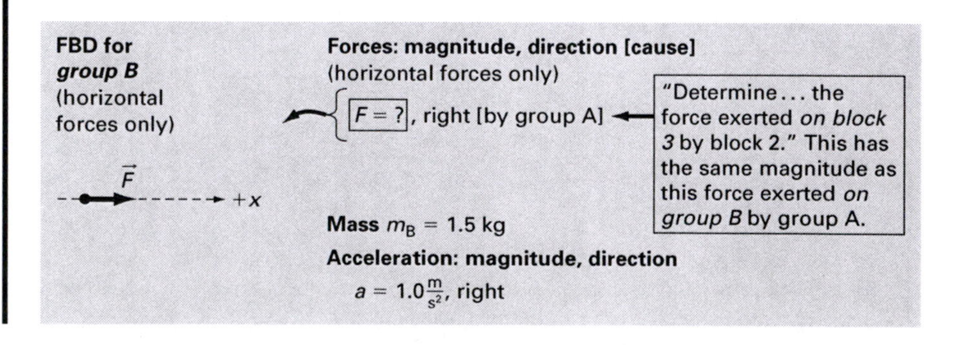

Equation 5.1a for group B:

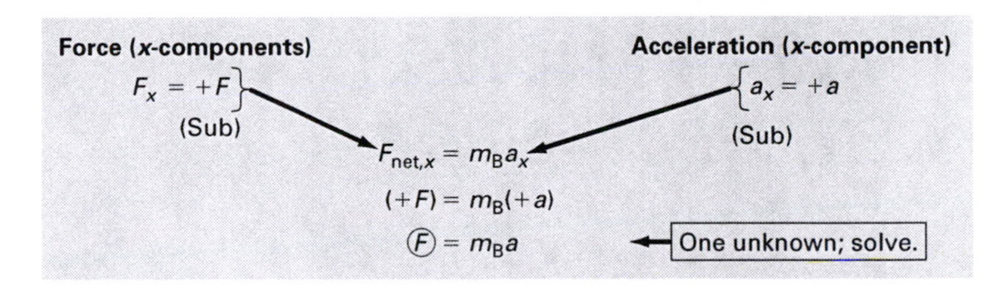

DOESN'T IT MATTER THAT GROUP B PUSHES BACK ON GROUP A?

Newton's third law tells us that since group A pushes *on group B* (magnitude F, direction *right*), then group B also pushes *on group A* with an equal and opposite force (same magnitude, F, direction *left*).

- The force acting *on group A* affects only group A, and so it goes only in the FBD for group A if we made this FBD (we won't).
- The force acting *on group B* affects only group B, and so it goes only in the FBD for group B. *This* is the FBD we made!

Answers $P = 2.5$ N, $F = 1.5$ N ■

CHAPTER 6

CIRCULAR MOTION AND CENTRIPETAL FORCE

IN THIS CHAPTER, we focus on circular motion and how it is related to centripetal force. We first look at circular motion, without thinking about force yet.

6.1 TANGENTIAL SPEED AND CENTRIPETAL ACCELERATION

The tangential velocity vector, $\vec{v}$, of an object in circular motion is perpendicular to the radius of the circle, *tangent* to the curved path. The equation:

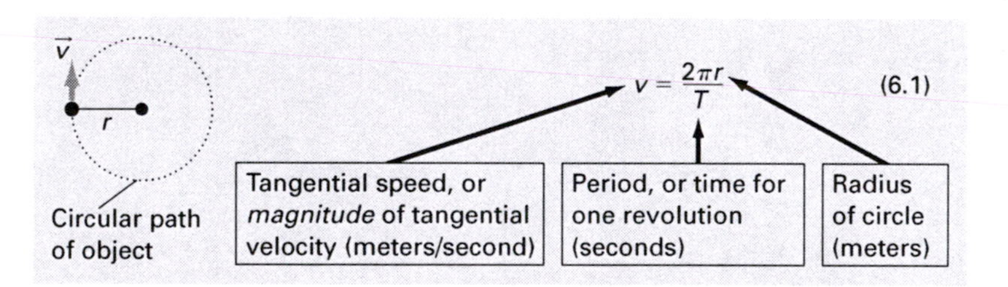

Is the object *accelerating?* YES! Even if its speed is constant? Yes, because the *direction* of its velocity vector is always *changing*. We call it *centripetal* acceleration because $\vec{a}$ is directed *toward the center* of the circle when the speed is constant. The word *centripetal* means "toward the center." Two equations we use:

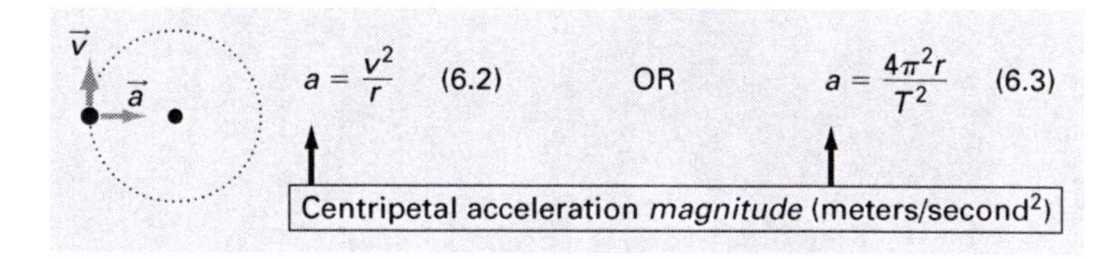

The quantities v, r, and T mean the same thing as in Equation 6.1. Equation 6.3 comes from substituting Equation 6.1 into Equation 6.2.

6.2 COMPARING CIRCULAR MOTION AT TWO DIFFERENT RADII

EXERCISE 6.1

A merry-go-round has an outer radius of 2.0 m. While it is spinning, the centripetal acceleration at the outer edge is 3.0 m/s^2. What is the acceleration at a point 1.5 m from the center?

Solution

(1) Type of problem

Circular motion, comparing two objects/circles: Use Equation 6.2 or 6.3, and 6.1 if needed.

(2) Sort by object and/or circle

We are comparing the circular motion at two different radii, so we sort information for two different circles. *For each circle,* we keep track of a, v, r, and T, the quantities in Equations 6.2 and 6.3:

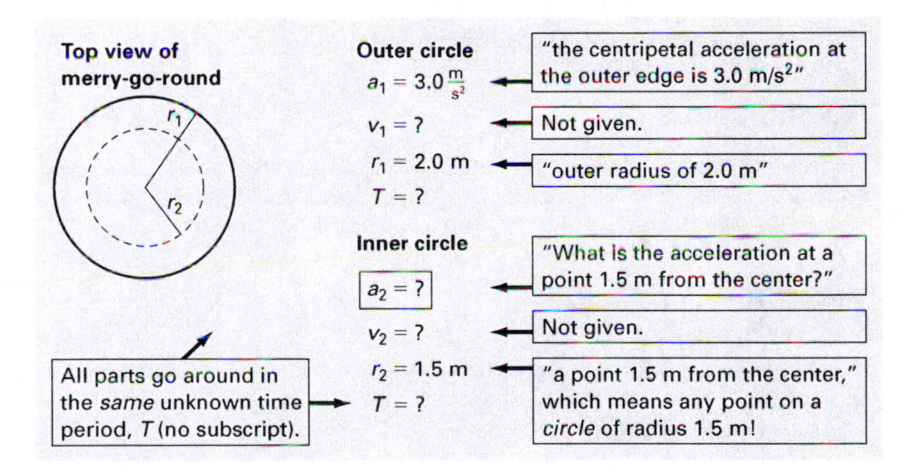

(3) Equations & unknowns and (4) Outline solution

Because we are *comparing* circles, we want equations *relating the circles* (there are none in this problem) and equations *for each circle.*

For each circle, we decide between Equations 6.2 and 6.3. Because the circles have the same unknown period, T, we pick the one with T in it: Equation 6.3. We write it out for each circle, and outline the solution:

Outer circle $$a_1 = \frac{4\pi^2 r_1}{(T)^2}$$ (1) One unknown; solve for T.

Inner circle $$(a_2) = \frac{4\pi^2 r_2}{(T)^2}$$ (2) Sub in T; solve for a_2.

Or you can divide these two equations, but be careful with the algebra!

IN COMPARISON OR RATIO PROBLEMS, LOOK FOR THE QUANTITY THAT IS THE SAME!

Here, the two circles have the *same* period, *T*. Even though *T* is *unknown*, this fact reduces the number of unknowns and leads to our solution.

Answer $a_2 \cong 2.25 \cong 2.3\frac{\text{m}}{\text{s}^2}$ ■

Write what you need to, but focus on THINKING this way!

6.3 COMPARING CIRCULAR MOTION AT TWO DIFFERENT SPEEDS

Here is a different kind of comparison problem.

EXERCISE 6.2

In rounding a circular curve on a horizontal, unbanked road, your speed is 15 m/s, and you experience a certain centripetal acceleration. At what speed should you drive the same curve to reduce the centripetal acceleration to one-third of the original value?

Solution

(1) Type of problem

Circular motion, comparing two objects/circles: Use Equation 6.2 or 6.3, and 6.1 if needed.

(2) Sort by object and/or circle

We are to compare the circular motion for two different speeds, so we will sort all of the quantities in Equations 6.2 and 6.3 as if for two different objects:

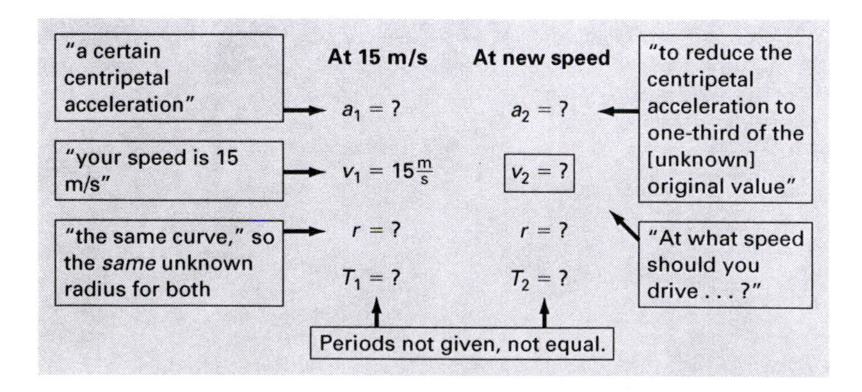

(3) Equations & unknowns and (4) Outline solution

An equation *relating the objects* comes from the statement that the reduced centripetal acceleration is *one-third of the original value:* $a_2 = \frac{1}{3}a_1$.

For each object, should we use Equation 6.2 or 6.3? Since we want to solve for a speed, v_2, we use the equation with v in it, or Equation 6.2:

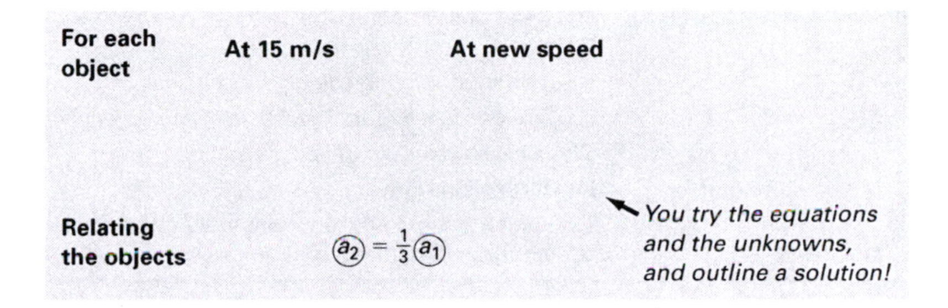

Answers for (3) Equations & unknowns and (4) Outline solution

At first it may seem like there are too many unknowns, but some cancel out:

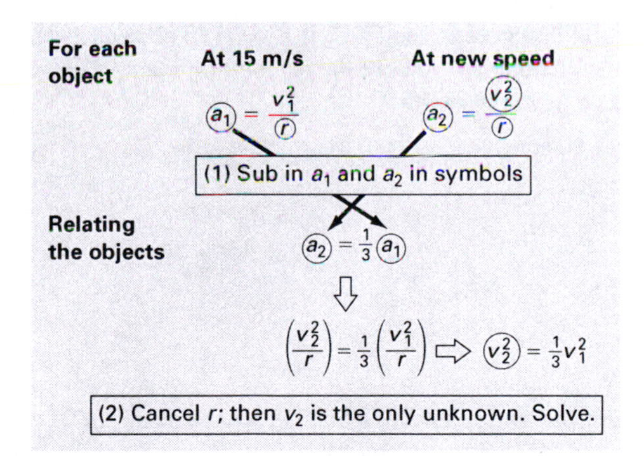

Answer $v_2 \cong 8.66 \cong 8.7\,\frac{\text{m}}{\text{s}}$ ■

6.4 HOW TO SET UP CIRCULAR MOTION COMPARISON PROBLEMS

Here is a summary of the setup steps we followed in the previous two exercises.

Circular Motion Comparison Problems—Mental and Written Steps

Mental→	**(1) Type of problem** Circular motion, comparing two objects/circles: Use Equations 6.1, 6.2, and/or 6.3.
Mental and written→	**(2) Sort *by object* and/or *circle*** For each object/circle, sort out known and unknown values of *a*, *v*, *r*, and *T*.
Mental and written→	**(3) Equations & unknowns** Write: • Equation(s) *relating the objects/circles* • Equations 6.1, 6.2, and/or 6.3 *for each object/circle* Circle each unknown in each equation.
Mental→	**(4) Outline solution** The usual! If there are too many unknowns at first, try substituting something, and an unknown might cancel!

6.5 HOW TO THINK ABOUT CENTRIPETAL FORCE PROBLEMS

For the next several sections, we will focus on problems with *centripetal force.* We follow steps very similar to those we used in Chapter 5.

The term *centripetal force* is often MISUNDERSTOOD to be a force like tension or friction that goes in the FBD. It is NOT! The centripetal force is the *net,* or sum, of force components *toward the center.* For example, here we make the horizontal x-axis *toward the center:*

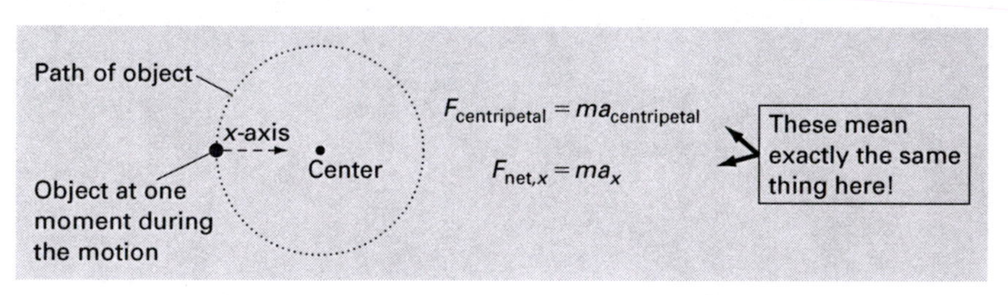

To set up a solution for this situation, we first draw a FBD with the usual *long-range* and *contact* forces acting on the object. Then we get the left and right sides of Equation 5.1a ($F_{net,x} = ma_x$) in the same way we did in Chapter 5:

- Left side: Add all the force x-components and substitute this sum in for $F_{net,x}$. This is the *centripetal force.*
- Right side: Substitute in the mass and the x-component of the acceleration. Now we use Equation 6.2 or 6.3 for the acceleration magnitude. For the previous figure, the acceleration x-component would be:

$$a_x = +a = +\frac{v^2}{r} \qquad \text{OR} \qquad a_x = +a = +\frac{4\pi^2 r}{T^2}$$

We will illustrate this process in the next several sections.

6.6 CIRCULAR MOTION WITH A HORIZONTAL STRING

EXERCISE 6.3

A rock of mass 1.0 kg is tied to the end of a horizontal string and is twirled in a horizontal circular path of radius 0.50 m. At what speed is the rock moving when the string breaks if the breaking strength of the string is 12 N?

Solution

It helps to picture a snapshot of the situation:

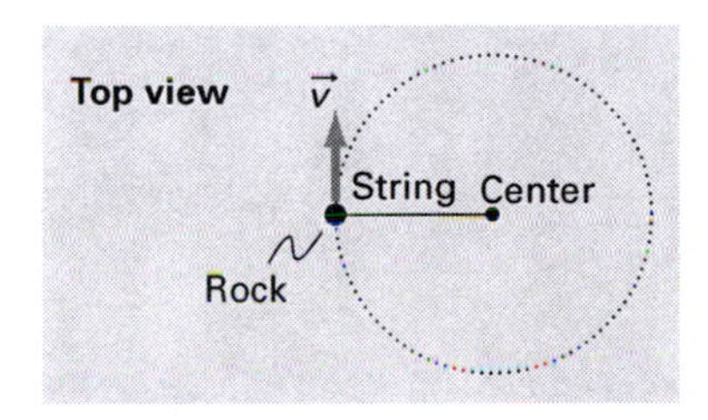

(1) Type of problem

Force (tension is mentioned) and circular motion: Use Equations 5.1a and/or 5.1b, along with Equation 6.2 or 6.3, and 6.1 if needed.

(2) Sort by object

The tension in the string acts *on the rock,* and the *rock* is the object in circular motion at a known radius. So we will look at *force, mass,* and *acceleration* (the quantities in Equations 5.1a and 5.1b) in relation to *the rock.* Since we have centripetal acceleration, we also consider *speed, radius,* and *period* (quantities in Equations 6.2 and/or 6.3).

Subtle point: The problem says that the string is horizontal. However, if you try twirling a rock on a string in a horizontal circle, the string is always angled somewhat downward. For the string to be horizontal, imagine the rock sliding in a horizontal circle on a frictionless tabletop. The two vertical forces (the weight and the normal) would cancel out, so we *completely ignore them.*

Now we identify all the force vectors acting *on the rock* (horizontal forces only):

Long-range forces: We are ignoring the weight, a vertical force.

Contact forces: The rock is touching the string:

- Tension: Magnitude $F_T = 12$ N (just before breaking), direction *toward center of circle (along string, away from rock)* [caused by *the string*].

DON'T CONFUSE TENSION WITH PERIOD!

We use F_T for *tension* to keep from confusing it with the *period, T.*

Next we draw a FBD and list the other relevant quantities:

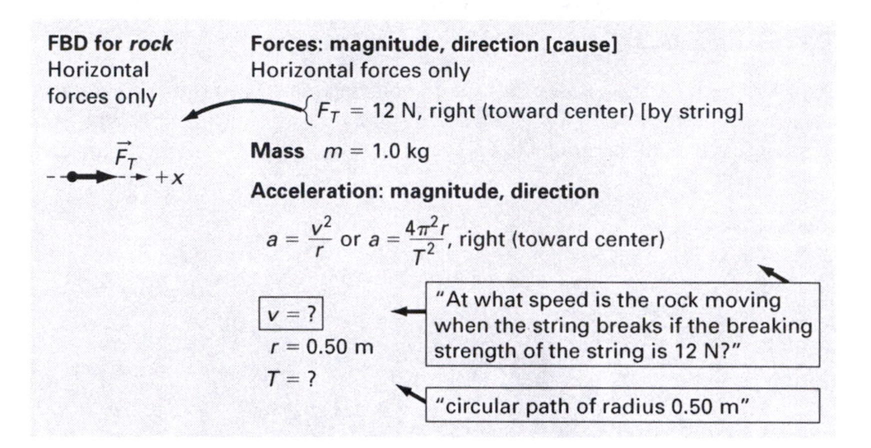

This shows us that the net (centripetal) force, $F_{net,x}$, is caused by the tension force.

BUT ISN'T THERE AN OUTWARD FORCE?!

Not acting *on the rock!* The string pulls *inward on the rock.* Newton's third law tells us that the rock pulls *outward on the string,* but this force would go in a FBD for *the string* if we made one (we won't)!

(3) Equations & unknowns and (4) Outline solution

Next, we find the x-components and put them into Equation 5.1a:

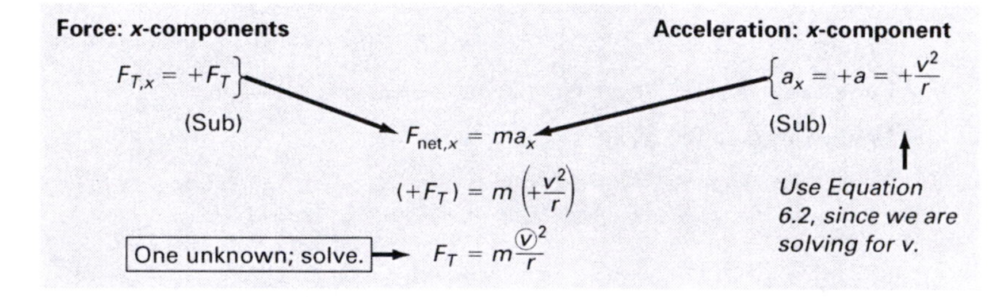

Answer $v \cong 2.449 \cong 2.4\ \frac{\text{m}}{\text{s}}$ ■

6.7 HOW TO SET UP CENTRIPETAL FORCE PROBLEMS

Here is a summary of what we did in the previous exercise, which combined circular motion with the setup steps from Chapter 5 for force problems.

Centripetal Force Problems—Mental and Written Steps

Mental→ **(1) Type of problem**

For force and circular motion, use:

- Right triangle trig for vectors at angles
- Equations 5.1a and/or 5.1b
- Equations 6.1, 6.2, and/or 6.3

Mental and written→ **(2) Sort *by object***

For *each object* for which we know (or want to know) about forces, we sort out:

- **Forces vectors:** magnitudes and directions; draw a FBD
- **Mass**
- **Acceleration vector:**
  - Magnitude (also need: v, r, and T)
  - Direction (toward center)

Mental and written→ **(3) Equations & unknowns**

For each object:

- **Force components:** Determine from the FBD; then add and sub into *left side* of Equation 5.1a and/or 5.1b.
- **Mass:** Sub into *right side* of Equation 5.1a and/or 5.1b.
- **Acceleration components:**
  - Use Equation 6.2 or 6.3, and 6.1 if needed, to get acceleration magnitude; then . . .
  - Use axes from the FBD to get components; then sub into *right side* of Equation 5.1a and/or 5.1b.

Simplify and then circle unknowns.

Mental→ **(4) Outline solution**

We will use these steps throughout the rest of this chapter.

6.8 CIRCULAR MOTION WITH A STRING AT AN ANGLE

In the following exercise, the circular motion is horizontal but the string is not.

EXERCISE 6.4

Your speedometer is broken, so you hang a rock from a string tied to the ceiling of your car in order to determine your speed. (Doesn't *everyone* do this?) While you are driving straight at a constant speed, the rock hangs straight down. As you drive around an unbanked curve of radius 75 m, the string makes an angle of 15 degrees with the vertical as the rock swings away from the center of the curved path. What is your speed?

Solution

Here is a snapshot of the situation:

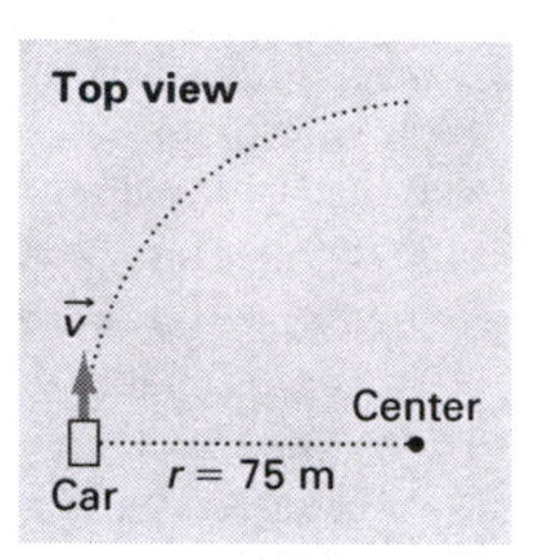

Or:

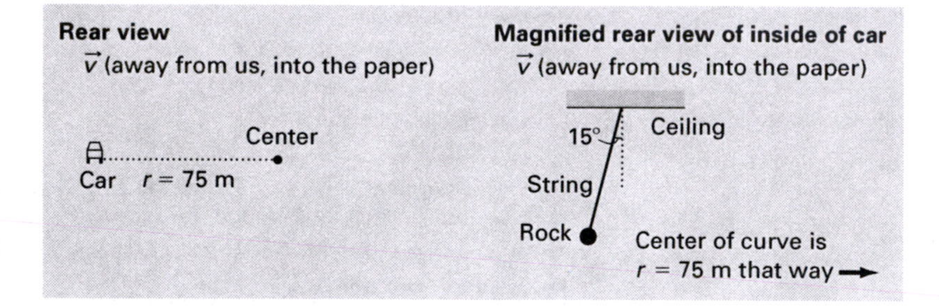

(1) Type of problem

Force (tension is never mentioned but is *implied* because there is a *string*) and circular motion: Use Equations 5.1a and/or 5.1b, along with Equation 6.2 or 6.3, and 6.1 if needed.

(2) Sort by object

We focus *on the rock,* looking at *force, mass,* and *acceleration.* For centripetal acceleration, we also need *speed, radius,* and/or *period.*

Before drawing the FBD:

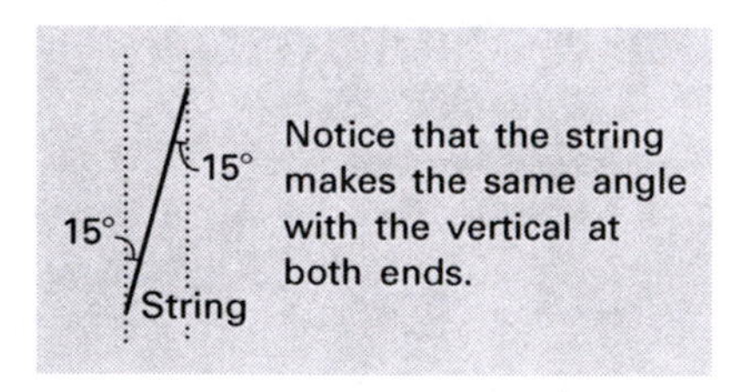

In addition to the *long-range* weight force, there is the *contact* tension force acting on the rock, caused by the string:

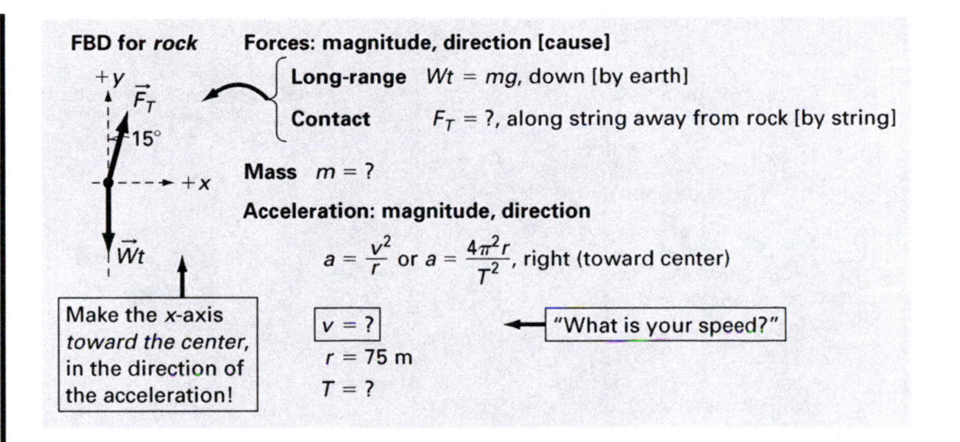

PUT ONLY LONG-RANGE AND CONTACT FORCES IN THE FBD!

You might be tempted to put some "outward" force in the FBD. But there is NO outward force *on the rock*. The rock *does* pull outward and downward *on the string*—Newton's third law. This force would go in a FBD *for the string*, not for the rock. Stick to the usual *long-range* and *contact* forces, even in circular motion!

(3) Equations & unknowns and (4) Outline solution

We use a right triangle to get the components of the tension vector (enlarged view):

$F_{T,x}$, $F_{T,y}$, F_T, 15°

$$\frac{F_{T,x}}{F_T} = \sin 15° \Rightarrow F_{T,x} = +F_T \sin 15°$$

Make this *positive*, since it is *right*.

$$\frac{F_{T,y}}{F_T} = \cos 15° \Rightarrow F_{T,y} = +F_T \cos 15°$$

Make this *positive*, since it is *up*.

The x-components and Equation 5.1a:

Force: *x*-components **Acceleration: *x*-component**

$$F_{T,x} = +F_T \sin 15° \quad \text{(Sub)} \rightarrow F_{\text{net},x} = ma_x \leftarrow a_x = +a = +\frac{v^2}{r} \quad \text{(Sub)}$$

Use Equation 6.2, since we are solving for v.

$$(+F_T \sin 15°) = m\left(+\frac{v^2}{r}\right)$$

$$\textcircled{F_T} \sin 15° = \textcircled{m}\frac{\textcircled{v}^2}{r}$$

This means that the x-component of the tension ($F_T \sin 15°$) is the net (centripetal) force, $F_{\text{net},x}$, which causes the centripetal acceleration of the rock.

The y-components and Equation 5.1b:

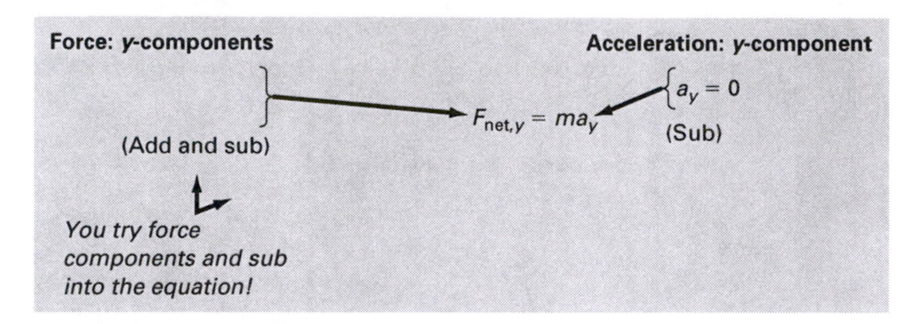

Now we put the equations together:

From x-components $F_T \sin 15° = m \frac{v^2}{r}$

From y-components

You finish the equations and the unknowns and outline a solution!

If it looks like there are too many unknowns, maybe something will cancel out!

Answers for . . .

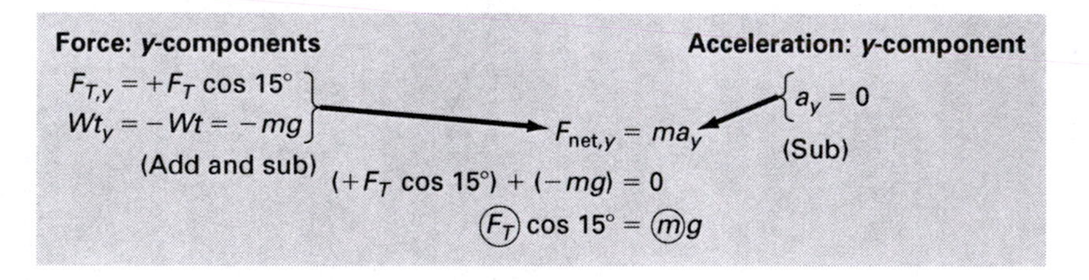

Meaning: the y-component of the tension ($F_T \cos 15°$) supports the rock's weight (mg).

Answers for (3) Equations & unknowns and (4) Outline solution

From x-components $F_T \sin 15° = m \frac{v^2}{r}$

(1) Solve for F_T in symbols and sub.

From y-components $F_T \cos 15° = mg$

(2) The unknown m cancels, so only one unknown remains. Solve for v.

Answer $v \cong 14.0 \cong 14 \frac{\text{m}}{\text{s}}$ ■

Remember, you don't need to write *all* of this when you work a problem. Much of it is mental. If you are not that comfortable with it yet, write out more until the mental part is more automatic for you!

6.9 CIRCULAR MOTION ON AN UNBANKED ROAD WITH FRICTION

EXERCISE 6.5

At a maximum speed of 18 m/s, a car can round the unbanked horizontal curve, radius 85 m, without skidding. Determine the coefficient of static friction between the car's tires and the road.

Solution

Here is a snapshot:

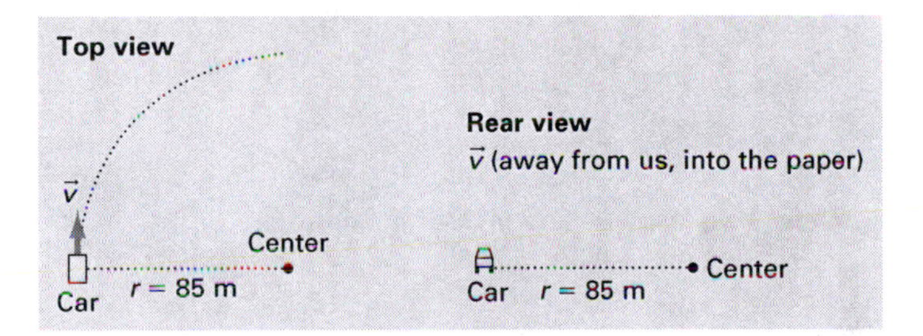

This situation is just like the previous exercise, except now we look at the motion of *the car* (caused by friction from the road surface) rather than the motion of the rock (caused by tension in the string). Even though the car is moving, the friction force is not kinetic friction. It is *static friction* because there is *no skidding* between the road and tires. Since the car is at its maximum speed without skidding, we use the *maximum* static friction force (magnitude: $f_{s,\max} = \mu_s n$).

What is the direction of $\vec{f}_{s,\max}$? The car's inertia is "trying" to go in a *straight line,* but the friction force makes it curve *to the right* (as in the snapshot). So $\vec{f}_{s,\max}$ must be *to the right,* or *inward.*

But isn't there an outward force acting *on the car?* NO! There *is* an outward friction force caused by the car, acting *on the road*—this force would go in a FBD *for the road* if we made one (we won't)!

(1) Type of problem

Force (friction is mentioned) and circular motion: Use Equations 5.1a and/or 5.1b, along with Equation 6.2 or 6.3, and 6.1 if needed.

(2) Sort by object

We look at *force, mass,* and *acceleration,* as well as *speed, radius,* and *period* for the car.

In addition to the weight as the long-range force, there is contact with the road surface, which causes the normal force *and* the maximum static friction force. We treat all of the tires as *one* contact point:

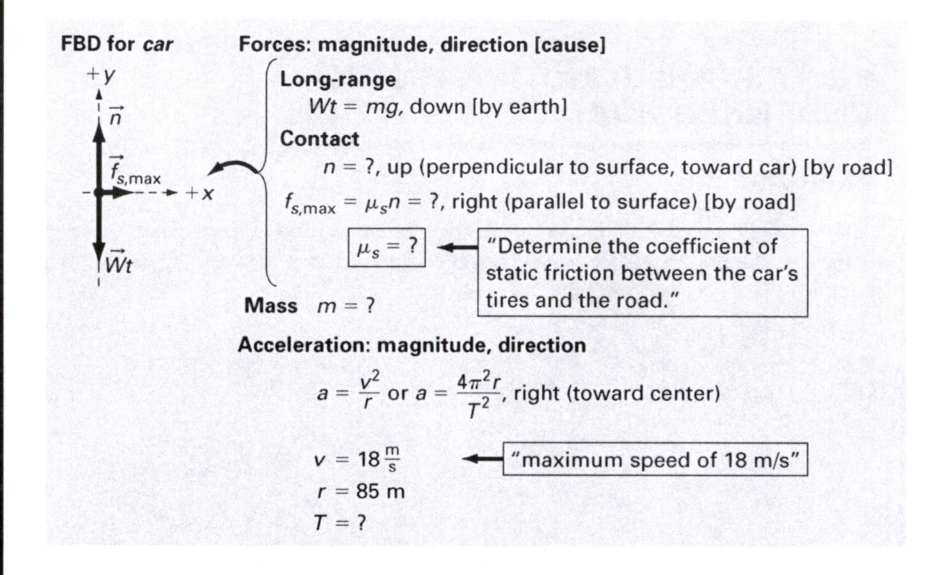

(3) Equations & unknowns and (4) Outline solution

The x-components and Equation 5.1a:

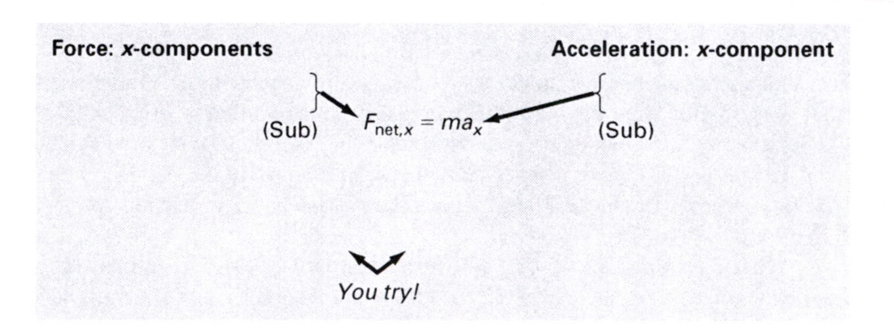

We also need the y-components and Equation 5.1b to determine an equation for n:

Force: y-components | **Acceleration: y-component**

$n_y = +n$

$Wt_y = -Wt = -mg$

(Add and sub) → $F_{net,y} = ma_y$ ← (Sub) $a_y = 0$

$$(+n) + (-mg) = 0$$

$$n = mg$$

Putting the equations together:

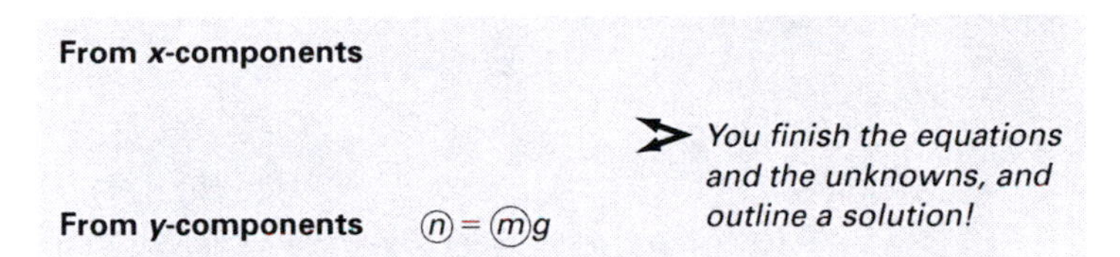

Answers for . . .

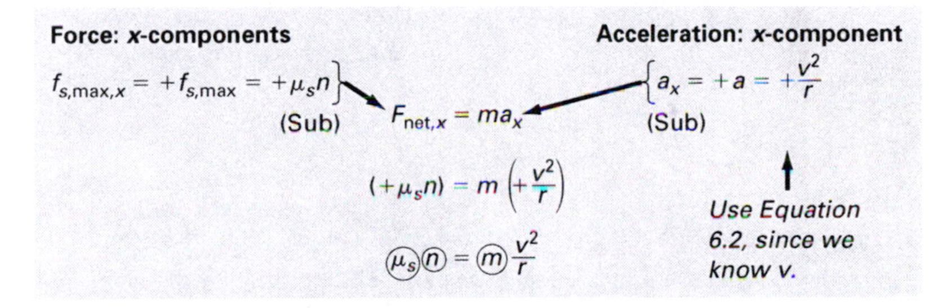

Answers for (3) Equations & unknowns and (4) Outline solution

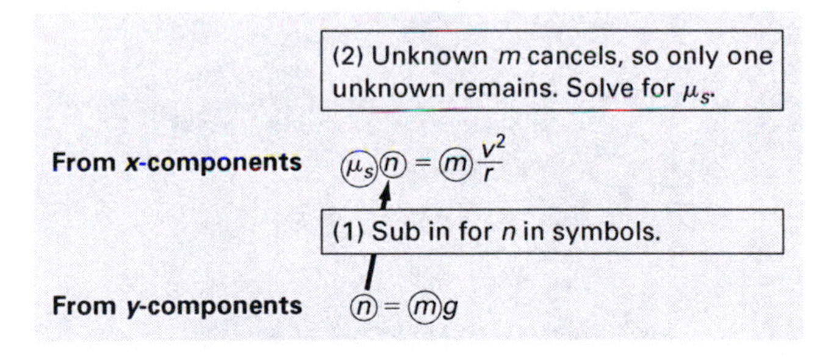

Answer $\mu_s \cong 0.389 \cong 0.39$ ■

6.10 CIRCULAR MOTION ON A BANKED ROAD WITHOUT FRICTION

In the next exercise, the road, or track, is banked at an angle, but the car is still moving in a *horizontal* circle.

EXERCISE 6.6

A car drives at 22 m/s on the banked track shown in the next figure. At what radius should the car drive in order not to rely on friction at all?

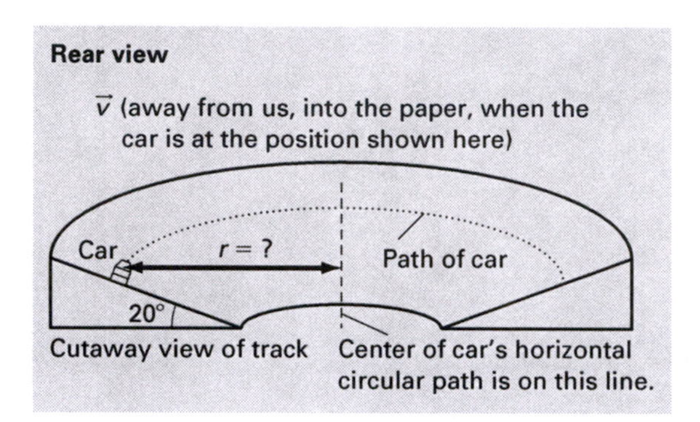

Solution

(1) Type of problem

Force (even though there is no friction, there must be a force keeping the car in circular motion on the track) and circular motion: Use Equations 5.1a and/or 5.1b, along with Equation 6.2 or 6.3, and 6.1 if needed.

(2) Sort by object

We look at *force, mass,* and *acceleration,* as well as *speed, radius,* and *period* for the car:

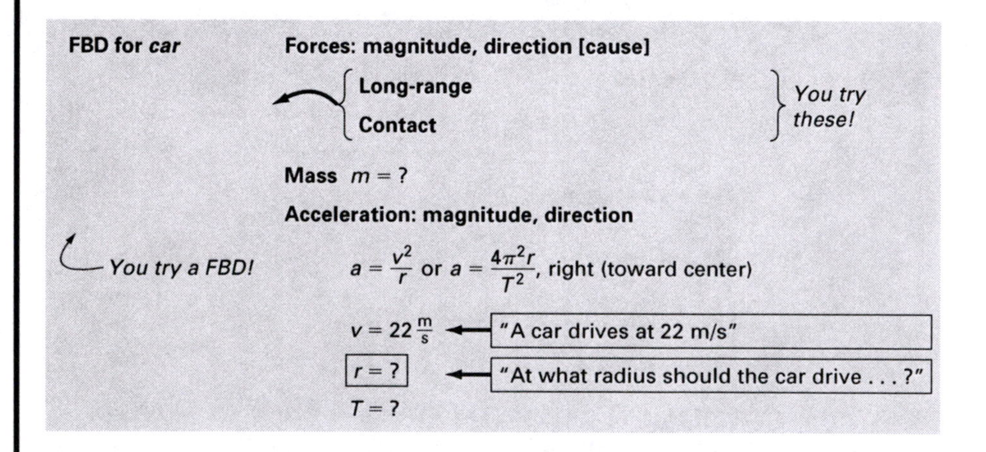

(3) Equations & unknowns

Which directions should we use for the axes?

USE AXES PARALLEL AND PERPENDICULAR TO THE CENTRIPETAL ACCELERATION!

In Chapter 5, we had an object accelerating up an incline. We chose axes *parallel and perpendicular to the acceleration,* which were also parallel and perpendicular to the track surface. That meant fewer force components to calculate, and made $a_x = +a$ and $a_y = 0$. Easier math! BUT . . .

Here, the acceleration of the car is NOT parallel to the incline but is *toward the center* of the circular path, or *horizontal* in this case. So we choose *horizontal* and *vertical axes*. This makes $a_x = +a$ and $a_y = 0$ here also. EASIER math! Hooray!

We use a right triangle to determine the components of $\vec{n}$ (enlarged view):

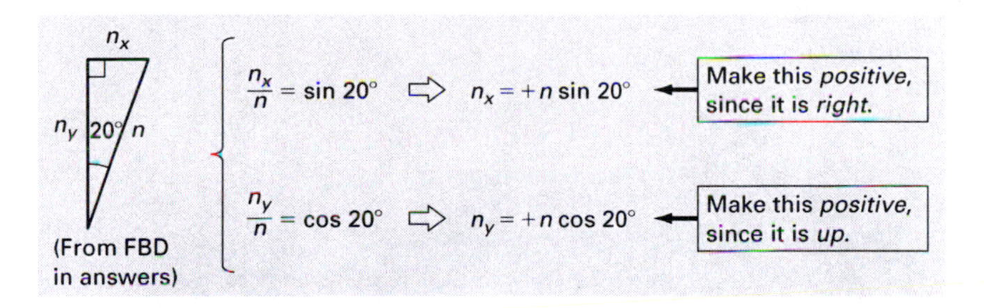

The component equations:

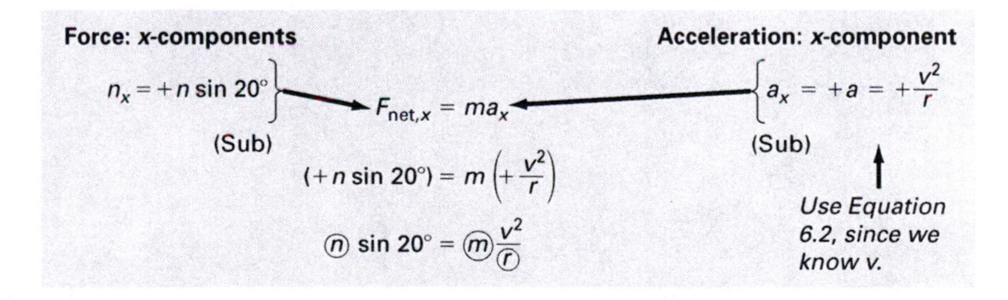

Force: *y*-components

$n_y = +n \cos 20°$

$Wt_y = -Wt = -mg$

(Add and sub)

Acceleration: *y*-component

$a_y = 0$

(Sub)

$$F_{net,y} = ma_y$$

$$(+n \cos 20°) + (-mg) = 0$$

$$Ⓝ \cos 20° = Ⓜ g$$

(4) Outline solution

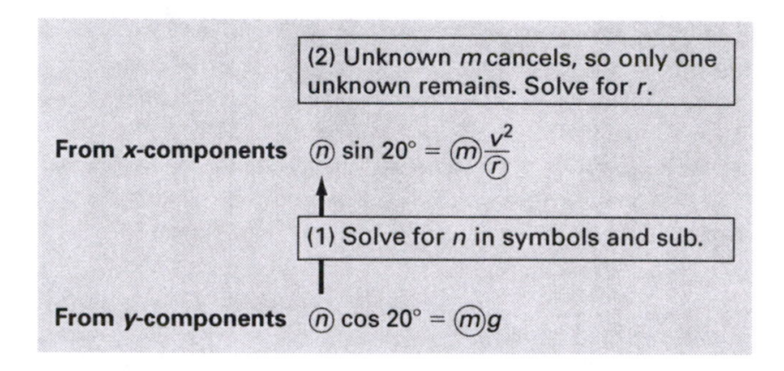

Answers for . . .

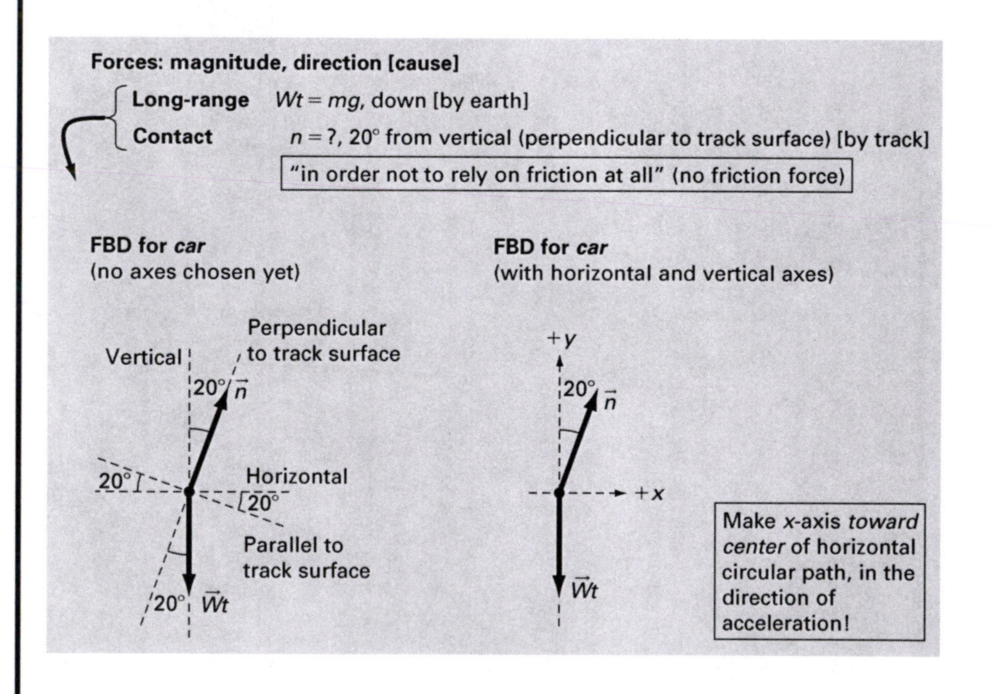

It is easy to make a mistake with the normal force angle. Be careful! The x-component of the normal force causes the centripetal acceleration. The y-component of the normal force supports the weight of the car.

Answer $r \cong 136\ \text{m} \cong 140\ \text{m}$ ■

6.11 VERTICAL CIRCULAR MOTION—LOWEST POINT

The next few exercises deal with objects moving in *vertical* circles.

EXERCISE 6.7

A 0.20-kg rock is tied to a string and moves in a vertical circle with radius 0.65 m. At the lowest point, the rock has speed 2.1 m/s. At this lowest point, determine the magnitudes of (a) the tension force acting on the rock and (b) the centripetal force acting on the rock.

Solution

A snapshot of the situation:

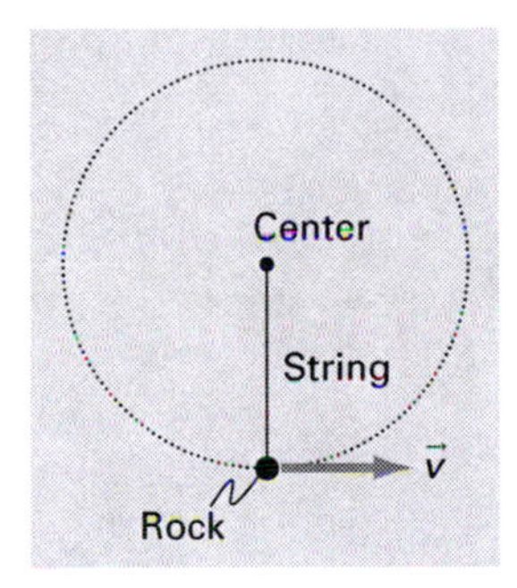

(1) Type of problem

Force (tension and centripetal force are mentioned) and circular motion: Use Equations 5.1a and/or 5.1b, along with Equation 6.2 or 6.3, and 6.1 if needed.

(2) Sort by object

Quantities and FBD for the rock:

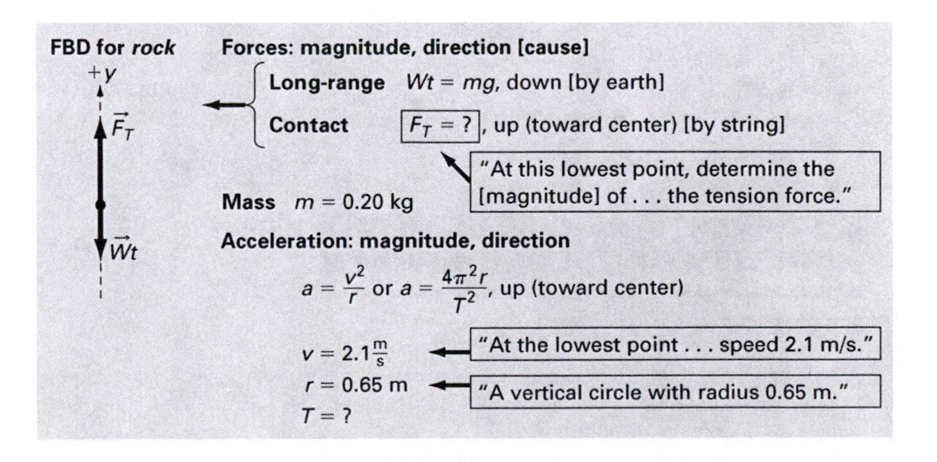

The tension has to be larger than the weight to make the net force toward the center.

(3) Equations & unknowns and (4) Outline solution

For part (a), we use Equation 5.1b with the y-components:

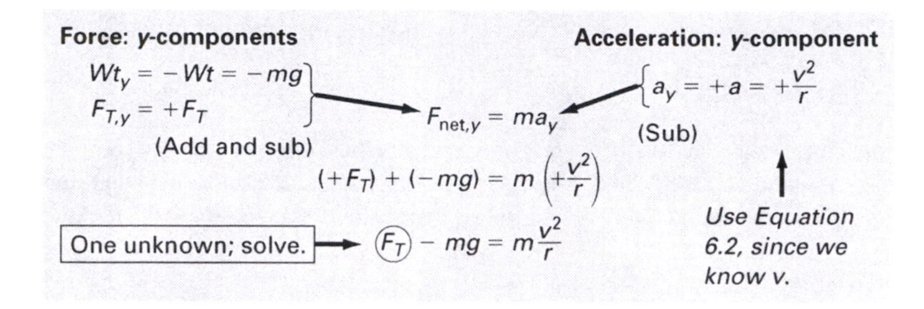

For part (b), there are two ways to think about it:

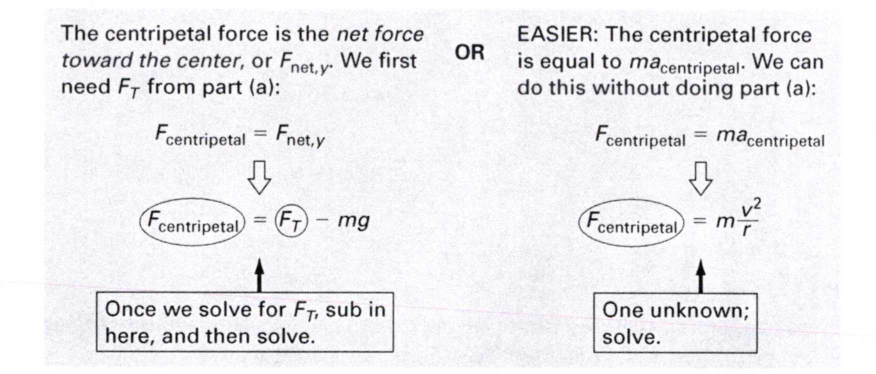

Answers $F_T \cong 3.32 \cong 3.3$ N, $F_{\text{centripetal}} \cong 1.36 \cong 1.4$ N ■

As a variation of the previous exercise, there might be a car at the lowest point of a circular dip in the road. We would set this up in exactly the same way except the tension force (by the string) would be replaced with a normal force (by the road).

6.12 VERTICAL CIRCULAR MOTION—HIGHEST POINT, UPSIDE-DOWN

EXERCISE 6.8

A woman of mass 50.0 kg riding in a roller-coaster car is upside-down at the very top of the loop-the-loop of radius 7.00 m, while moving at a speed of 9.50 m/s. (a) What is her apparent weight (as a multiple of her true weight)? (b) What is the minimum speed necessary to just barely make it through the loop without her coming out of the seat? (Ignore any friction forces, forces by the seat back, or by any passenger restraint bars/belts.)

Solution

(1) Type of problem

Force and circular motion: Use Equations 5.1a and/or 5.1b, along with Equation 6.2 or 6.3, and 6.1 if needed.

(2) Sort by object, (3) Equations & unknowns, and (4) Outline solution

We will do these steps separately for parts (a) and (b).

Part (a)

Besides the long-range weight force caused by the earth, the seat causes a normal force to act on the woman. At the top of the loop, the seat is *above* her and so pushes *vertically downward* on her. The note to "[i]gnore any friction forces, forces by the seat back, or by any passenger restraint bars/belts" means there are no other forces.

WHAT IS APPARENT WEIGHT?

This is *not necessarily equal* to her true weight: $Wt = mg = 490$ N. *Apparent weight* is what she FEELS: the NORMAL FORCE caused by *contact* with the seat.

The quantities and FBD:

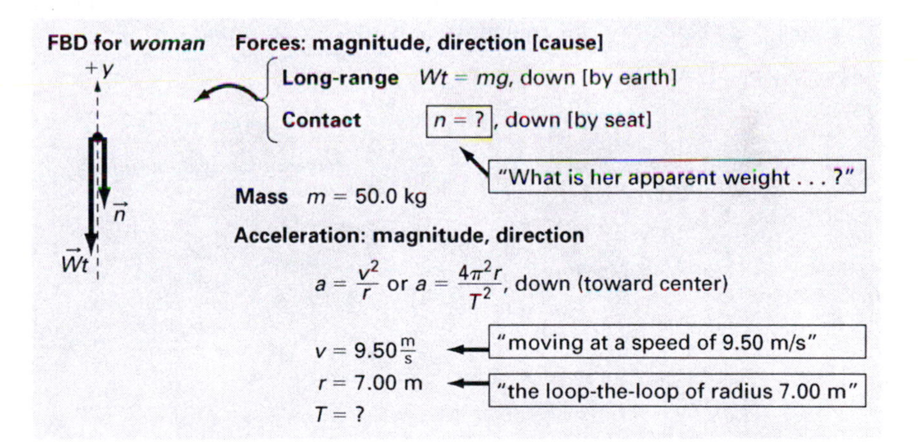

The y-components and Equation 5.1b:

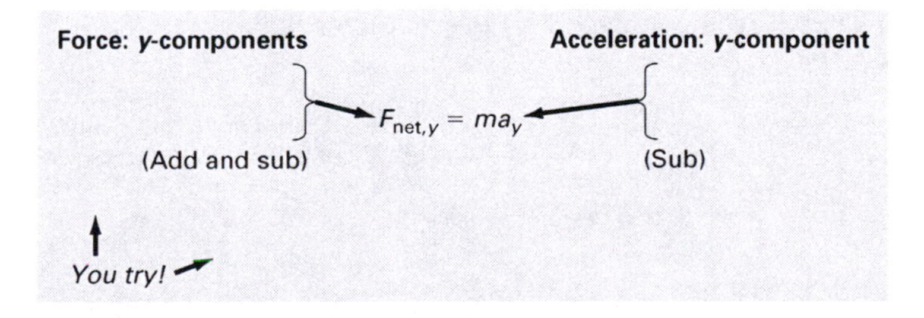

Part (b)

We want "the minimum speed necessary to just barely make it through the loop without her coming out of the seat." What does this mean?

> *LOSING CONTACT MEANS THE NORMAL FORCE IS ZERO!*
>
> At the *minimum speed,* the woman just barely begins to lose *contact* with the seat, and so the *normal force* (the *contact* force caused by the seat) acting on her is *zero.* So she *feels* weightless! Slower than this minimum speed: She falls away from the seat, or the car falls with her away from the track.

Everything is the same as in part (a), except now the normal force is known ($n = 0$), and the speed, v, is unknown. So we just reuse the equation from part (a), with a different unknown (check the answers if you did not yet get this equation):

$$n = m\frac{v^2}{r} - mg$$

← Sub in $n = 0$. One unknown; solve.

It was very useful to solve part (a) in symbols before putting in numbers, so we could reuse the equation in part (b) for a slightly different question.

Answers for . . .

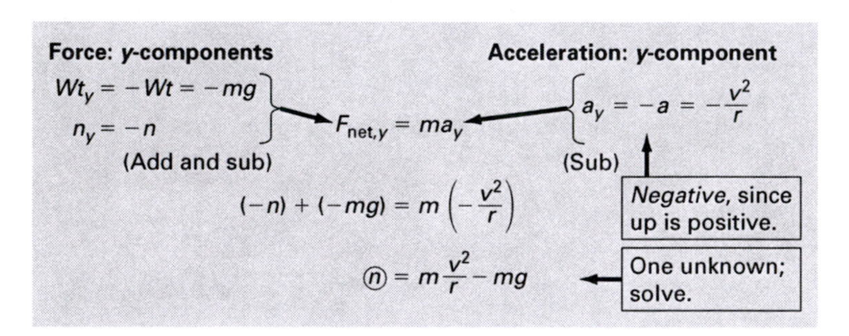

Answers

(a) Apparent weight: $n \cong 154.6 \cong 155$ N

To give the answer *as a multiple of her true weight,* we find the ratio:

$$\frac{n}{Wt} \cong \frac{(154.6 \text{ N})}{(490 \text{ N})} \cong 0.3155 \cong 0.316$$

So the apparent weight is 0.316 times the true weight, or $n = (0.316)Wt$.

(b) $v \cong 8.283 \cong 8.28 \frac{\text{m}}{\text{s}}$ ■

6.13 VERTICAL CIRCULAR MOTION—HIGHEST POINT, RIGHT-SIDE-UP

EXERCISE 6.9

The woman in the roller-coaster car from the previous exercise passes right-side-up over the top of a hill of radius 10.0 m. At the highest point, her apparent weight is one-third of her true weight. What is her speed?

Solution

(1) Type of problem

Force and circular motion: Use Equations 5.1a and/or 5.1b, along with Equation 6.2 or 6.3, and 6.1 if needed.

(2) Sort by object

Since her "apparent weight" (i.e., *normal force*) "is one-third of her true weight," then the normal force magnitude is $n = \frac{1}{3}Wt = \frac{1}{3}mg$. The FBD is different from in the previous exercise because now the seat is *below* her, and so the normal force is vertically *up:*

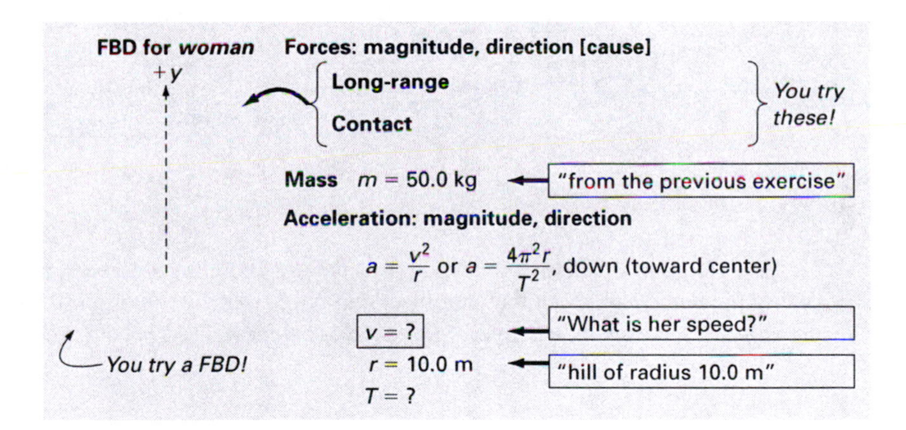

(3) Equations & unknowns and (4) Outline solution

The y-components and Equation 5.1b:

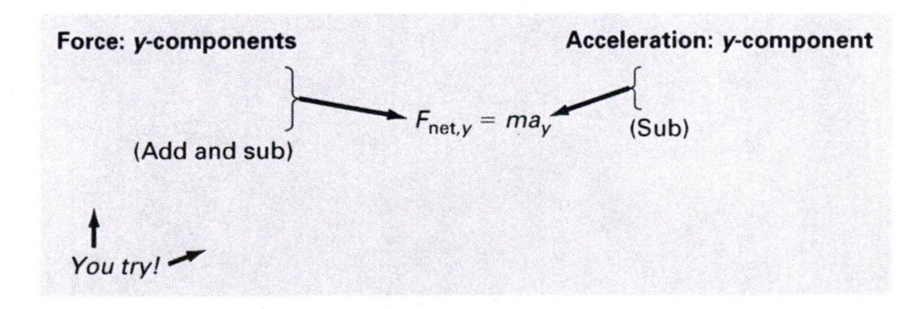

Answers for . . .

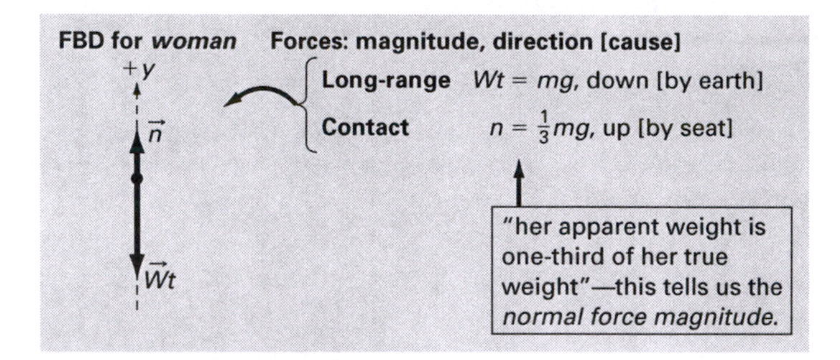

Answers for (3) Equations & unknowns and (4) Outline solution

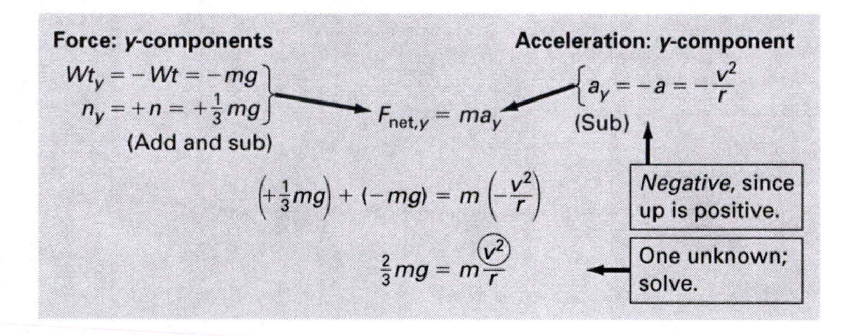

Answer $v \cong 8.083 \cong 8.08\,\frac{\mathrm{m}}{\mathrm{s}}$ ■

As a variation of the previous exercise, we are sometimes asked a question involving the *maximum speed* that an object can have over the top of a hill *without losing contact.* Here again, the *normal force* (or apparent weight) becomes *zero.*

CHAPTER 7

GRAVITATION AND ORBITS

IN THIS CHAPTER, we focus on gravitational force and circular orbit problems, using the methods from Chapters 5 and 6. There is one new fundamental equation:

Gravitational force *magnitude* (newtons)	$F_g = G\dfrac{m_1 m_2}{r^2}$	(7.1)

The quantities in this equation are defined like so:

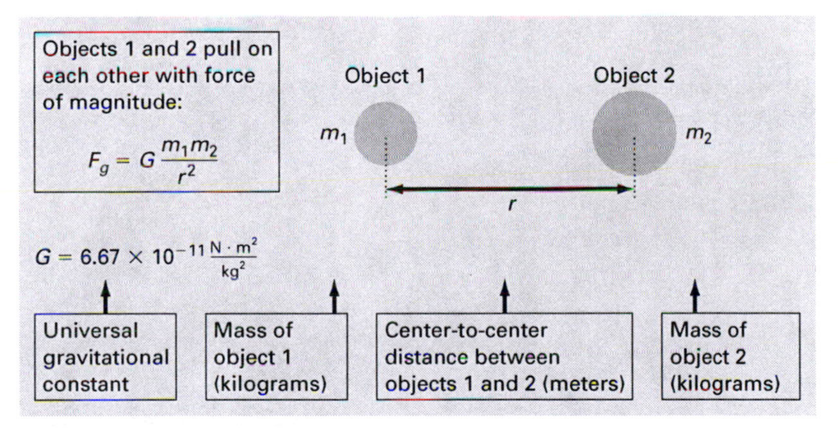

They each cause *exactly the same-sized force,* regardless of which mass is bigger! Here, the force acting on object 1 is to the right, caused by object 2. The force acting on object 2 is to the left, caused by object 1.

7.1 WEIGHT AND *g* AT A PLANET'S SURFACE

Here is a common type of gravitational force problem.

EXERCISE 7.1

A block of mass 1250 kg sits on the surface of a planet. The planet's mass is 4.10×10^{23} kg and its radius is 5.20×10^6 m. Determine (a) the weight of the block and (b) the value of g at the planet's surface.

Solution

(1) Type of problem

The *weight* of the block is the *gravitational force,* caused by the planet, acting on the block. Up to now, we have said that weight is directed vertically down, but it helps to see it more generally:

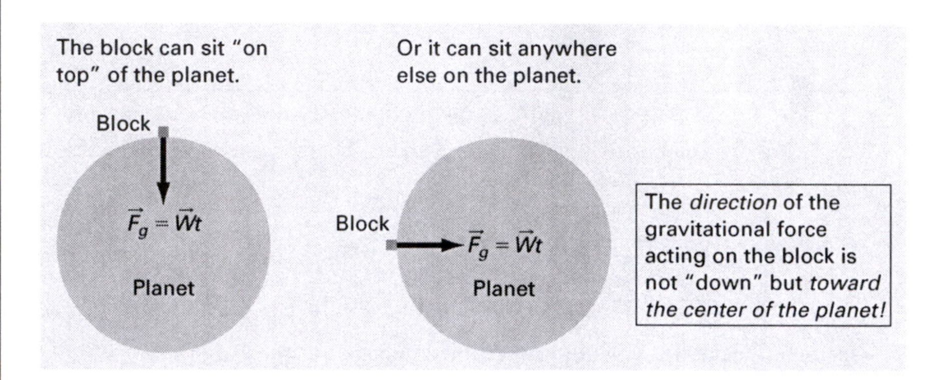

We use Equation 7.1, along with the "old" equation for weight ($Wt = mg$). However, g is NOT $9.8\frac{\text{m}}{\text{s}^2}$ because the planet is not the earth!

(2) Sort by object

We set up this situation for Equation 7.1:

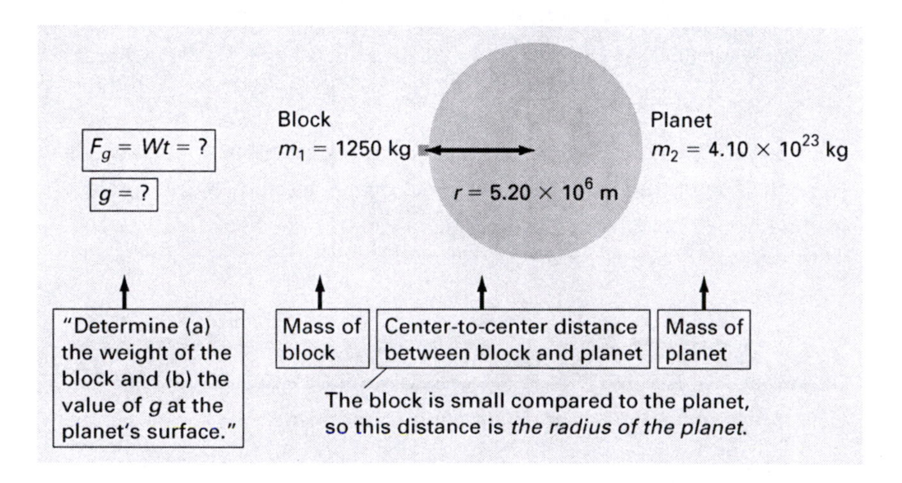

The block also pulls *on the planet,* with the same-sized force, but here we are looking at the force *on the block.*

(3) Equations & unknowns and (4) Outline solution

Our two equations:

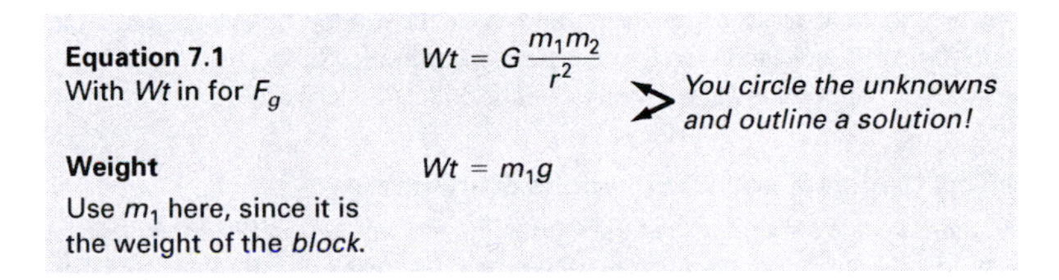

Answers for (3) Equations & unknowns and (4) Outline solution

Equation 7.1	$\textcircled{Wt} = G\dfrac{m_1 m_2}{r^2}$	(1) Solve for *Wt*.
Weight	$\textcircled{Wt} = m_1\textcircled{g}$	(2) Sub in *Wt*; solve for *g*.

Answers $Wt \cong 1264 \cong 1.26 \times 10^3$ N, $g \cong 1.011 \cong 1.01\frac{\text{m}}{\text{s}^2}$ ■

7.2 ADDING GRAVITATIONAL FORCE VECTORS

When more than one gravitational force pulls on an object, we add the forces as vectors.

EXERCISE 7.2

A satellite is somewhere directly between the earth and the moon. At what distance x from the earth's center is the net gravitational force on the satellite zero? Ignore gravitational effects due to other bodies such as the sun. The figure is not to scale.

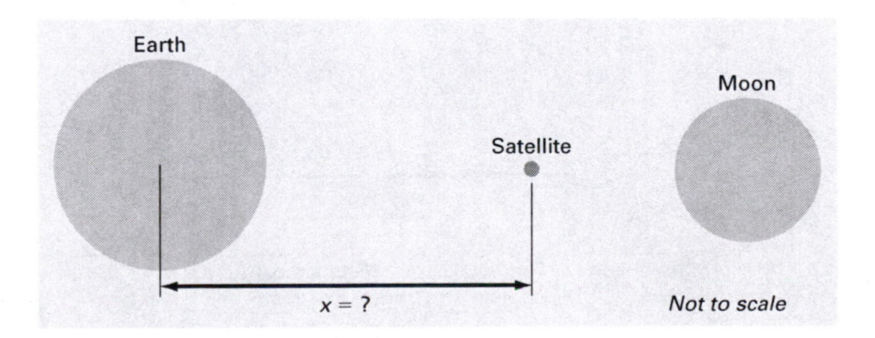

Solution

(1) Type of problem

Gravitational force: Use Equation 7.1.

Note: We can look up the earth's radius and mass, the moon's radius and mass, the distance between the earth and moon, and so on, in any textbook. So we treat quantities like these as *known.* We will give the values we need at the end of the exercise.

(2) Sort by object and (3) Equations & unknowns

The main idea: We want the "net gravitational force *on the satellite*" to be zero. We are looking at how the earth and moon each pull *on the satellite,* NOT at the earth and moon pulling on each other.

There are two long-range gravitational forces acting *on the satellite:* a force to the left caused by the earth ($\vec{F}_e$) and a force to the right caused by the moon ($\vec{F}_m$):

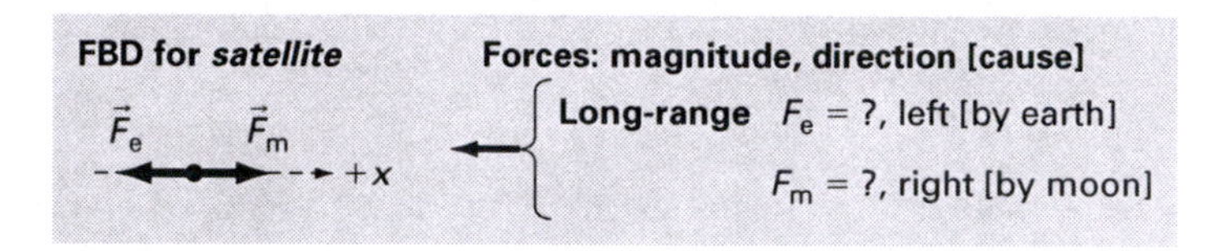

The x-components add to equal zero, so the magnitudes must be equal:

Equal magnitudes $F_e = F_m$

Now we use Equation 7.1 to get an equation for each of these unknown magnitudes.

Force caused by the earth

To calculate F_e, we use *the center-to-center distance between the satellite and the earth.* Looking only at the satellite and the earth (ignoring the moon for now), we set up for Equation 7.1:

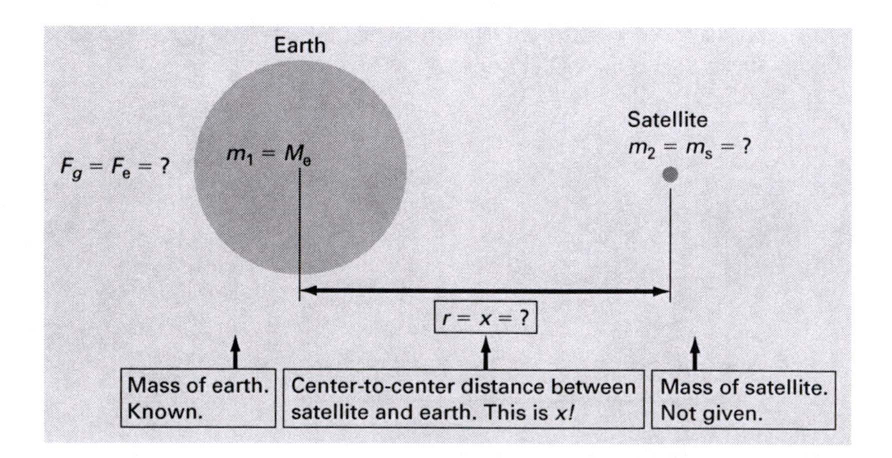

So:

Equation 7.1	becomes	
$F_g = G\frac{m_1 m_2}{r^2}$ ⇨	$F_e = G\frac{M_e m_s}{x^2}$	**Force caused by the earth**

USE THE CORRECT r IN EQUATION 7.1!

A common mistake here is to use the radius of the earth, R_e, to get F_e:

$$F_e = G\frac{M_e m_s}{R_e^2}$$ ← WRONG! Don't make this mistake!

Do you see what is wrong? In Equation 7.1, r is the *center-to-center distance* between the two objects. This wrong equation would be correct if the satellite were *on the earth's surface,* a distance R_e from the center of the earth.

Force caused by the moon

To get an equation for F_m, we need the center-to-center distance between the satellite and the moon. Since x is unknown, this distance is also unknown, but we can write it in terms of x:

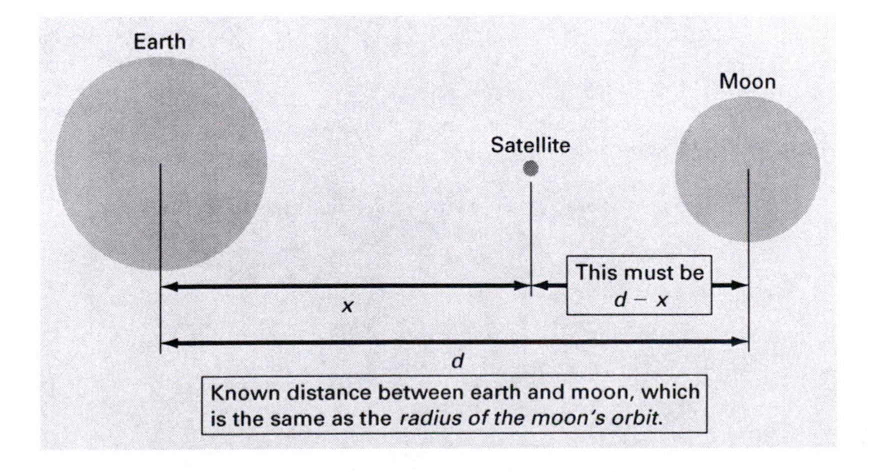

Now we look at only the satellite and the moon, and set up for Equation 7.1:

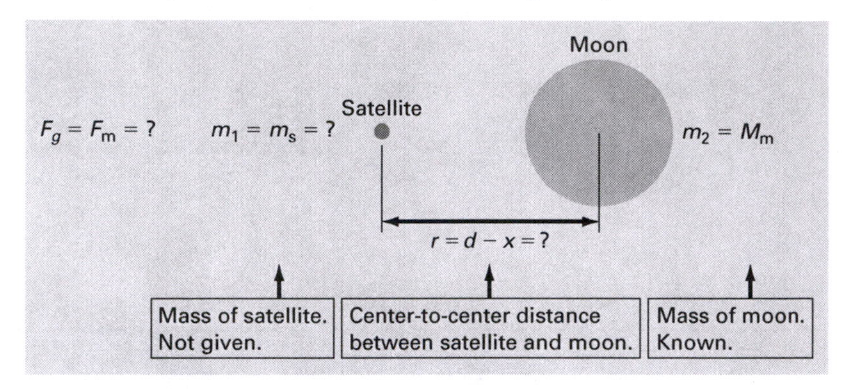

So:

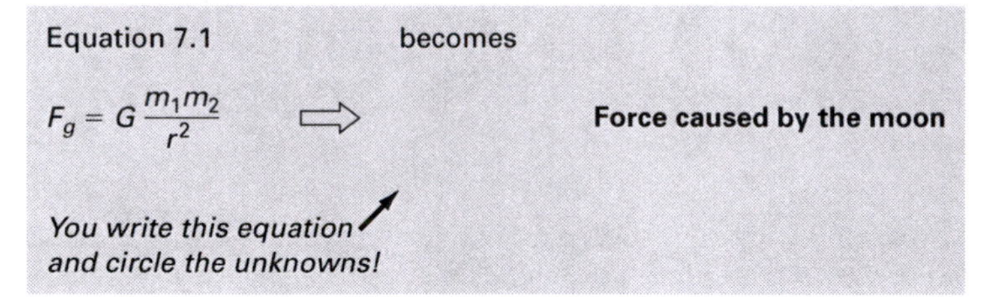

The answer is in the next part.

(4) Outline solution

Writing all of the equations together:

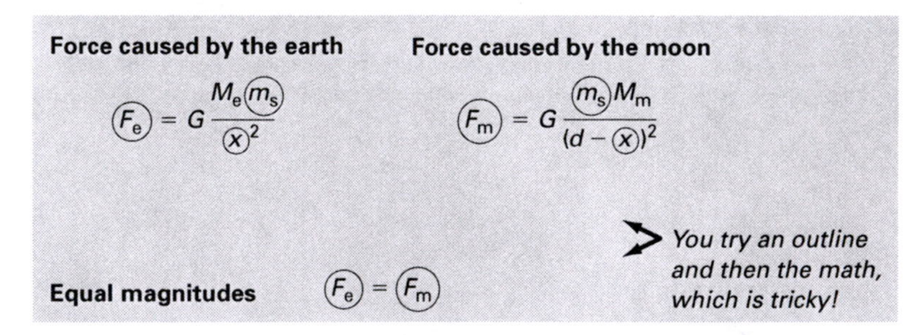

Answer for (4) Outline solution

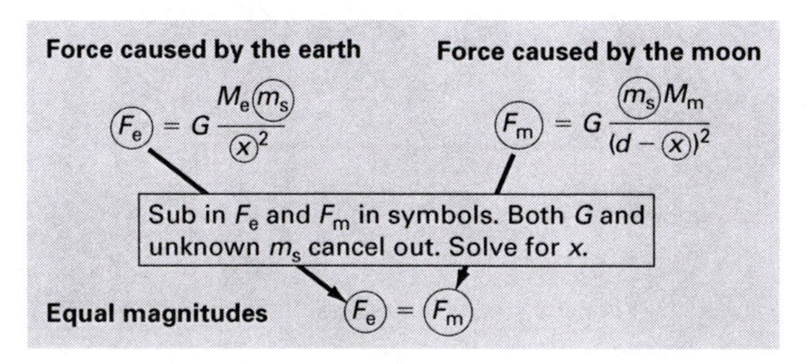

Answer for the math

The algebra here is harder than usual, and so we show the details:

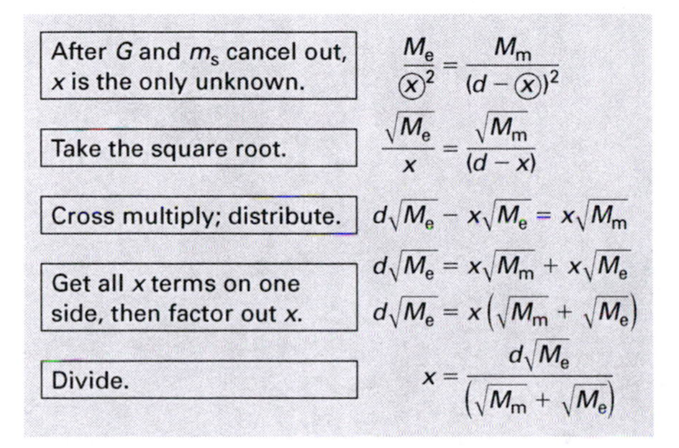

In the end, the three numbers we need to look up are the mass of the earth ($M_e = 5.98 \times 10^{24}$ kg), the mass of the moon ($M_m = 7.36 \times 10^{22}$ kg), and the center-to-center distance between the earth and the moon ($d = 3.82 \times 10^8$ m), which is the same as the radius of the moon's orbit. We never need the radius of either the earth or the moon for this problem.

Answer $x \cong 3.44 \times 10^8$ m ■

7.3 CIRCULAR ORBIT PROBLEMS

Why does a satellite move in a circular orbit around a planet, rather than in a straight line as its inertia would have it? As in Chapter 6, there must be a force *toward the center* of the circular path: It is the *gravitational force* caused by the planet, calculated with Equation 7.1.

This applies to a satellite orbiting a planet, a moon orbiting a planet, a planet orbiting the sun, and so on. Even though orbits are generally elliptical, we often approximate them to be circular.

The next exercise shows how to use all the main equations that might be needed for *any* circular orbit problem. Remember, most of the steps are *mental!*

EXERCISE 7.3

A satellite is in a circular, synchronous orbit at an altitude $h = 8.00 \times 10^6$ m above the surface of a planet. The radius R of the planet is 1.30×10^6 m, and the period of the planet's rotation is 16.0 hours (the planet's "day"). Determine (a) the orbital speed, v, of the satellite and (b) the mass of the planet.

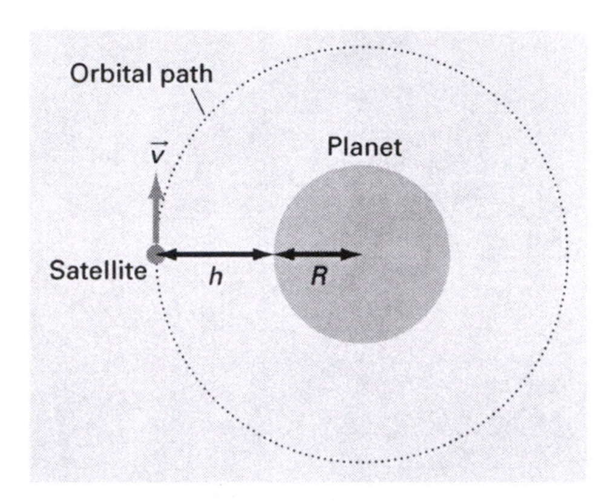

WHAT IS A SYNCHRONOUS ORBIT?

The word *synchronous* means that the orbital period, T, of the satellite is *equal to* the rotation period (the "day") of the planet. So here, $T = 16.0$ hours for the satellite.

Geosynchronous is the term used if the planet is the *earth,* and so the orbital period of a *geosynchronous* satellite is about 24 hours.

Solution

(1) Type of problem

The main ideas in this problem:

- **Gravitational force:** Use Equation 7.1.
- **Force and circular motion:** As outlined in Chapter 6, use Equations 5.1a and/or 5.1b, along with Equation 6.2 or 6.3, and 6.1 if needed.

(2) Sort by object and/or circle and (3) Equations & unknowns

Before doing the main setup, we need the center-to-center distance, which in this case is the radius of the satellite's orbit:

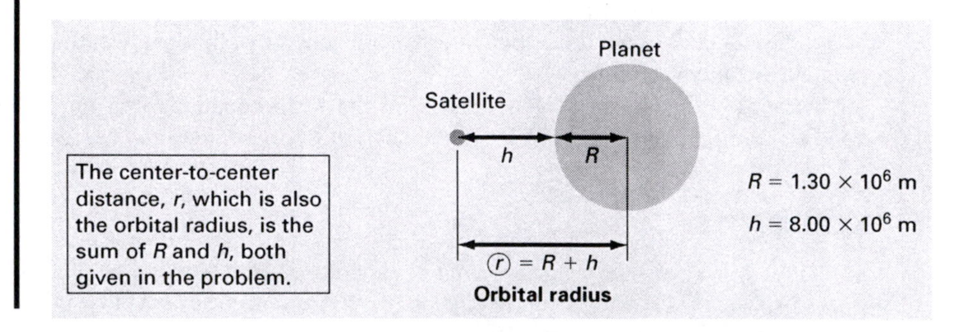

DON'T CONFUSE THE PLANET'S RADIUS, R, WITH THE ORBIT RADIUS, r!

They are related by the equation $r = R + h$, where h is the *altitude* above the surface.

Gravitational force

The setup for Equation 7.1:

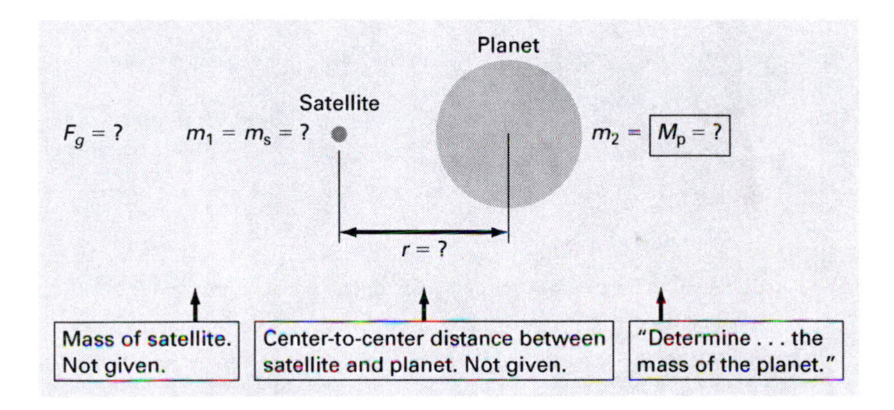

So:

Equation 7.1 becomes

$$F_g = G\frac{m_1 m_2}{r^2} \Rightarrow F_g = G\frac{m_s M_p}{r^2}$$ **Gravitational force**

Force and circular motion

First, unit conversions:

"[S]ynchronous orbit . . . and the period of the planet's rotation is 16.0 hours"—this is the orbital period of the satellite!

$$T = 16.0\ \text{hr} \times \left(\frac{3600\ \text{s}}{1\ \text{hr}}\right) = 57{,}600\ \text{s}$$

WATCH THE UNITS IN GRAVITATION PROBLEMS!

Convert masses to *kilograms,* distances to *meters,* and times to *seconds.* This is the case for most problems, but especially gravitation problems, because quantities are often given in other units.

The FBD and circular motion quantities *for the satellite:*

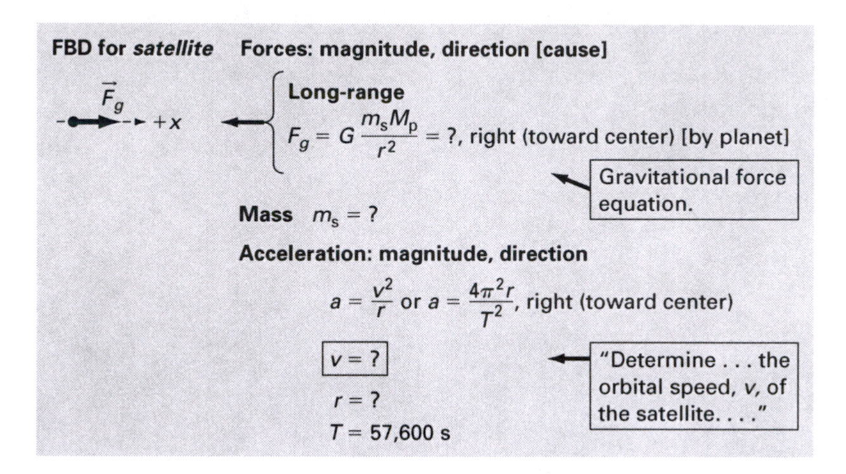

We can use Equation 5.1a with either of the centripetal acceleration equations:

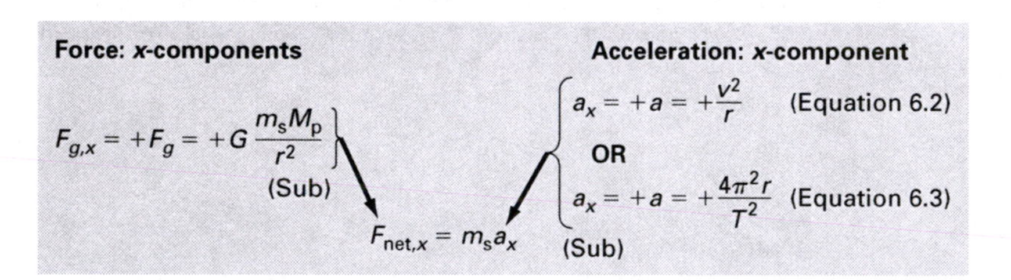

This gives two possible equations:

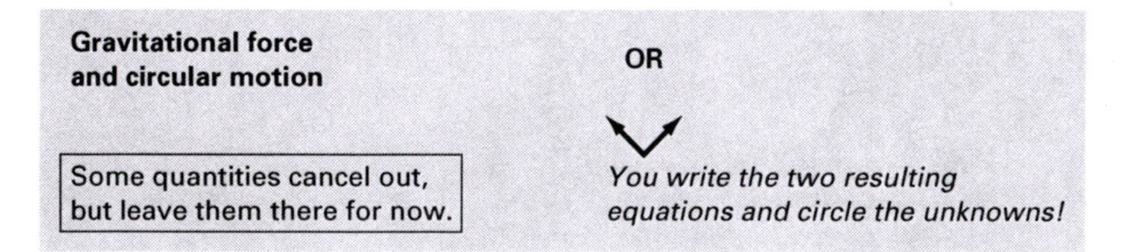

The answers are in the next part.

Finally, we know T and want to solve for v, so we also use Equation 6.1 for the tangential or orbital speed:

Orbital speed $(v) = \frac{2\pi (r)}{T}$

(4) Outline solution

We put all the equations together:

Gravitational force and circular motion $G\frac{m_s M_p}{r^2} = m_s \frac{v^2}{r}$ **OR** $G\frac{m_s M_p}{r^2} = m_s \frac{4\pi^2 r}{T^2}$

You outline a solution!

Orbital radius $r = R + h$ **Orbital speed** $v = \frac{2\pi r}{T}$

Answers for (4) Outline solution

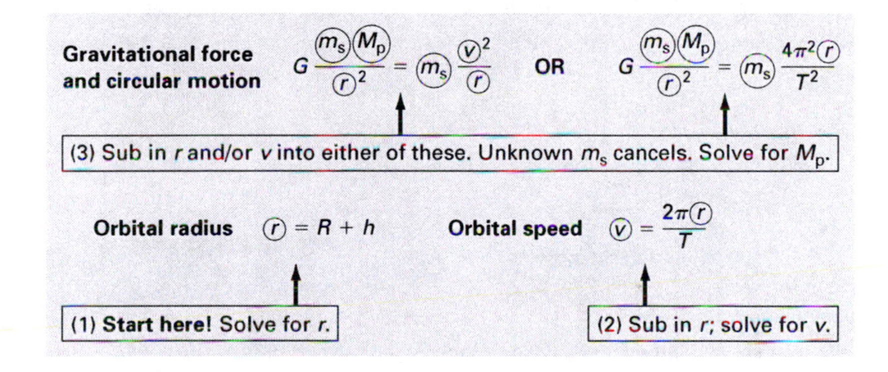

Answers $v \cong 1.01 \times 10^3\ \frac{\text{m}}{\text{s}}$, $M_p \cong 1.43 \times 10^{23}$ kg ■

7.4 CIRCULAR ORBIT EQUATIONS

Here is a summary of the equations from the previous exercise, which can be used for nearly all problems involving gravitational force with circular orbits:

Gravitational force and circular motion $G\frac{m_s M_p}{r^2} = m_s \frac{v^2}{r}$ **OR** $G\frac{m_s M_p}{r^2} = m_s \frac{4\pi^2 r}{T^2}$

Orbital radius $r = R + h$ **Orbital speed** $v = \frac{2\pi r}{T}$

Go through the previous exercise to make sure you understand what they mean!

7.5 COMPARING ORBITS AT TWO DIFFERENT RADII

Now we use these circular orbit equations in an exercise comparing two orbits.

EXERCISE 7.4

Satellites 1 and 2 both orbit a planet in circular orbits at different radii. Satellite 1 has twice the speed of satellite 2. Determine the ratio, r_2/r_1, of the orbital radii.

Solution

(1) Type of problem

These are circular orbits, so the setup and equations for each orbit are nearly identical to those in the previous exercise.

(2) Sort by object and/or circle and (3) Equations & unknowns

When comparing two objects, we need equations *relating the objects* and equations *for each object.*

Relating the objects
The problem statement gives us this equation:

Relating the objects $v_1 = 2v_2$ "Satellite 1 has twice the speed of satellite 2."

We won't bother circling unknowns, because nearly everything is "unknown."

For each object
The setup for each satellite is nearly identical to that in the previous exercise. So we use the same circular orbit equations, separately for each satellite (except we don't need $r = R + h$ since altitude is not mentioned).

Which circular orbit equation(s) should we use? We know something about *speeds* ($v_1 = 2v_2$) and we *want* to know something about the *radii* (r_2/r_1). However, we know nothing about *periods* (T_1 or T_2). So:

$$G\frac{m_s M_p}{r^2} = m_s\frac{v^2}{r}$$

The best equation has in it both *speed, v,* and *radius, r,* but NOT *period, T.*

Here it is with subscripts for the mass, speed, and radius of each satellite:

Satellite 1 $G\dfrac{m_1 M_p}{r_1^2} = m_1\dfrac{v_1^2}{r_1}$ **Satellite 2** $G\dfrac{m_2 M_p}{r_2^2} = m_2\dfrac{v_2^2}{r_2}$

We can simplify these by canceling some terms.

(4) Outline solution

Putting the equations together:

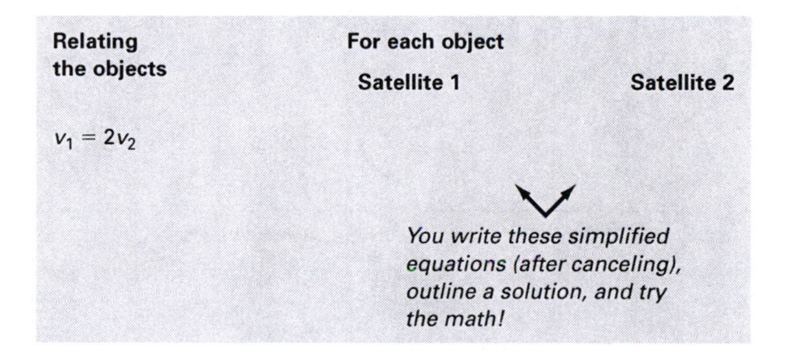

Answers for (4) Outline solution

One possible outline and the math:

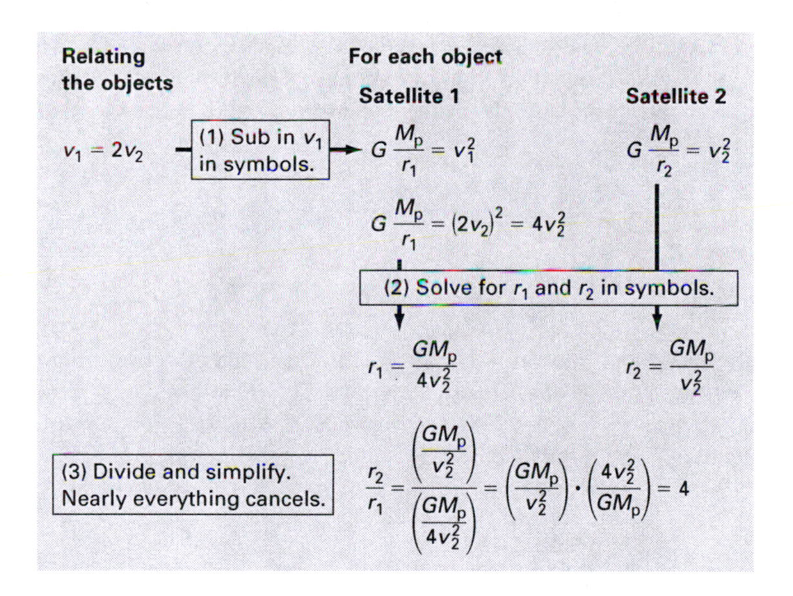

Answer $\frac{r_2}{r_1} = 4$ ■

CHAPTER **8**

WORK AND ENERGY

IN THIS CHAPTER, we show how to approach problems that involve the quantity *work,* the quantities *work* and *energy* together, and *conservation of energy.*

8.1 WORK DONE BY A CONSTANT/ AVERAGE FORCE

Say a block is being dragged across a tabletop surface during a time *interval.* Beginning at the *initial* instant (I), it moves through a displacement, $\vec{d}$, ending at the *final* instant (F). We can picture the displacement vector and the interval like so:

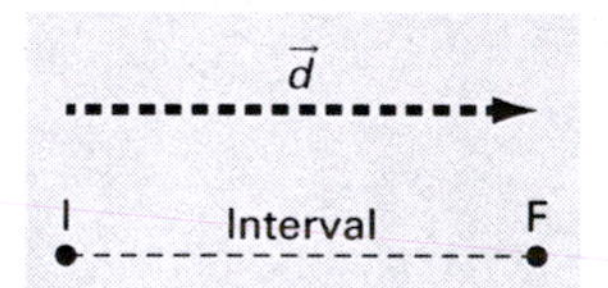

During this entire interval, there are forces (like weight, tension, normal, friction, etc.) acting on the block. To calculate the work done by *one* of these forces (call it $\vec{F}$), we need *the angle θ between the vectors* $\vec{F}$ and $\vec{d}$. One possible situation is shown here:

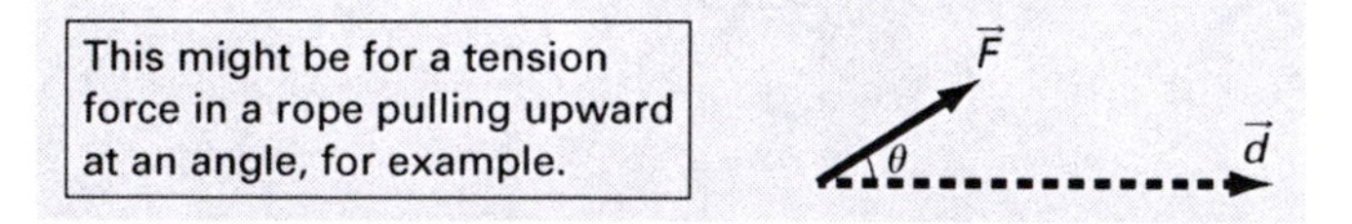

The equation to calculate work:

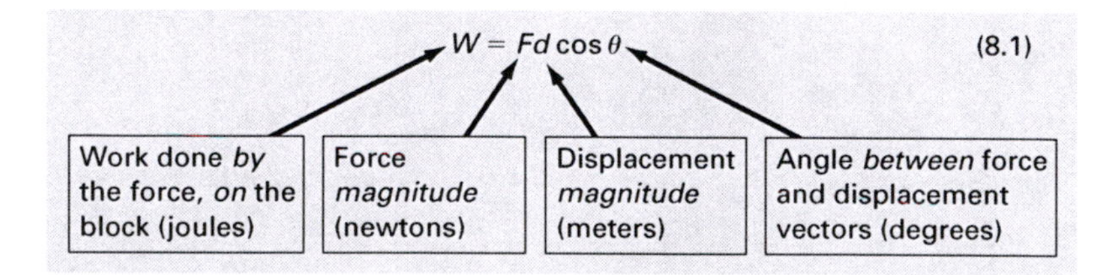

UNDERSTAND EACH TERM IN THE EQUATION!

"Work" is an everyday term that physics has borrowed and redefined with a very specific meaning. Pay attention to the details in the equations and definitions!

Although calculated by using two vectors, work is a *scalar* and so has *no direction,* but it does have a *sign:*

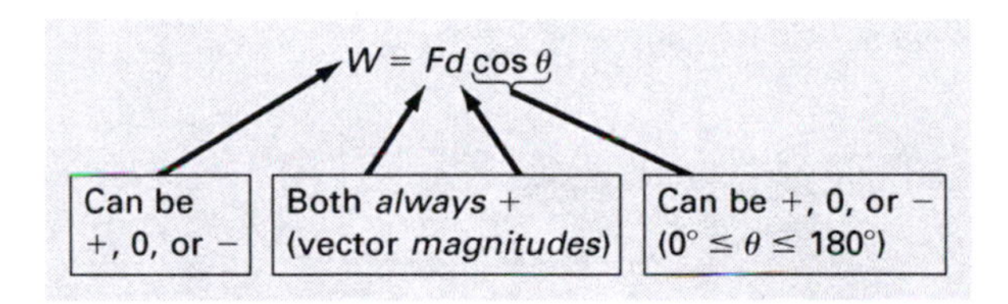

8.2 WORK PROBLEMS—WITH TWO OR MORE FORCES

As usual, we detail all the mental and written steps, and so the exercises shown here are much longer than what you would ever write on your own paper.

EXERCISE 8.1

Two boys pull a block along the floor for a distance of 4.0 m toward the east. One boy pulls with force $\vec{F}_1$ (magnitude 75 N) and the other pulls with $\vec{F}_2$ (also magnitude 75 N). A friction force, $\vec{f}_k$ (magnitude 25 N), opposes the motion. (a) How much work is done on the block by each of these forces? (b) What is the total work done on the block by all three forces?

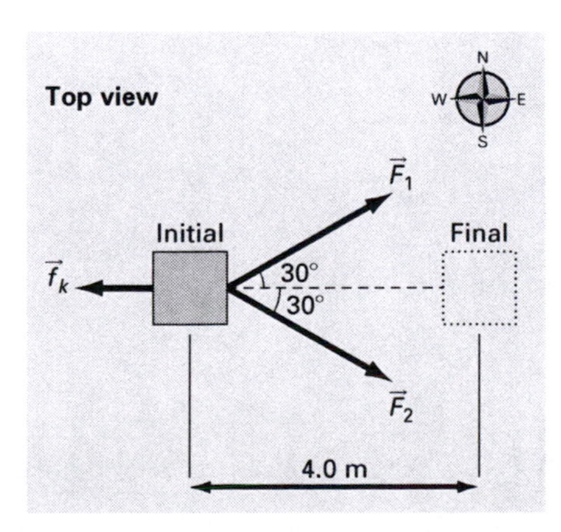

Solution

(1) Type of problem

Work: Use Equation 8.1 for each force individually; then add to get total work.

(2) Sort by displacement and force vectors

To calculate work with Equation 8.1, we sort out information for each of the following:

- **Displacement vector:** *magnitude* and *direction*
- **Force vectors:** *magnitude* and *direction* for each, and *work done* by each, and then the *total work* done by all the force vectors

Displacement vector

This goes from I to F of the interval:

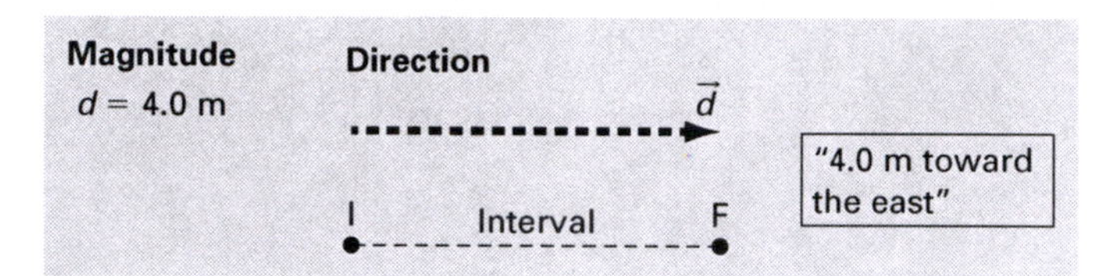

Force vectors

These are given in the problem:

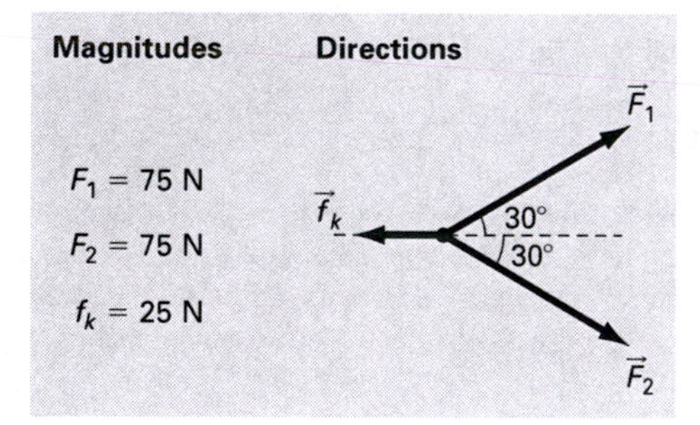

And the work:

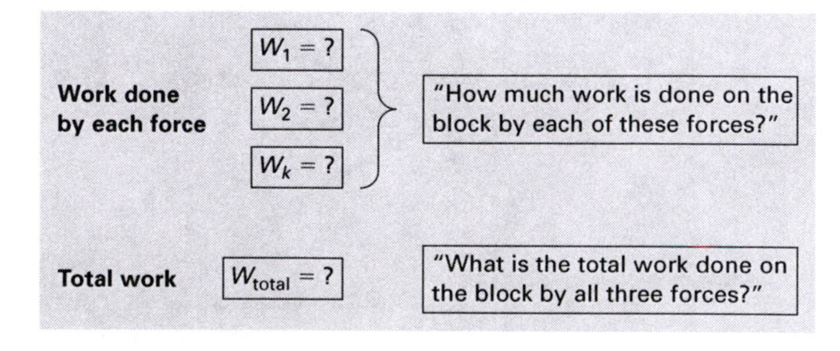

(3) Equations & unknowns

Two parts here:

- **Work done by each force:** Equation 8.1 for each
- **Total work:** Just add

Work done by each force

For *each* force vector, we do the mental and written steps illustrated here for $\vec{F}_1$:

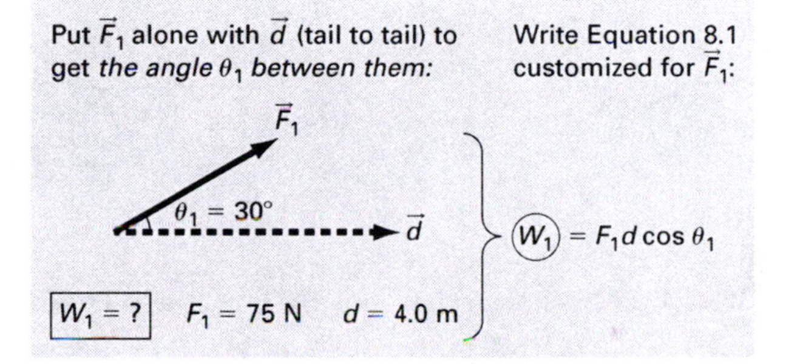

Now we do the same thing for $\vec{F}_2$:

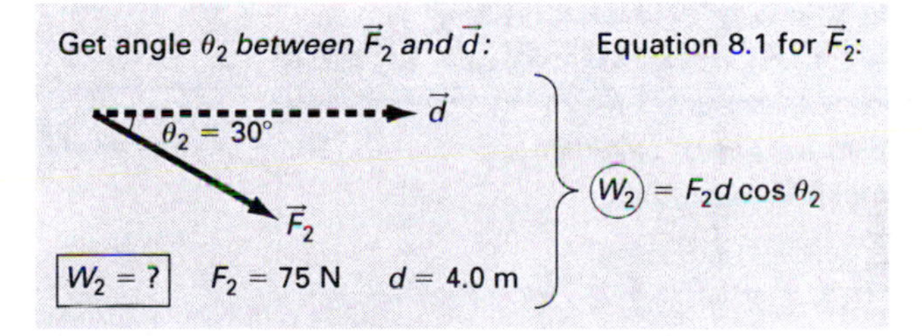

And, finally, for $\vec{f}_k$, which is in the opposite direction of $\vec{d}$*:*

Get angle θ_k *between* $\vec{f}_k$ *and* $\vec{d}$*:*

Equation 8.1 for $\vec{f}_k$*:*

$\theta_k = 180°$

$\vec{f}_k$ $\vec{d}$

$$W_k = f_k d \cos\theta_k$$

$W_k = ?$ $f_k = 25$ N $d = 4.0$ m

Total work

Work is a scalar, which can be positive or negative but has *no components*. So, to get the total work, we *just add* the work values from the individual forces:

$$W_{total} = W_1 + W_2 + W_k$$

(4) Outline solution

The outline is very simple:

Work done by each force

$(W_1) = F_1 d \cos \theta_1$

$(W_2) = F_2 d \cos \theta_2$

$(W_k) = f_k d \cos \theta_k$

(1) One unknown in each; solve.

Total work $(W_{total}) = (W_1) + (W_2) + (W_k)$

(2) Sub in the work done by each force, and solve for W_{total}.

Answers $W_1 = W_2 \cong 259.8 \cong 260$ J, $W_k = -100$ J, $W_{total} \cong 419.6 \cong 420$ J ■

8.3 WORK PROBLEMS—WHEN FORCES ARE NOT GIVEN

If a problem asks us to calculate work but the force vectors are NOT given, we can often use a FBD to get them.

EXERCISE 8.2

A 2.0-kg block slides a distance of 3.0 m down a frictionless ramp angled at 30° from the horizontal. (a) Determine the work done by each force acting on the block. (b) Determine the work done by the net force on the block.

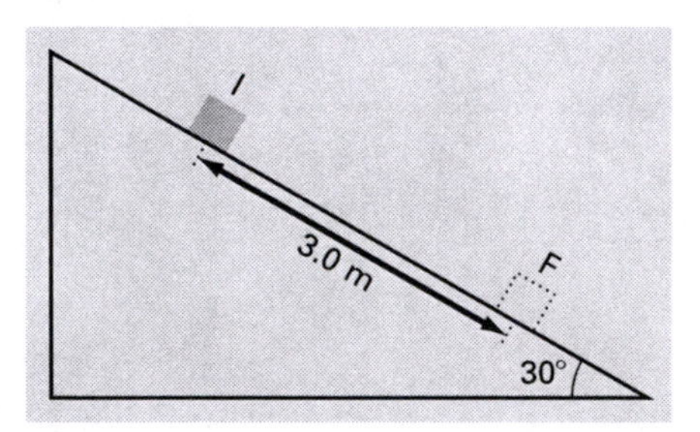

Solution

(1) Type of problem

Work: No forces are given, so we use a FBD to help get magnitudes and directions of forces on the block. Otherwise, the setup is just like that in the previous exercise.

(2) Sort by displacement and force vectors

Displacement vector

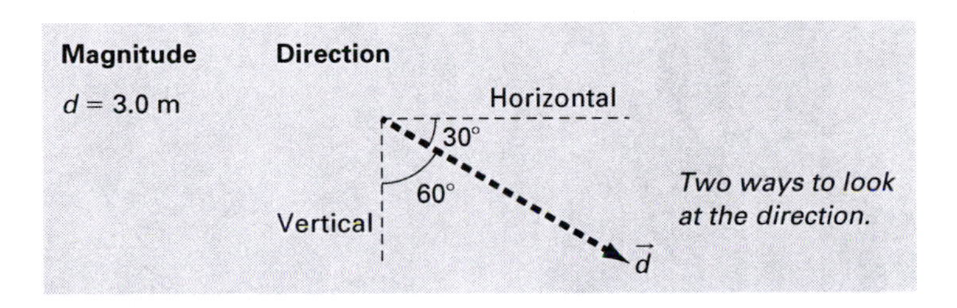

Force vectors

Because the forces acting on the block are not specifically given in the problem, we identify them and make a FBD (see Section 5.1 for a review):

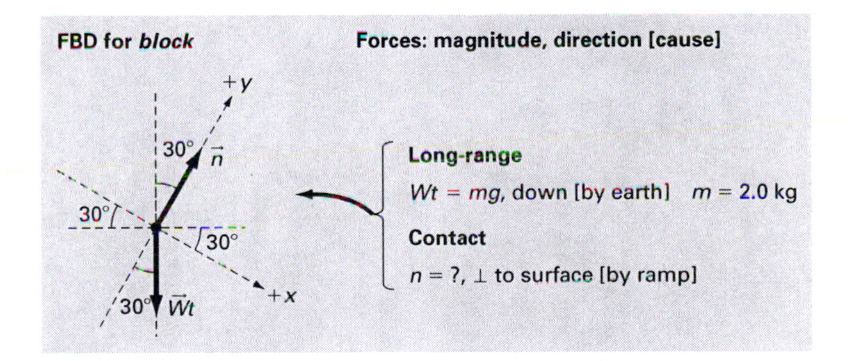

We don't need the x- and y-axes (to calculate components) here, but we will in the next exercise!

Sorting out each of the work quantities:

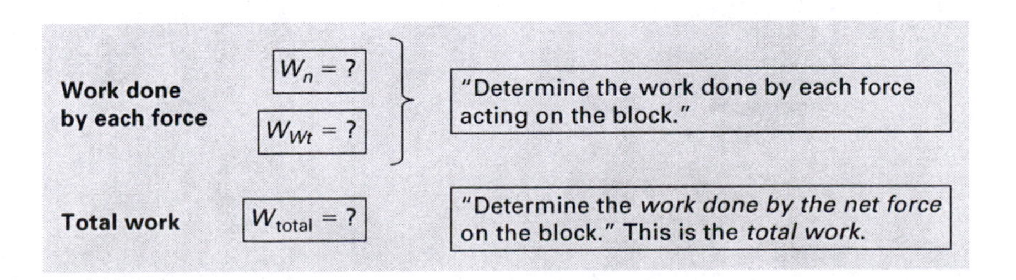

(3) Equations & unknowns

Work done by each force

First, $\vec{n}$:

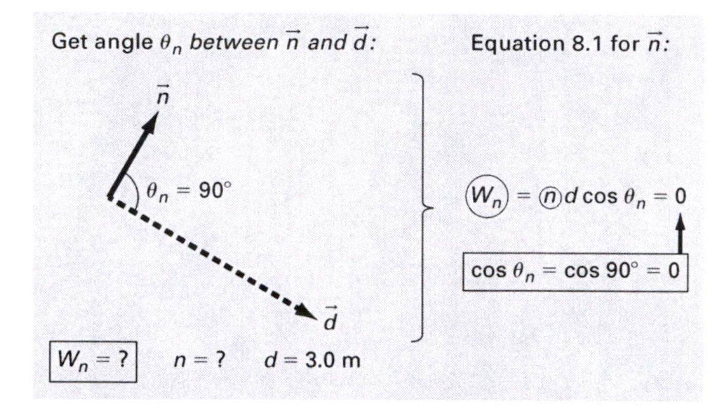

Since $\vec{n}$ is perpendicular to $\vec{d}$, then $W_n = 0$. It does not matter that n is unknown. Hooray!

For $\vec{Wt}$:

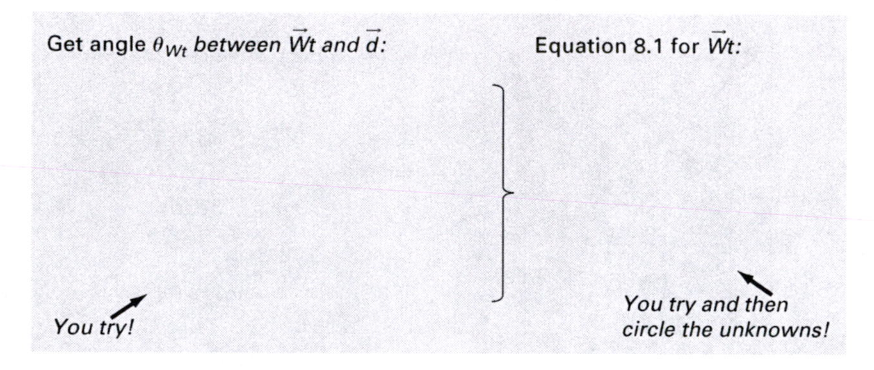

Total work

We add the individual work values to get the total:

$$W_{total} = W_n + W_{Wt}$$

(4) Outline solution

Putting the equations together:

Work done by each force

$$W_n = n d \cos\theta_n = 0$$

$$W_{Wt} = mgd \cos\theta_{Wt}$$

You try an outline!

Total work

$$W_{total} = W_n + W_{Wt}$$

Answers for . . .

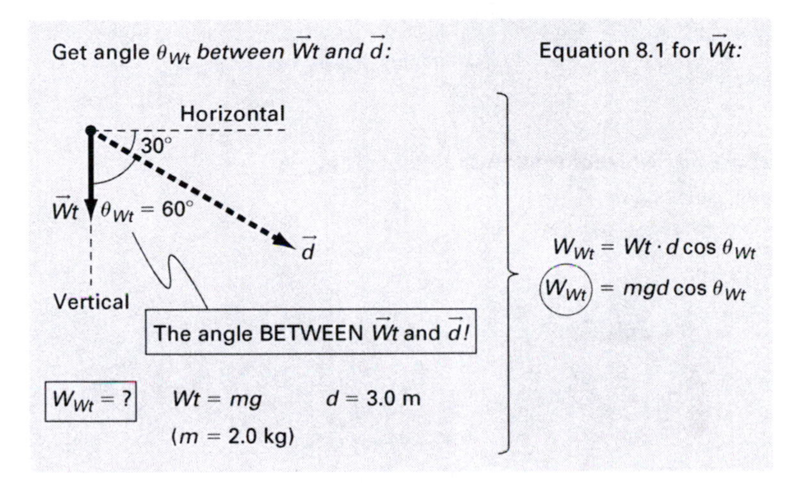

Answers for (4) Outline solution

Work done by each force	$W_n = n d \cos \theta_n = 0$	(1) Already solved! Zero!
	$W_{Wt} = mgd \cos \theta_{Wt}$	(2) One unknown; solve.
Total work	$W_{\text{total}} = W_n + W_{Wt}$	(2) Sub in the work done by each force, and solve for W_{total}.

Answers $W_n = 0$, $W_{Wt} \cong 29.4 \cong 29 \text{ J}$, $W_{\text{total}} \cong 29.4 \cong 29 \text{ J}$ ■

In the previous exercise, we never solved for n, but in the next exercise we will need to.

EXERCISE 8.3

(This is like the previous exercise, but with friction and different unknowns.) A 2.0-kg block slides a distance of 3.0 m down a ramp angled at 30° from the horizontal. Kinetic friction is present. The total work done on the block by all forces is +17 J. (a) Determine the work done by the kinetic friction force. (b) Determine the coefficient of kinetic friction between the block and the ramp surface.

Solution

(1) Type of problem

Work: Set up just like in the previous exercise!

USE THE SAME SETUP REGARDLESS OF THE QUESTION!

We now have different unknowns from the previous exercise, but both problems involve *work*. We use the same setup, no matter what the unknowns are.

(2) Sort by displacement and force vectors

Displacement vector

Just like in the previous exercise:

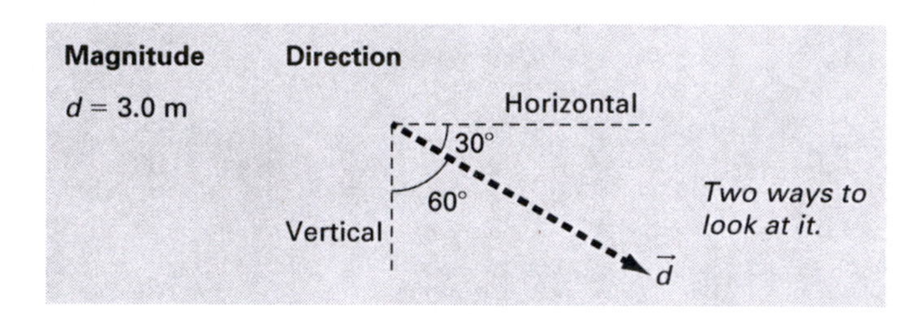

Force vectors

Now the FBD includes kinetic friction:

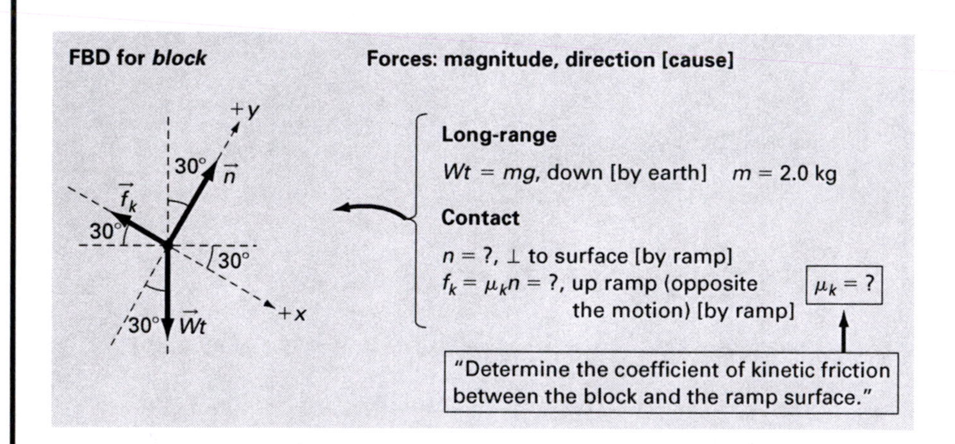

HOW IS THE COEFFICIENT OF KINETIC FRICTION RELATED TO WORK?

The coefficient of friction, μ_k, is related to the friction force ($f_k = \mu_k n$), which does work, W_k, on the block.

The work quantities:

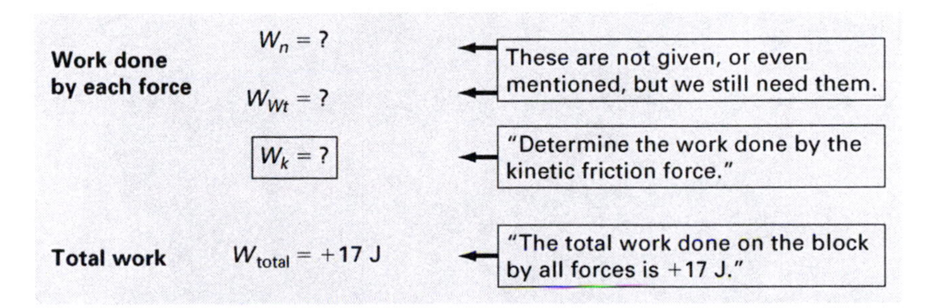

(3) Equations & unknowns

In some work problems, we get an equation from the FBD in addition to our other equations.

Equation from FBD

Because there is kinetic friction, we need an equation for n (since $f_k = \mu_k n$). To do this, we get y-components from the FBD and use Equation 5.1b. See Chapter 5 for a detailed review of how to do this for an object on an incline. We first use a right triangle to get the y-component of the weight vector (enlarged view):

Wt_y, 30°, Wt, Wt_x

$$\frac{Wt_y}{Wt} = \cos 30° \Rightarrow Wt_y = -Wt \cos 30°$$

Make this *negative,* since it is in negative y-direction.

For review, you try the y-components and Equation 5.1b:

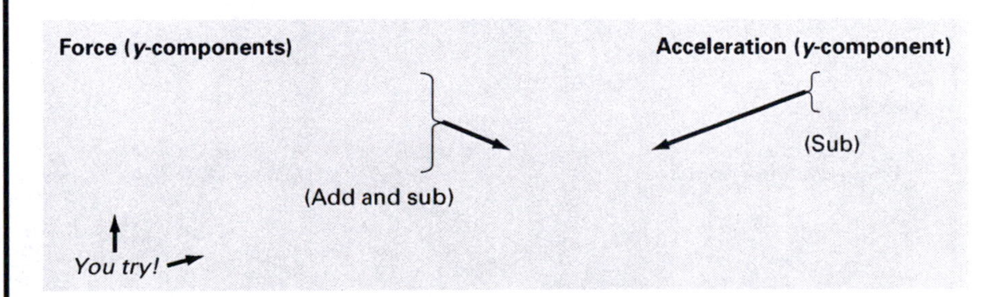

Hint: If there is acceleration, it is in the x-direction along the ramp surface. That means $a_y = 0$, and so *we do not need to know the acceleration magnitude.*

Work done by each force

The work equations for $\vec{n}$ and $\vec{Wt}$ are exactly the same as in the previous exercise.

Now for $\vec{f}_k$:

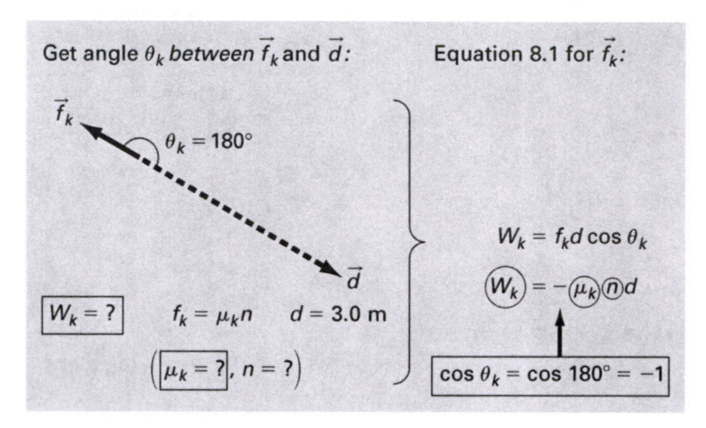

Total work

Add the individual work values:

$$W_{total} = W_n + W_{Wt} + W_k$$

(4) Outline solution

Finally, we put together all the equations, including those for W_n and W_{Wt}, from the previous exercise:

Equation from FBD

$$n = mg\cos 30°$$

Work done by each force

$$W_n = n d \cos\theta_n = 0$$

$$W_{Wt} = mgd\cos\theta_{Wt}$$

$$W_k = -\mu_k n d$$

Total work

$$W_{total} = W_n + W_{Wt} + W_k$$

You try an outline!

Answers for . . .

Force (y-components)

$$n_y = +n = ?$$

$$Wt_y = -Wt\cos 30° = -mg\cos 30°$$

(Add and sub)

Acceleration (y-component)

$$a_y = 0$$

(Sub)

$$F_{net,y} = ma_y$$

$$(+n) + (-mg\cos 30°) = m(0)$$

$$n = mg\cos 30°$$

Answers for (4) Outline solution

One possible outline:

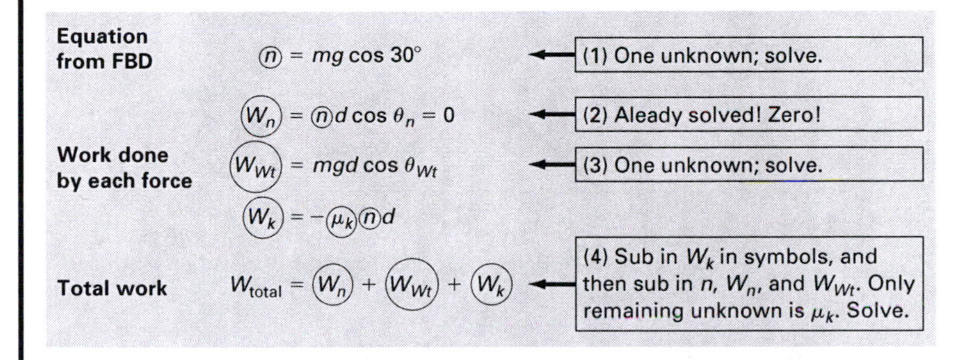

Answers $W_k \cong -12.4 \cong -12$ J, $\mu_k \cong 0.244 \cong 0.24$ (watch the signs!) ■

8.4 HOW TO SET UP WORK PROBLEMS

Here is a summary of what we have done in the last few exercises:

Work Problems—Mental and Written Steps

Mental →	**(1) Type of problem** Work: Use Equation 8.1 for each force individually, and then add to get total work. Use Equation 5.1a or 5.1b if needed.

Mental and written →	**(2) Sort by displacement and force vectors**
	Sort the following known or unknown quantities:
	• **Displacement vector:** *magnitude, direction*
	• **Force vectors:**
	  • For forces not given: Use FBD if needed.
	  • *Magnitude, direction,* and *work done* (for each force vector)
	  • *Total work* done by all the force vectors

Mental and written →	**(3) Equations & unknowns** • **Equation from FBD:** Use Equation 5.1a or 5.1b in direction of *zero acceleration.* • **Work done by each force:** Find the angle between $\vec{d}$ and each force vector, and then customize Equation 8.1 for each force vector. • **Total work:** Just add. Simplify and then circle the unknowns!
Mental and written →	**(4) Outline solution**

8.5 THE WORK–ENERGY THEOREM—*KE* ONLY

Here we discuss the first of two versions of the *work–energy theorem,* which gives us equations to relate work and energy. This version involves only *kinetic energy,* which can be calculated for an object with this equation:

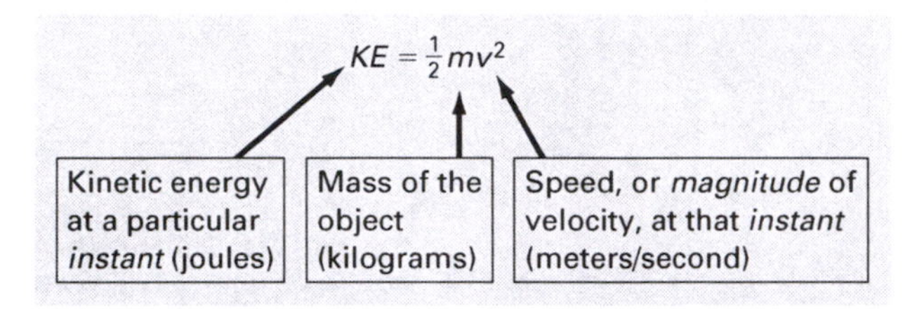

Say a block has kinetic energy and speed—KE_0 and v_0, respectively—at an *initial instant.* During an *interval,* forces do total work, W_{total}, on the block. This changes the motion so that the block has new values, KE and v, at the *final instant.* We can picture the interval and write the equation in terms of either kinetic energy or speed:

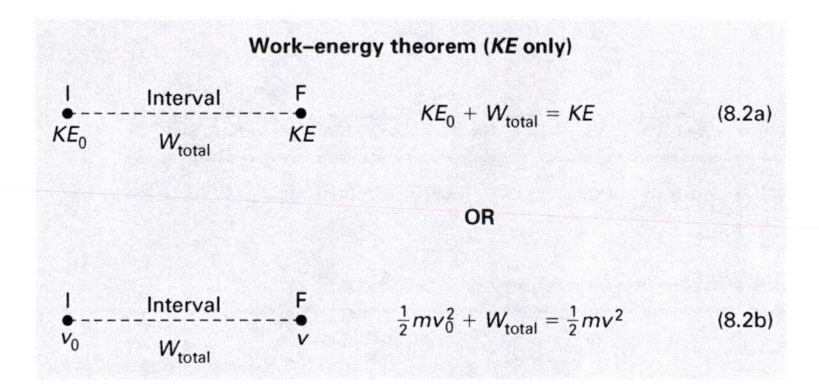

We use whichever equation best fits the problem.

EXERCISE 8.4

(This is just like Exercise 8.2, but now with kinetic energy thrown in.) A 2.0-kg block slides a distance of 3.0 m down a frictionless ramp angled at 30° from the horizontal. The block's final kinetic energy is 35 J. Determine its initial kinetic energy.

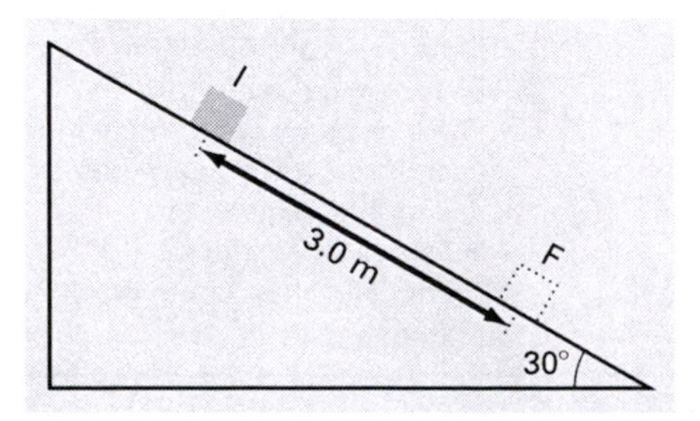

THINK INTERVALS!

When reading a problem, words like *initial, final, initially,* and so on are clues to set up an interval with initial/final instants. This is just like what we did in earlier chapters, but here we use quantities in Equation 8.2a or 8.2b.

Solution

(1) Type of problem

There is a motion *interval,* and *force* is implied because the block's weight pulls it: Use the work–energy theorem (*KE* only), specifically Equation 8.2a, since *kinetic energies* are mentioned.

(2) Sort by . . . and (3) Equations & unknowns

We combine these steps together in two parts:

- **Total work:** Determine W_{total} as outlined in Section 8.4.
- **Work–energy (*KE* only):** Set up a motion interval for Equation 8.2a.

Total work

This was all done in Exercise 8.2: $W_{total} \cong 29.4 \cong 29$ J. We would need to do all of the setup and calculations from Exercise 8.2 here if we had not already done them.

Work–energy (KE only)

The motion interval for Equation 8.2a:

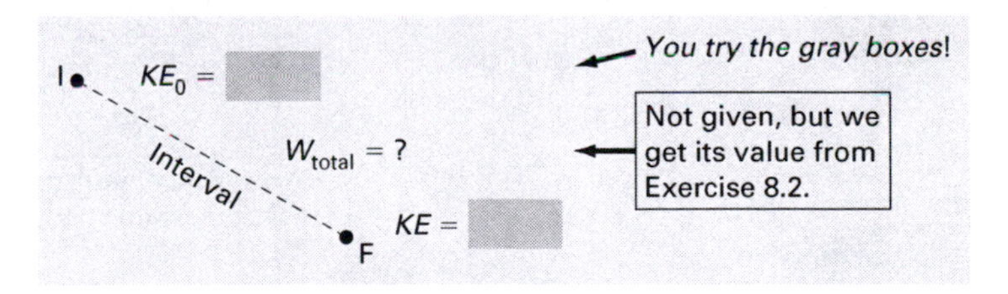

Equation 8.2a for this interval:

$$KE_0 + W_{total} = KE$$

(4) Outline solution

We put this equation together with those from Exercise 8.2:

Work done by each force	$W_n = (n) d \cos \theta_n = 0$ $(W_{Wt}) = mgd \cos \theta_{Wt}$	
Total work	$(W_{total}) = (W_n) + (W_{Wt})$	➤ *You try an outline!*
Work–energy (*KE* only)	$(KE_0) + (W_{total}) = KE$	

Answers for gray boxes

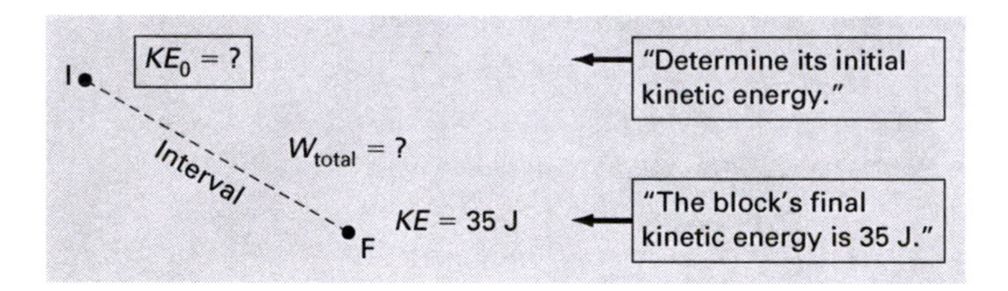

Answers for (4) Outline solution
One possible outline:

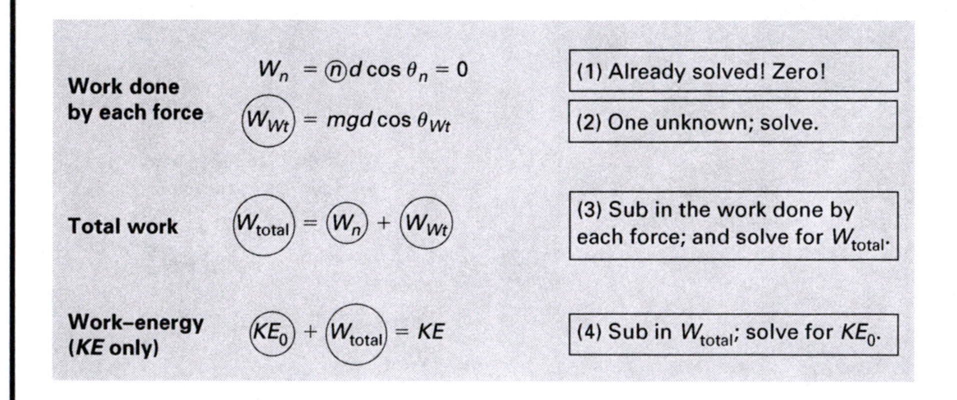

Answer $KE_0 \cong 5.6$ J ■

EXERCISE 8.5

A bullet of mass 4.0 grams initially moving at 280 m/s hits a large wood post and penetrates a distance of 6.0 cm into the wood before it stops. What is the average stopping force acting on the bullet?

Solution

(1) Type of problem

It helps to picture the situation and displacement vector like so:

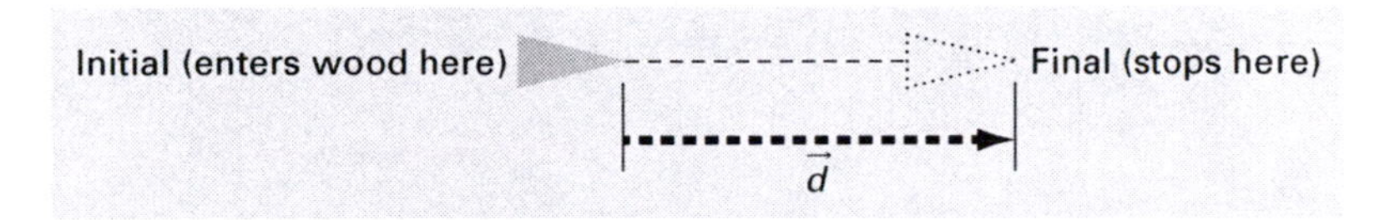

There is a motion *interval* with *force:* Use the work–energy theorem (*KE* only), specifically Equation 8.2b, since *speeds* are mentioned. (We could also solve this problem with methods from Chapter 5.)

(2) Sort by displacement and force vectors, and interval, and
(3) Equations & unknowns

First, some unit conversions:

"penetrates a distance of 6.0 cm into the wood"	$d = 6.0\text{ cm} \times \left(\frac{1\text{ m}}{100\text{ cm}}\right) = 0.060\text{ m}$
"A bullet of mass 4.0 grams"	$m = 4.0\text{ grams} \times \left(\frac{1\text{ kg}}{1000\text{ grams}}\right) = 0.0040\text{ kg}$

Total work

We follow the setup outline in Section 8.4 to get W_{total}. First, we need an equation for the work, W_s, done by the stopping force, $\vec{F}_s$. For a diplacement to the right, the stopping force must be *opposite* this, or to the left:

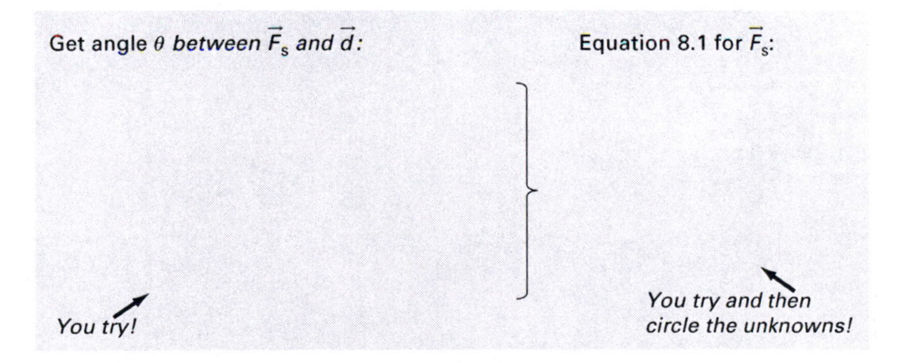

This is the only force vector doing work, so this is also the total work:

$$W_{\text{total}} = W_s \quad \Rightarrow \quad \textcircled{W_{\text{total}}} = -\textcircled{F_s}\, d$$

***Work–energy* (KE *only*)**
The motion interval, set up for Equation 8.2b:

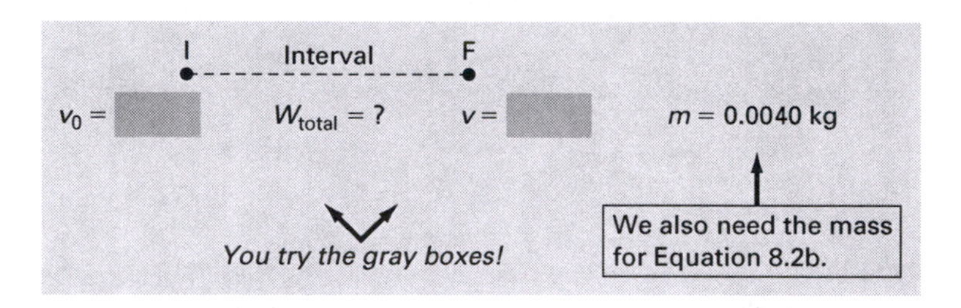

Equation 8.2b for this situation:

$$\frac{1}{2}mv_0^2 + W_{total} = \frac{1}{2}mv^2$$

(4) Outline solution
Finally, we put the equations together:

Total work $W_{total} = -F_s d$

Work–energy (*KE* only) $\frac{1}{2}mv_0^2 + W_{total} = \frac{1}{2}mv^2$

You try an outline!

Answers for . . .

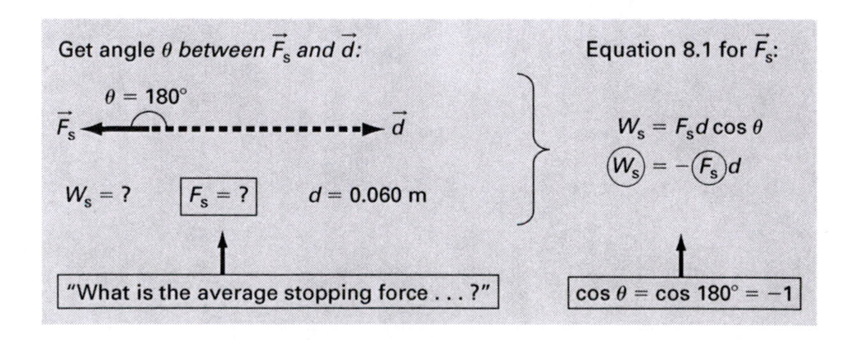

Answers for gray boxes

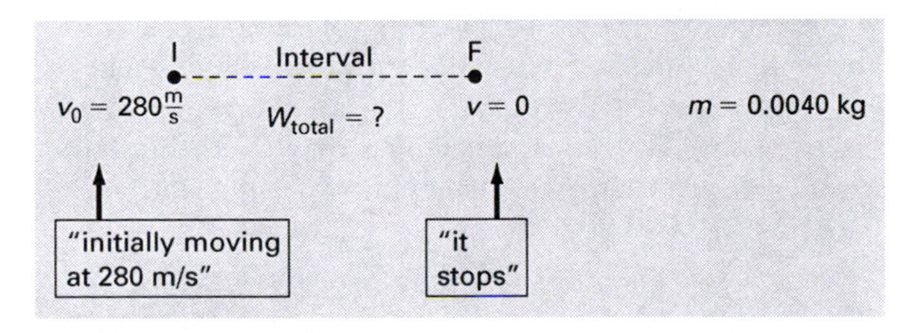

Answers for (4) Outline solution

One possible outline:

Total work	$W_{\text{total}} = -F_s d$	(2) Sub in W_{total}; solve for F_s.
Work–energy (*KE* only)	$\frac{1}{2}mv_0^2 + W_{\text{total}} = \frac{1}{2}mv^2$	(1) One unknown; solve for W_{total}.

Answer $F_s \cong 2613 \cong 2600$ N (*magnitude* should come out *positive!*)

Intermediate answer: $W_{\text{total}} \cong -157$ J. *Negative* total work means that it *takes away kinetic energy!* ■

8.6 HOW TO SET UP WORK-ENERGY PROBLEMS—*KE* ONLY

Here is a summary of what we did in the previous two exercises:

Work-Energy (*KE* Only) Problems—Mental and Written Steps

Mental →	**(1) Type of problem** Motion *interval* and constant/average *forces:* Use the "*KE* only" version of the work-energy theorem, Equation 8.2a or 8.2b.
Mental and written →	**(2) Sort by . . . and (3) Equations & unknowns** We combine these steps in two parts: • **Total work:** Sort by displacement and force vectors; get equation or value for W_{total} (as outlined in Section 8.4). • **Work–energy (*KE* only):** Sort quantities by interval for either Equation 8.2a or 8.2b.
Mental and written →	**(4) Outline solution**

8.7 POTENTIAL ENERGY, CONSERVATIVE AND NONCONSERVATIVE FORCES

Before moving on to the next version of the work–energy theorem, we need to discuss *potential energy* as well as *conservative* and *nonconservative forces.*

There are several types of potential energy. The main kind we deal with is *gravitational* potential energy of an object:

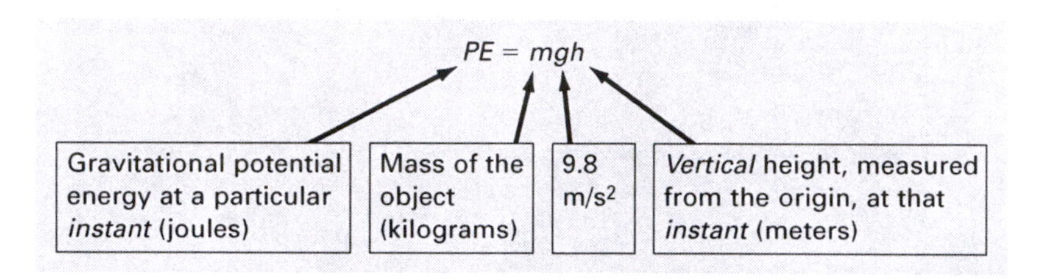

Conservative Forces and Potential Energy

Gravitational force, or weight, is one example of a *conservative force* for which the following is always true:

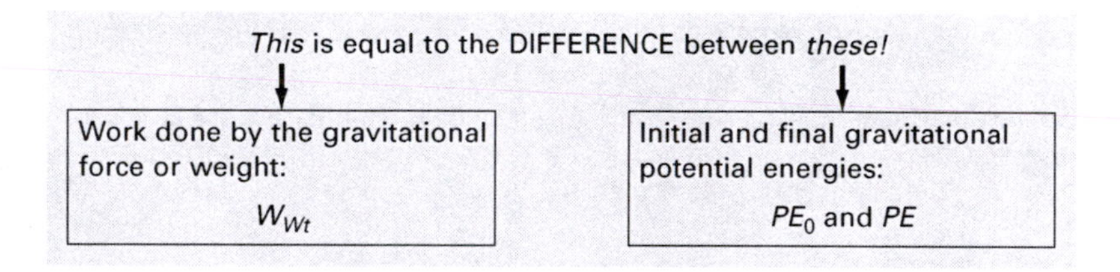

This makes calculations *easier:*

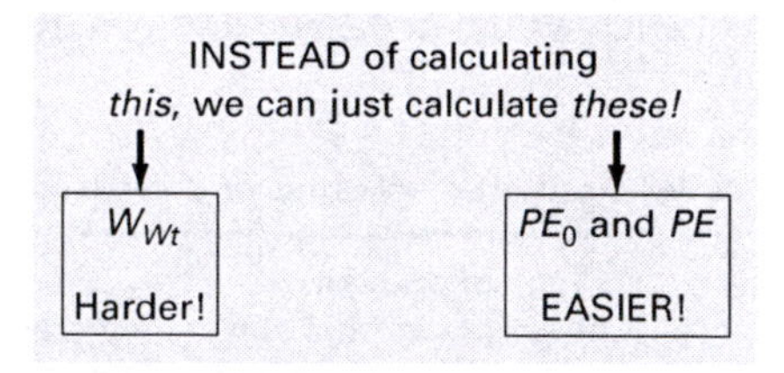

This leads to the "*KE* and *PE*" version of the work–energy theorem, discussed in the next section.

This idea works for ANY *conservative* force having a *related potential energy,* such as elastic or spring forces, electric forces, magnetic forces, and others. For example, *instead* of calculating the *work done by a spring,* we can just calculate the *initial and final elastic potential energies* of the spring.

Nonconservative Forces

All other forces we have discussed up to now are *nonconservative:* They have *no related potential energy!* This includes friction, normal, tension, person-caused forces, motor-caused forces, and so on.

8.8 THE WORK–ENERGY THEOREM—*KE* AND *PE*

This is the *better, easier-to-do-problems-with,* version of the work–energy theorem. The main difference between the two versions is in the *work calculations.* The total work, W_{total}, done on an object includes the work, W_c, done by *conservative* forces like weight, and work, W_{nc}, done by *nonconservative* forces. The "*KE* and *PE*" version requires fewer work calculations:

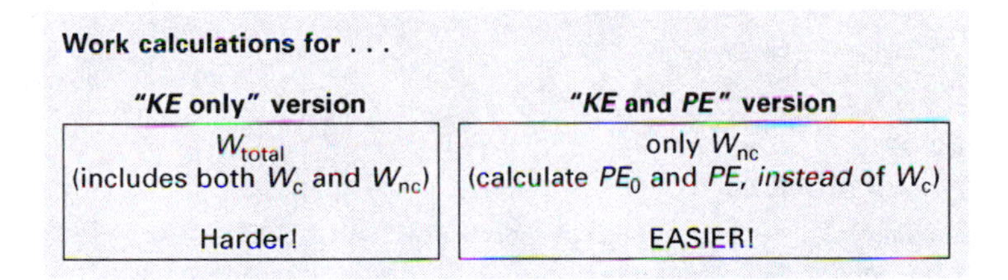

We can picture an interval that involves a height change, and write the equations in terms of either kinetic and potential energy or speed and height:

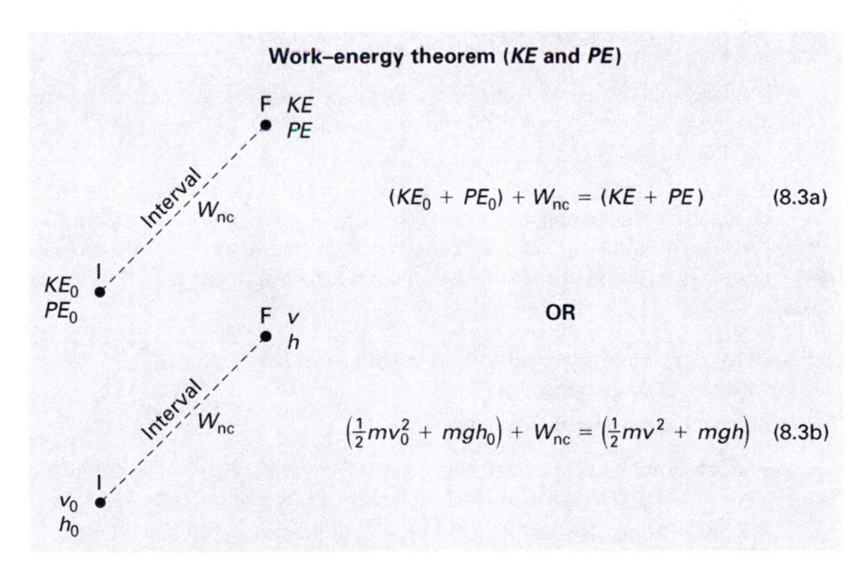

Our solution outline for this "*KE* and *PE*" version is very similar to what we used earlier for the "*KE* only" version.

EXERCISE 8.6

A girl of mass 25 kg stands on a scale on the floor of an elevator that is initially at rest. The elevator starts moving downward so that the scale reads 210 N. What is the elevator's velocity after it has traveled 1.0 m?

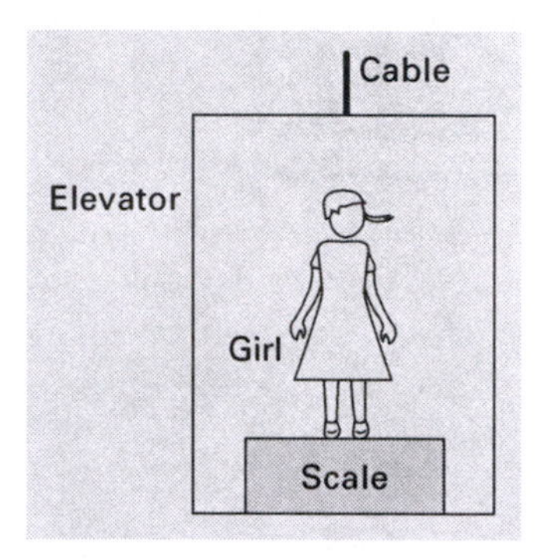

Solution

(1) Type of problem

Motion *interval:* Use the work–energy theorem. We will use the "*KE* and *PE*" version, specifically Equation 8.3b, because *speed* (magnitude of *velocity*) is mentioned.

> *FOR STRAIGHT-LINE MOTION, USE EITHER THE "KE ONLY" OR THE "KE AND PE" VERSION!*
>
> When the motion of an object is in a *straight line,* we can use either version of the work–energy theorem, but the "*KE* and *PE*" version has *fewer work calculations.*

The problem asks about the motion of *the elevator* but gives information about forces acting *on the girl:* Her mass tells us her *weight,* and the scale reading (which is the magnitude of the *normal force*) is given. Since the motion of the girl is identical to that for the elevator, we just focus *on the girl.*

(2) Sort by displacement and force vectors, and interval, and (3) Equations & unknowns

We combine these steps together in two parts:

- **Work done by *nonconservative* forces:** Determine W_{nc}, just as outlined in Section 8.4, EXCEPT we *ignore work done by the weight* since it is a *conservative* force.
- **Work–energy (*KE* and *PE*):** Set up a motion interval for Equation 8.3b.

Work done by* nonconservative *forces

Two forces act on the girl. The (long-range) gravitational force, $\vec{Wt}$, is *conservative.* The (contact) normal force, $\vec{n}$, is directed up and is *nonconservative.*

We only calculate the work done by the *nonconservative* force, $\vec{n}$. As outlined in Section 8.4, we put this *force vector* alone with the *displacement vector:*

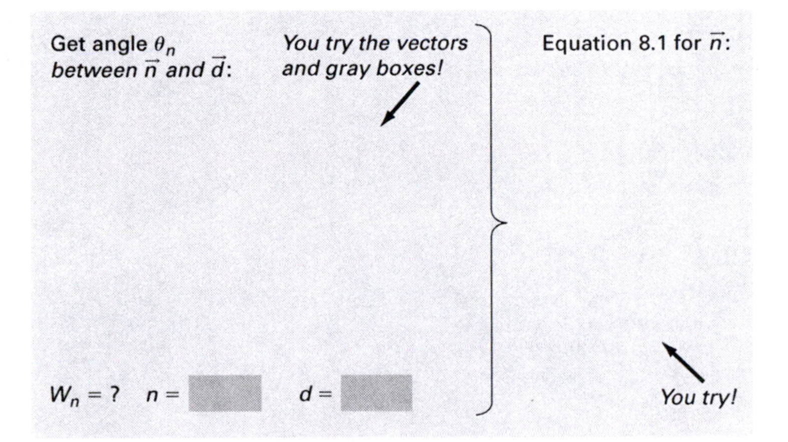

Since this is the only *nonconservative* force vector that does work:

$$W_{nc} = W_n \Rightarrow W_{nc} = -nd$$

***Work–energy* (KE *and* PE)**

The motion interval for Equation 8.3b:

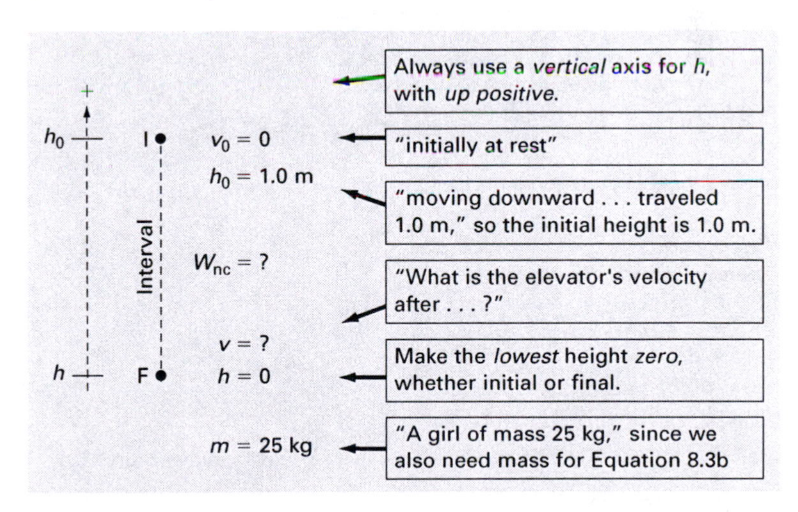

Equation 8.3b for this interval:

$$\left(\tfrac{1}{2}mv_0^2 + mgh_0\right) + W_{nc} = \left(\tfrac{1}{2}mv^2 + mgh\right) \leftarrow \textit{You circle the unknowns!}$$

Check the next figure for answers.

(4) Outline solution

We put all the equations together:

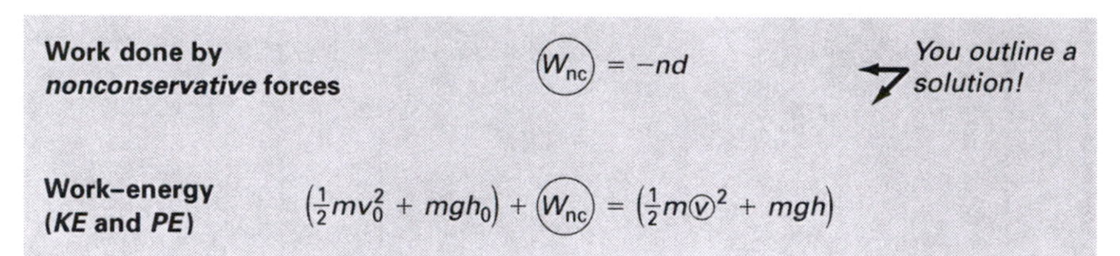

Answers for . . .

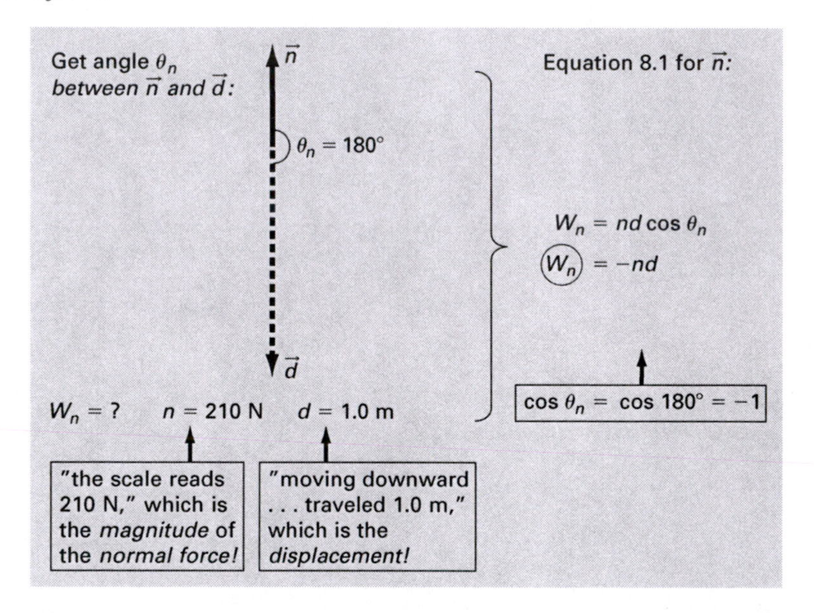

Answers for (4) Outline solution

Work done by *nonconservative* forces	$W_{nc} = -nd$	(1) Solve.
Work–energy (*KE* and *PE*)	$\left(\frac{1}{2}mv_0^2 + mgh_0\right) + W_{nc} = \left(\frac{1}{2}mv^2 + mgh\right)$	(2) Sub in W_{nc}; solve for v.

Answer $v \cong 1.67 \cong 1.7\frac{\text{m}}{\text{s}}$ (speed does not include sign to indicate direction)

Intermediate answer: $W_{nc} = -210$ J ■

A NEW WAY TO SOLVE AN OLD PROBLEM!

This exercise illustrates a new way to tackle an old problem: Exercise 5.2, part (c).

The next exercise involves *nonstraight motion,* and so we can use *only* the "*KE* and *PE*" version.

EXERCISE 8.7

At the bottom of the hill, the 1200-kg car moves at 25 m/s. It eventually slows to 15 m/s at the top of the hill, which is 45 m higher than the bottom. During the trip, the car's engine force (nonconservative) does $+5.0 \times 10^5$ J of work. How much work is done by all other nonconservative forces (friction, air resistance, etc.) combined?

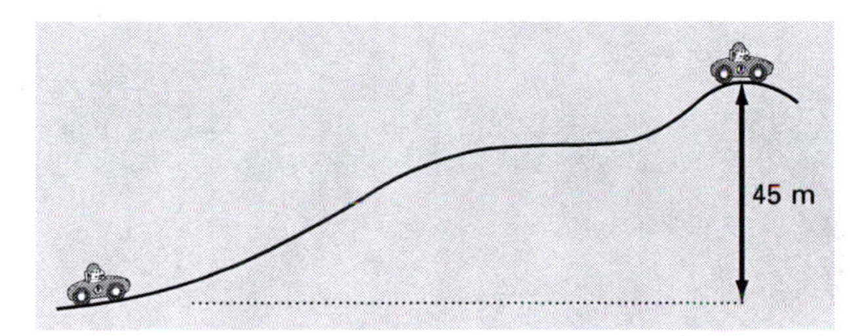

Solution

(1) Type of problem

Motion *interval:* Use the work–energy theorem. We will use the "*KE* and *PE*" version, specifically Equation 8.3b, because *speeds* and *heights* are mentioned.

FOR MOTION NOT IN A STRAIGHT LINE, USE ONLY THE "KE AND PE" VERSION!

When the motion of an object is *not in a straight line,* the work calculations in the "*KE* only" version of the work–energy theorem are often *very difficult* or *impossible!* Use the "*KE* and *PE*" version.

(2) Sort by displacement and force vectors, and interval, and
(3) Equations & unknowns

Work done by* nonconservative *forces

The total work done by *nonconservative* forces, W_{nc}, is unknown, but we know it is the sum of the work terms for all nonconservative forces:

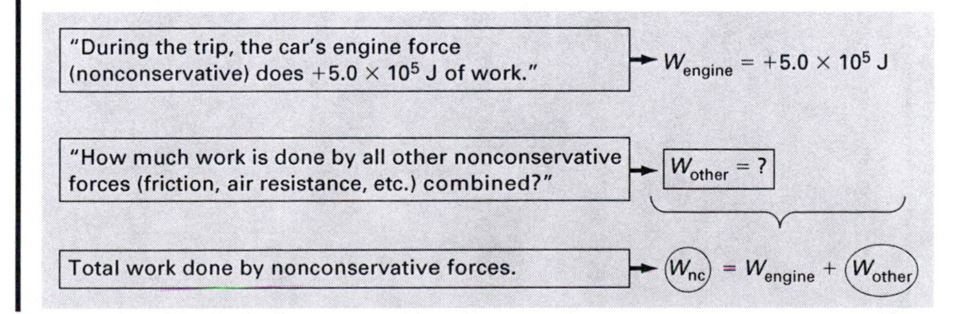

***Work–energy* (KE *and* PE)**

The motion interval for Equation 8.3b:

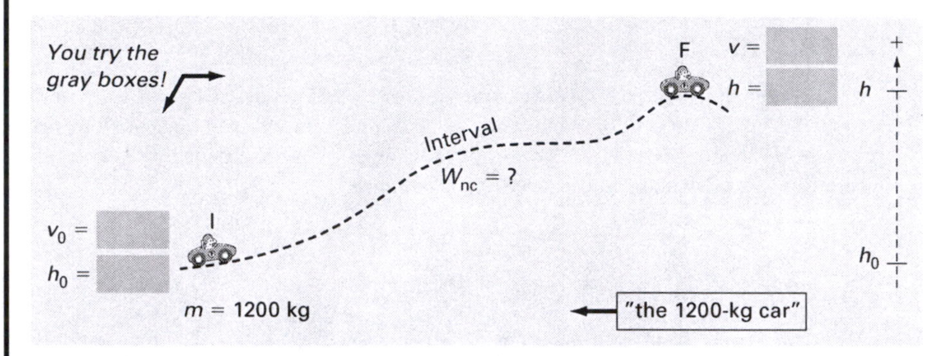

Equation 8.3b for this interval:

$$\left(\tfrac{1}{2}mv_0^2 + mgh_0\right) + W_{nc} = \left(\tfrac{1}{2}mv^2 + mgh\right)$$

(4) Outline solution

We put the equations together:

Work done by *nonconservative* forces

$$W_{nc} = W_{engine} + W_{other}$$

You outline a solution!

Work–energy (*KE* and *PE*)

$$\left(\tfrac{1}{2}mv_0^2 + mgh_0\right) + W_{nc} = \left(\tfrac{1}{2}mv^2 + mgh\right)$$

Answers for gray boxes

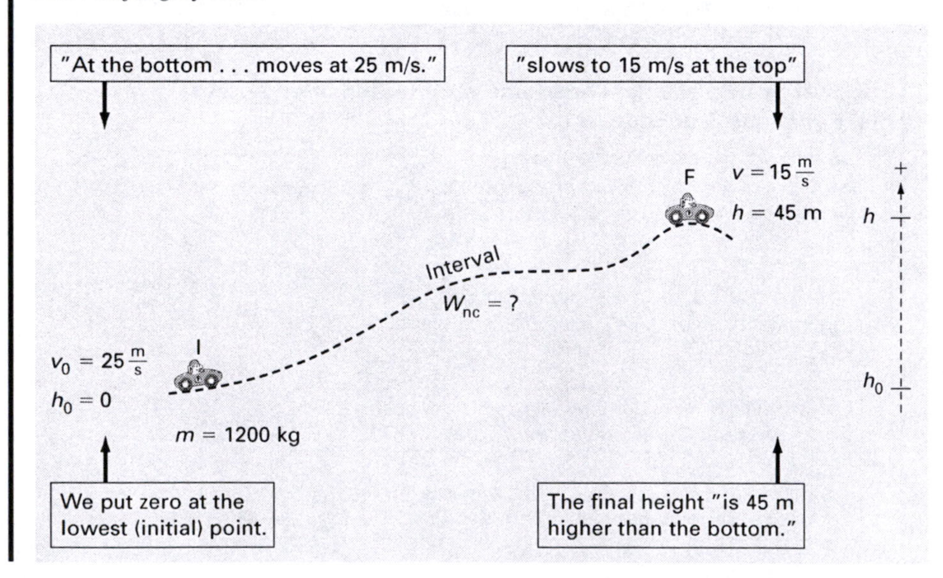

Answers for (4) Outline solution

Work done by *nonconservative* forces	$(W_{nc}) = W_{engine} + (W_{other})$	(2) Sub in W_{nc}; solve for W_{other}.
Work–energy (*KE* and *PE*)	$\left(\frac{1}{2}mv_0^2 + mgh_0\right) + (W_{nc}) = \left(\frac{1}{2}mv^2 + mgh\right)$	(1) Solve.

Answer $W_{other} \cong -2.11 \times 10^5 \cong -2.1 \times 10^5$ J (Negative! Watch signs!)

Intermediate answer: $W_{nc} \cong +2.89 \times 10^5$ J ■

8.9 HOW TO SET UP WORK–ENERGY PROBLEMS—*KE* AND *PE*

Here is a summary of what we did in the previous two exercises. This outline has a few differences compared to the outline for the "*KE* only" version.

Work–Energy (*KE* and *PE*) Problems—Mental and Written Steps

Mental →	**(1) Type of problem** Motion *interval:* Use the "*KE* and *PE*" version of the work–energy theorem, Equation 8.3a or 8.3b, in order to have fewer work calculations than the "*KE* only" version.
Mental and written →	**(2) Sort by . . . , & (3) Equations & unknowns** We combine these steps in two parts: • **Work done by *nonconservative* forces:** Sort by displacement and force vectors, and get equation or value for W_{nc} (as outlined in Section 8.4, *except ignore* work done by *conservative* forces like *weight*). • **Work–energy (*KE* and *PE*):** Sort quantities by interval for either Equation 8.3a or 8.3b. Simplify and then circle the unknowns.
Mental and written →	**(4) Outline solution**

CALCULATE W_{nc} IF THERE ARE MOTORS, ENGINES, PEOPLE, AND SO ON, OR KINETIC FRICTION!

These kinds of forces in a problem are clues that we need to calculate W_{nc}. They *add to* or *take away from* the total mechanical energy (*KE* + *PE*) of an object. For example:

- Although never mentioned in Exercise 8.6, a *motor* (or something like it) controls the motion of the girl/elevator. This *motor* is the *indirect cause* of the *nonconservative work* done on the girl by the normal force. The motor takes away mechanical energy because it does not allow the girl/elevator to free-fall while going down.

(*continued*)

- In Exercise 8.7, the *engine* adds mechanical energy to the car, while the other nonconservative forces (*friction, air resistance,* and so on) take away mechanical energy. These all cause W_{nc}.

8.10 CONSERVATION OF ENERGY—WHEN $W_{nc} = 0$

If NO forces from *motors, engines, people,* and so on, or *kinetic friction* (or similar nonconservative forces) directly or indirectly cause work, then $W_{nc} = 0$. In this situation, we have *conservation of mechanical energy,* or just *conservation of energy,* because the *initial total mechanical energy* ($KE_0 + PE_0$) is equal to the *final total mechanical energy* ($KE + PE$):

Since it is *zero,* we just leave W_{nc} out of Equations 8.3a and 8.3b.

Conservation of energy!

$$KE_0 + PE_0 = KE + PE \quad (8.4a)$$

OR

$$\tfrac{1}{2}mv_0^2 + mgh_0 = \tfrac{1}{2}mv^2 + mgh \quad (8.4b)$$

This means no work calculations at all! Hooray! The following is just like Exercise 8.4, but now we use *conservation of energy* to solve it.

EXERCISE 8.8

A 2.0-kg block slides a distance of 3.0 m down a frictionless ramp angled at 30° from the horizontal. The block's final kinetic energy is 35 J. Determine its initial kinetic energy.

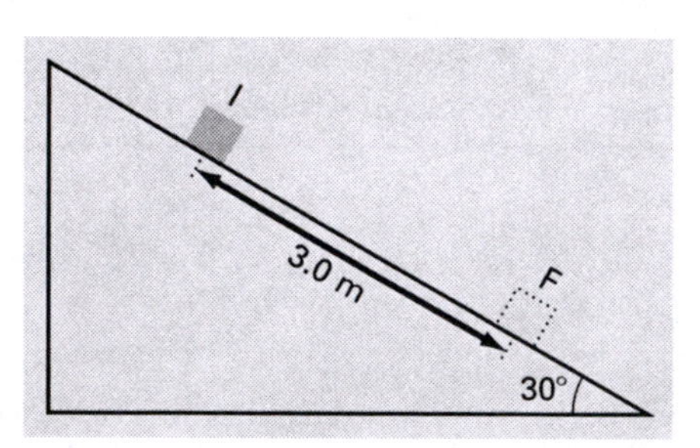

Solution

(1) Type of problem

Motion *interval,* and NO forces from *motors, engines, people,* and so on, or *kinetic friction* (or similar nonconservative forces) do work on the block: Use *conservation of energy,* specifically Equation 8.4a, since *kinetic energies* are mentioned.

If you don't like that explanation, then from Exercise 8.2 we know that only the weight, a *conservative* force, does work on the block, so $W_{nc} = 0$.

(2) Sort by interval, (3) Equations & unknowns, and (4) Outline solution

For potential energy, we need *vertical* heights, h_0 and h. We put zero at the lowest (final) point, so $h = 0$, and use a right triangle to get h_0:

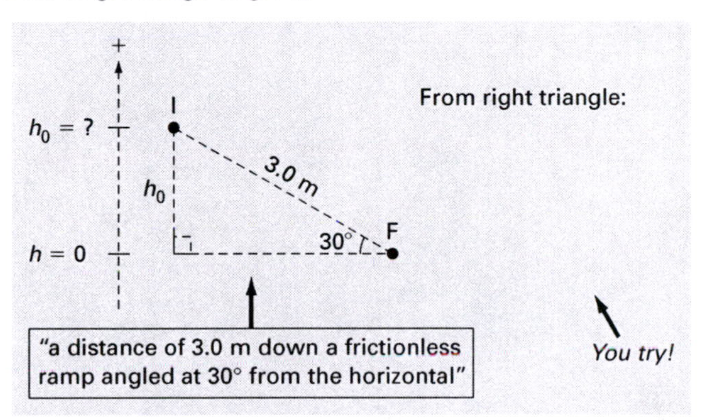

The interval for Equation 8.4a:

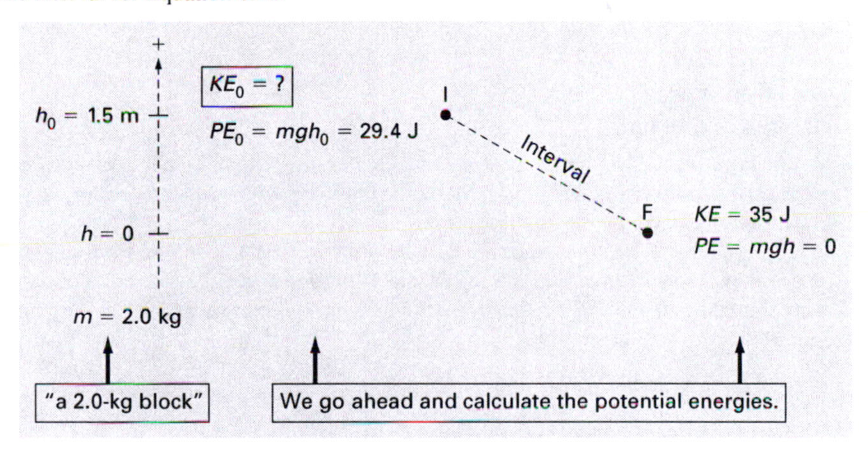

Equation 8.4a:

$$KE_0 + PE_0 = KE + PE \quad \boxed{\text{Solve.}}$$

Answer for . . .

From right triangle:

$$\sin 30° = \frac{h_0}{(3.0\text{ m})}$$

$$h_0 = 1.5\text{ m}$$

Answer $KE_0 \cong 5.6$ J ■

EXERCISE 8.9

A ball swings on a cable of length 2.5 m, attached to the ceiling. Initially, it is at its lowest point moving horizontally to the right at 5.0 m/s. What is its speed when its angle is 55° from the vertical?

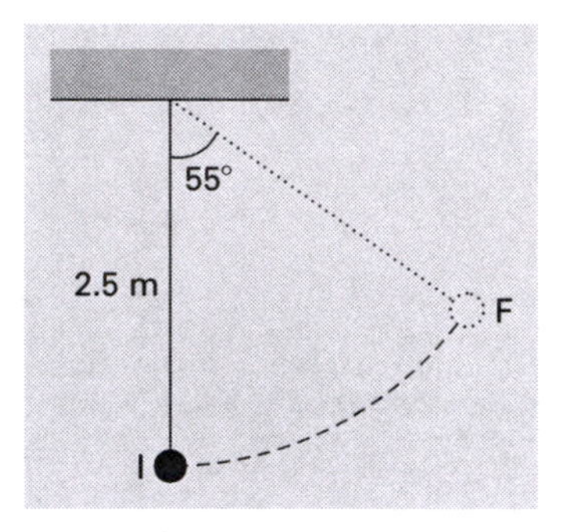

Solution

(1) Type of problem

Motion *interval,* and NO forces from *motors, engines, people,* and so on, or *kinetic friction* (or similar nonconservative forces) do work on the ball: Use *conservation of energy,* specifically Equation 8.4b, since *speeds* are mentioned.

If you don't like that explanation, there are two forces acting on the ball: The (long-range) gravitational force, $\vec{Wt}$, and the (contact) tension force, $\vec{T}$, directed along the string away from the ball. Only $\vec{T}$ is *nonconservative.* Because $\vec{T}$ is always at 90° to the path, it does *zero work:* $W_{nc} = 0$! Hooray!

(2) Sort by interval and (3) Equations & unknowns

On the vertical height axis, we put zero at the *lowest* point in the motion, so $h_0 = 0$. The hardest part of this solution is to determine a value for h, which we need for the final potential energy ($PE = mgh$). The next few figures show how to get h:

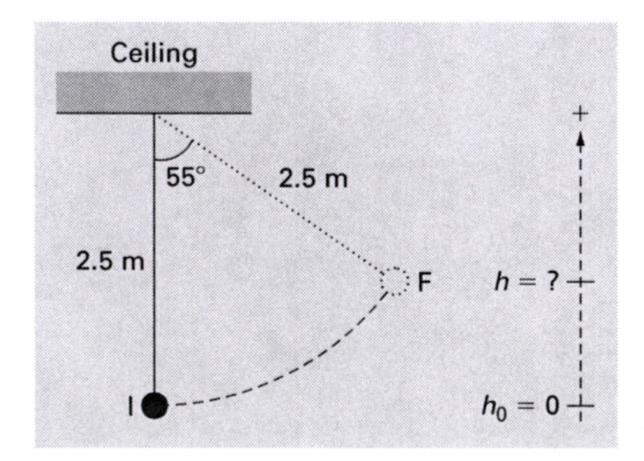

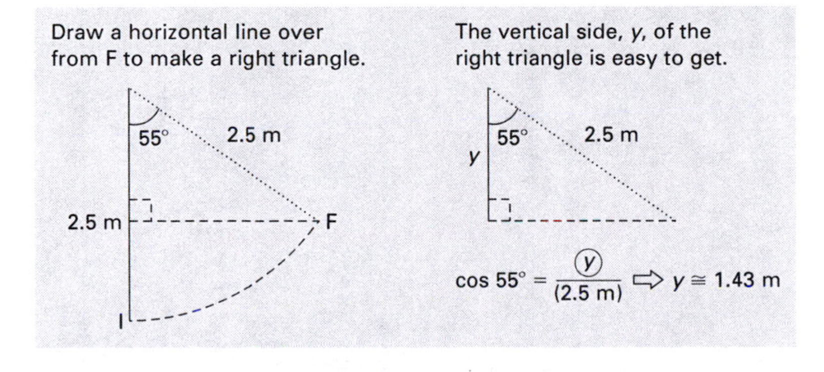

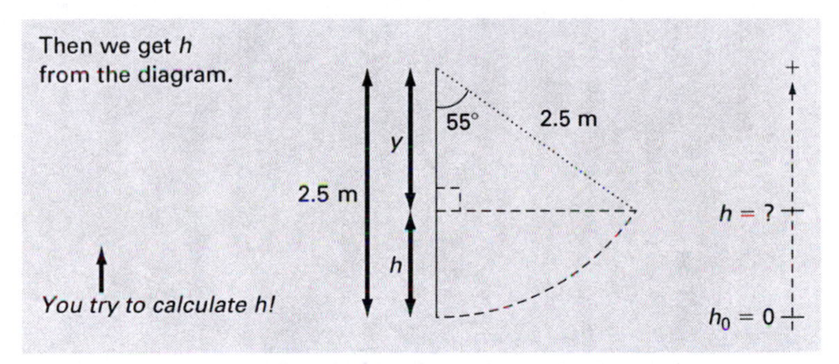

That was the hard part! Now we know h for the motion interval for Equation 8.4b:

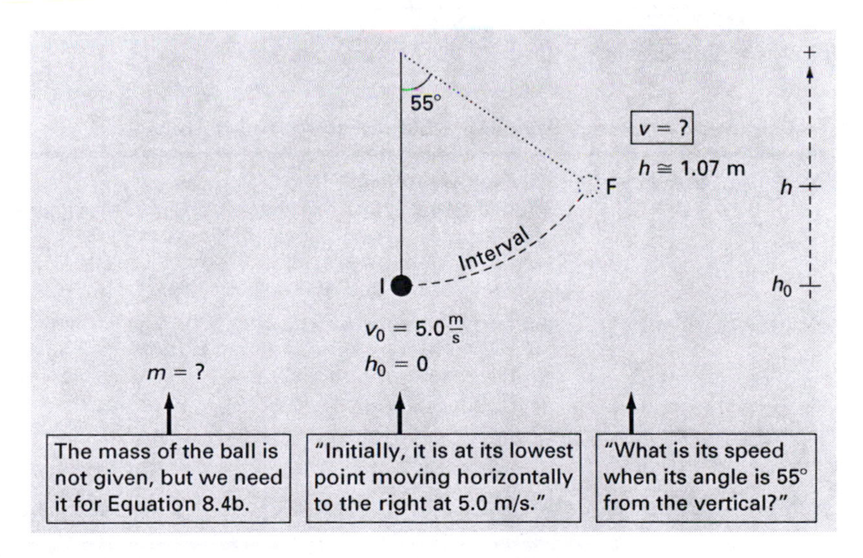

Equation 8.4b for this interval:

$$\frac{1}{2}mv_0^2 + mgh_0 = \frac{1}{2}mv^2 + mgh$$

(4) Outline solution

Since *every term* in the previous equation has m in it, we divide both sides by m to get rid of it. Then we have only one unknown:

$$\frac{1}{2}v_0^2 + gh_0 = \frac{1}{2}\text{ⓥ}^2 + gh \quad \boxed{\text{Solve.}}$$

Answers for . . .

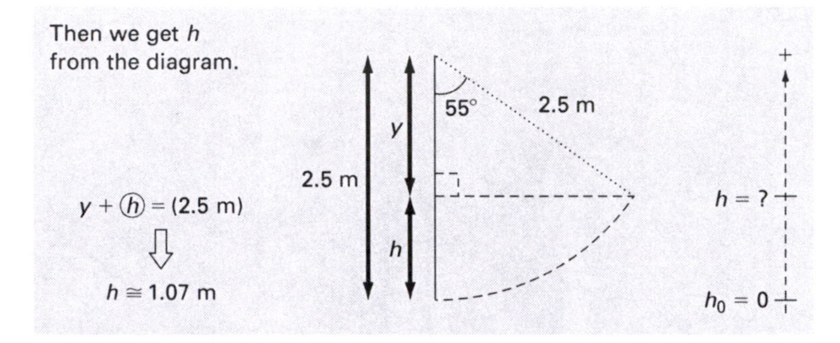

Answer $v \cong 2.0\frac{\text{m}}{\text{s}}$ ■

8.11 HOW TO SET UP CONSERVATION OF ENERGY PROBLEMS

Here is a summary of what we did in the previous two exercises:

Conservation of Energy Problems—Mental and Written Steps

Mental →	**(1) Type of problem** Motion *interval,* and NO forces from *motors, engines, people,* and so on, or *kinetic friction* (or similar nonconservative forces) do work on the object: Use *conservation of energy,* Equation 8.4a or 8.4b.
Mental and written →	**(2) Sort by interval and (3) Equations & unknowns** Sort quantities by interval for either Equation 8.4a or 8.4b, write the equation, simplify, and circle the unknowns.
Mental and written →	**(4) Outline solution**

8.12 HOW TO SPLIT UP A DIFFICULT PROBLEM

EXERCISE 8.10

A 1.0-kg ball is launched vertically upward from ground level and reaches a height of 2.5 m if there is no air resistance. If the ball is fired at the same initial speed, but air resistance is acting, it only reaches 2.2 m. What is the average air resistance force?

Solution

(1) Type of problem

This problem is much easier to swallow if we see it as *two separate problems:*

- **(A) Without air resistance:** The ball reaches 2.5 m maximum height. This is a motion interval with NO *nonconservative* forces that do work: Use conservation of energy, specifically Equation 8.4b, since *speeds* and *heights* are mentioned.
- **(B) With air resistance:** The ball reaches 2.2 m maximum height. Unknown: average air resistance force. This is a motion interval with the *nonconservative* air resistance force doing work: Use work–energy (*KE* and *PE*), Equation 8.3b, since *speeds* and *heights* are mentioned.

Both (A) and (B) have the *same initial speed,* v_0, just after launch. That means we try to solve for v_0 in one part and substitute it into the other part.

(2) Sort by . . . and (3) Equations & unknowns

Work done by* nonconservative *forces for (B)

We don't need to do this for (A), since there is no air resistance.

We look only at the motion *during flight,* beginning *just after* launch, so we completely *ignore the launch force* that got the ball going in the first place. For (B), the *nonconservative* force is the unknown air resistance force, $\vec{F}$, which opposes the motion and so must be down. Following the outline of Section 8.4, we put this force vector with the displacement vector:

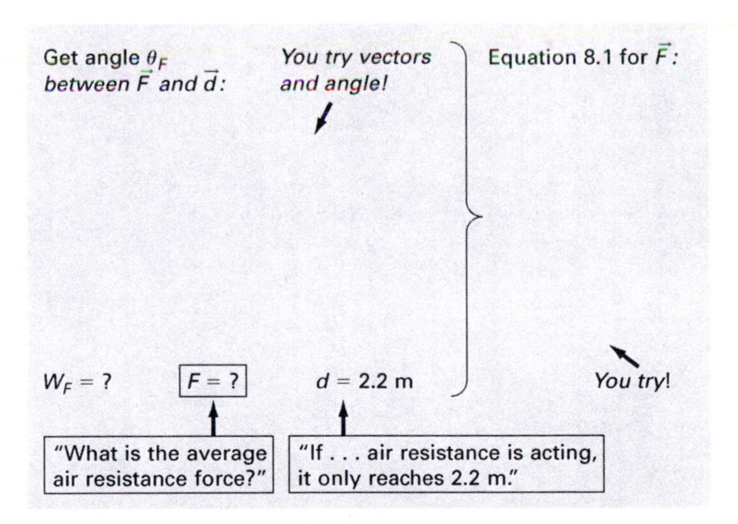

Since $\vec{F}$ is the only nonconservative force in (B), then:

$$W_{nc,B} = W_F \Rightarrow W_{nc,B} = -F d$$

Conservation of energy for (A), and work–energy* (KE *and* PE) *for (B)
The motion intervals:

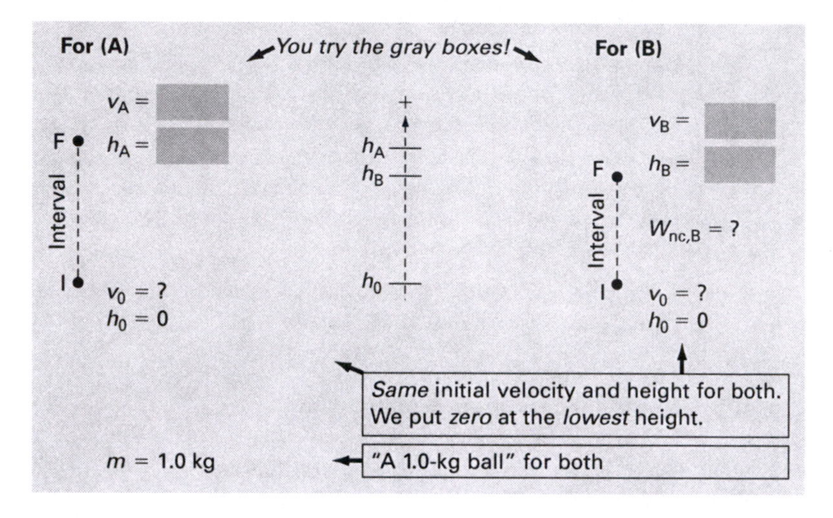

(4) Outline solution

We write the equations for each interval, together with the work equation for (B):

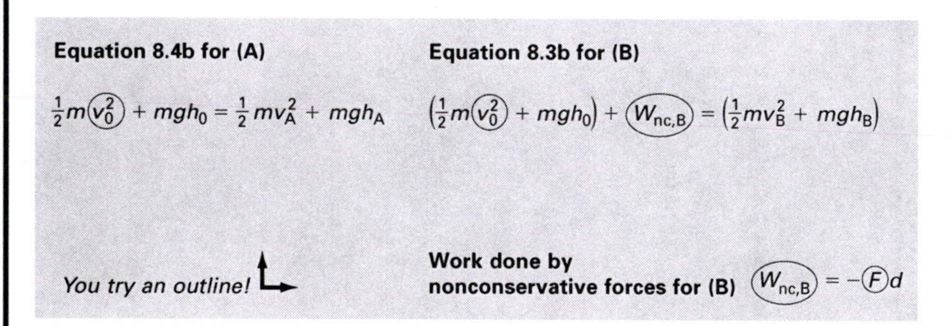

Answers for . . .

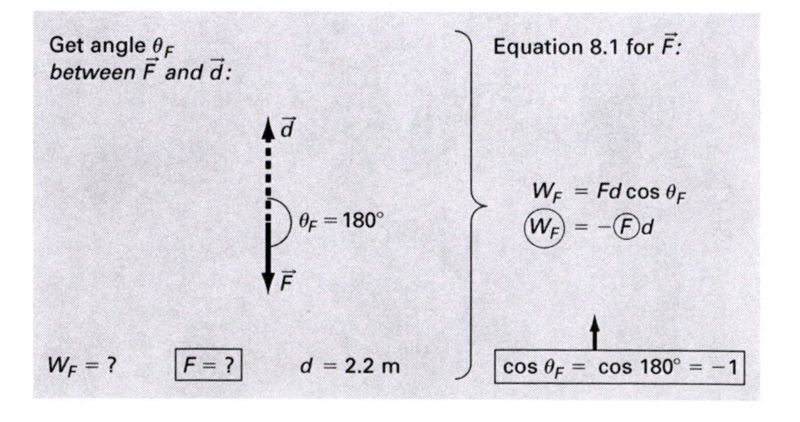

Answers for gray boxes

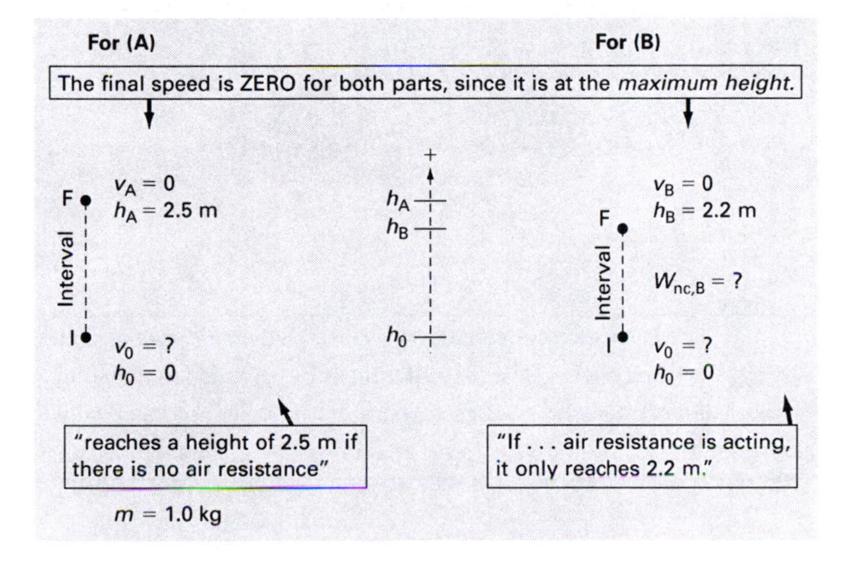

Answers for (4) Outline solution

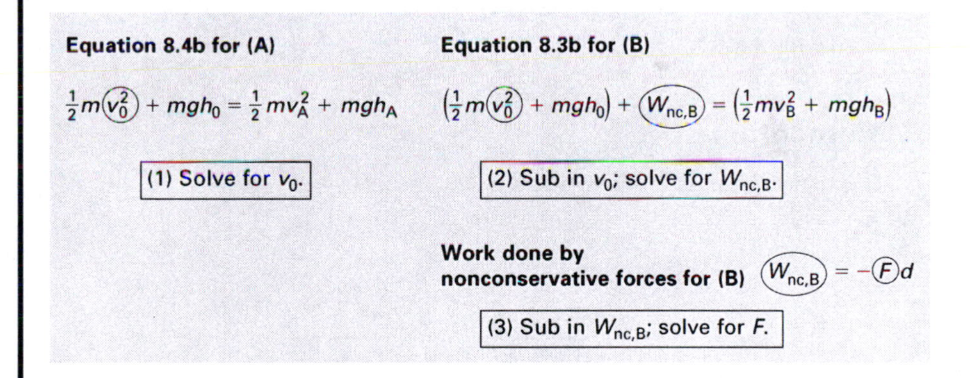

Answer $F \cong 1.34 \cong 1.3$ N (a vector magnitude should come out *positive!*)

Intermediate answers: $v_0 = 7.0\frac{\text{m}}{\text{s}}$, $W_{nc,B} = -2.94$ J (watch signs!) ■

CHAPTER 9

IMPULSE, MOMENTUM, AND CENTER OF MASS

THE RELATIONSHIP BETWEEN *impulse* and *momentum* is similar to that between work and energy. The big difference: Impulse and momentum are *vectors,* but work and energy are scalars. You can review how to deal with vectors in Chapter 3.

9.1 THE IMPULSE–MOMENTUM THEOREM

First, a few basics:

Momentum

The vector equation for momentum of an object:

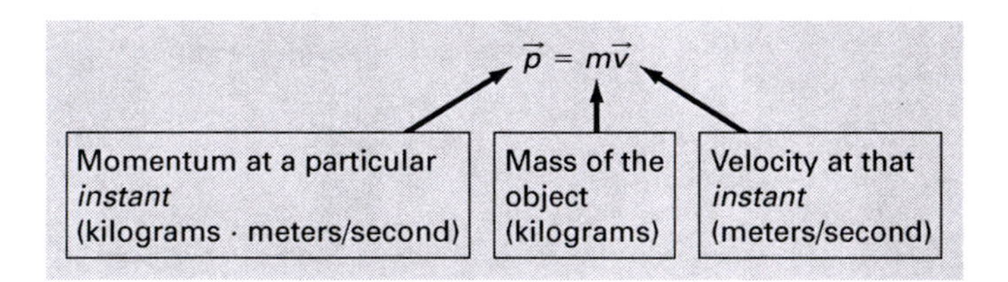

For calculations, we don't use the vector equation but instead the following versions of it:

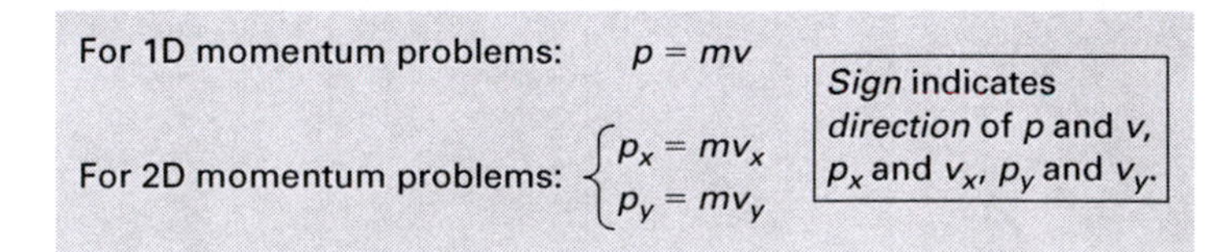

Impulse

The vector equation for the impulse caused by a constant/average force acting on an object:

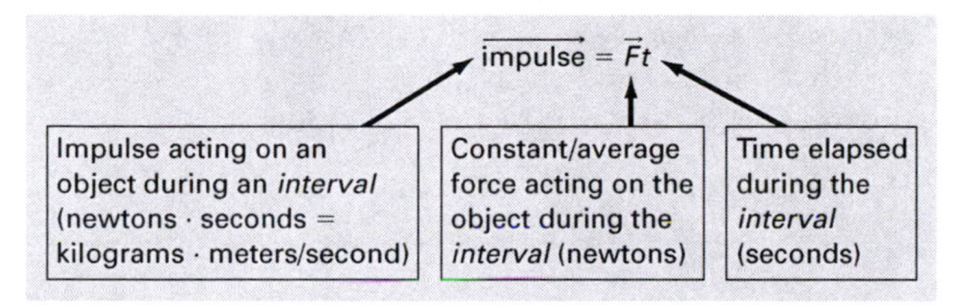

The versions of this equation that we actually use for calculations:

For 1D impulse problems: $\text{impulse} = Ft$

For 2D impulse problems: $\begin{cases} \text{impulse}_x = F_x t \\ \text{impulse}_y = F_y t \end{cases}$

Sign indicates *direction* of impulse and F, and their x and y-components.

Impulse = Change in Momentum

Think intervals and instants! Say a block moves in 1D and has momentum and velocity, p_0 and v_0, at an *initial instant.* During an *interval*, an impulse is caused by a force, F, acting on the block for an elapsed time, t. This changes the motion so that the block has new values, p and v, at the *final instant.* We can picture the interval as follows and write the equation relating these quantities in a few different ways:

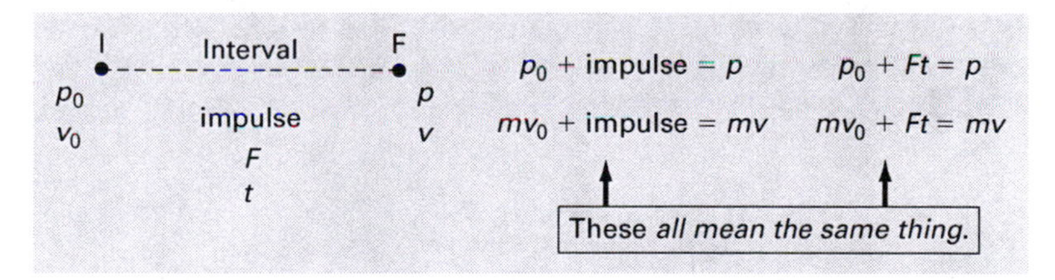

Rearranging the equations a bit, we write them all together as one big equation:

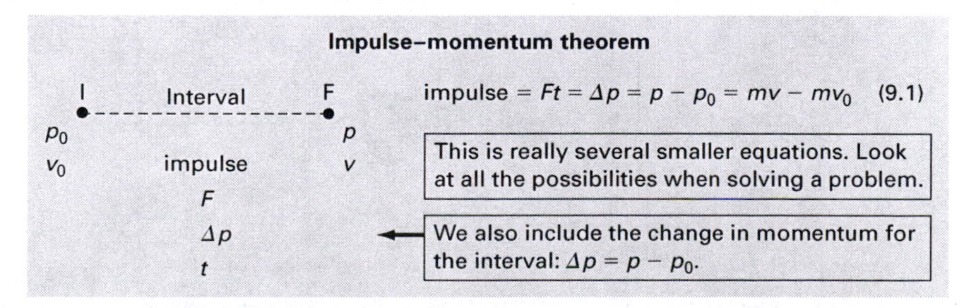

Equation 9.1 is the 1D version of a vector equation. For 2D problems, we write it with x- and/or y-components:

Impulse–momentum theorem—for 2D problems

x-components: $\text{impulse}_x = F_x t = \Delta p_x = p_x - p_{0x} = mv_x - mv_{0x}$ (9.1a)

y-components: $\text{impulse}_y = F_y t = \Delta p_y = p_y - p_{0y} = mv_y - mv_{0y}$ (9.1b)

Because m and t are not vectors, they do NOT have components.

9.2 1D IMPULSE AND MOMENTUM

Remember that we show much more detail, the *mental* and *written* steps, than you would write on paper when solving these problems on your own.

EXERCISE 9.1

A 2.0-kg ball travels at 12 m/s straight toward a wall. It hits the wall and then bounces straight back at 8.0 m/s. The impact lasts for 5.0×10^{-3} s. Find the (a) change in momentum of the ball, (b) impulse that acts on the ball, (c) average force that the wall exerts on the ball, and (d) average rate of change of momentum of the ball.

Solution

(1) Type of problem

Momentum, impulse, force, and time are all mentioned and there is a *motion-changing event* (the ball bounces off of the wall): Use Equation 9.1.

(2) Sort by interval

The collision happens during an *interval* of time (NOT at a single instant). The *initial instant* is *just before* the ball hits the wall, and the *final instant* is *just after* the ball leaves the wall. Assuming the ball is initially moving to the right, we can visualize the collision interval like so:

I • Interval
F •

This leads to the following setup diagram for Equation 9.1:

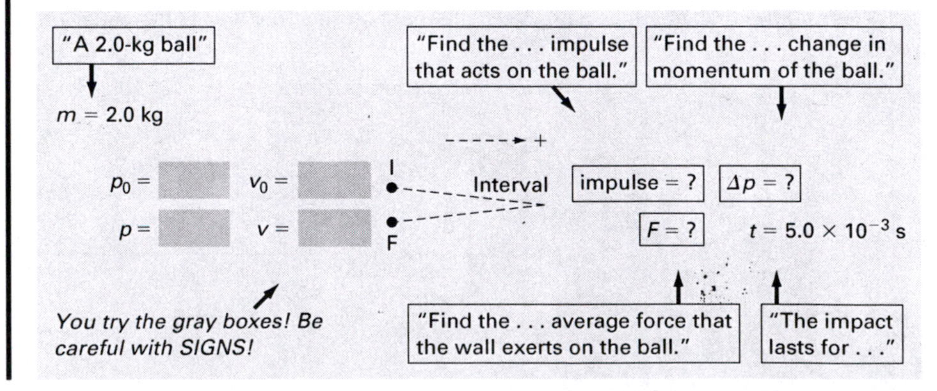

The force, F, in Equation 9.1 is actually the *net* force. But we use *only the force from the wall* and *completely ignore other forces* like weight. Why? Because any other forces either *cancel out* or they are *very small compared to* the large impact force from the wall!

(3) Equations & unknowns

We write Equation 9.1 for the interval and make an equation for part (d):

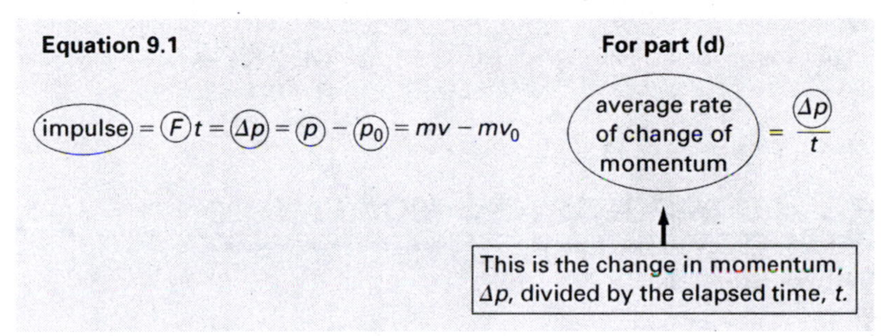

(4) Outline solution

You try to outline a solution on the previous figure, and then check the answers.

Answers for gray boxes

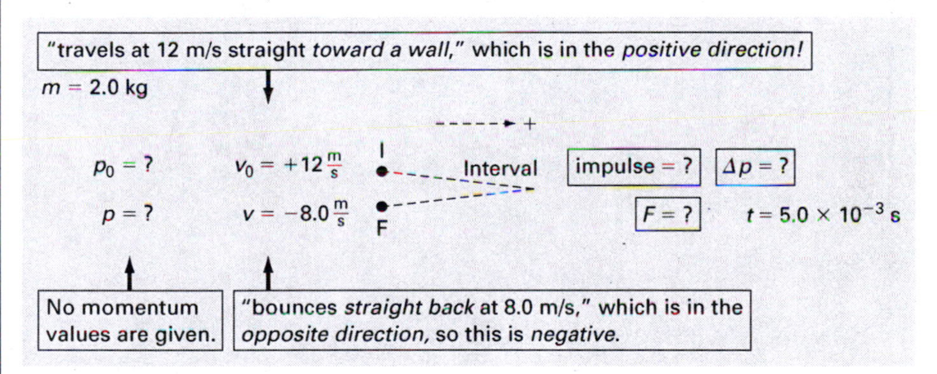

Answers for (4) Outline solution

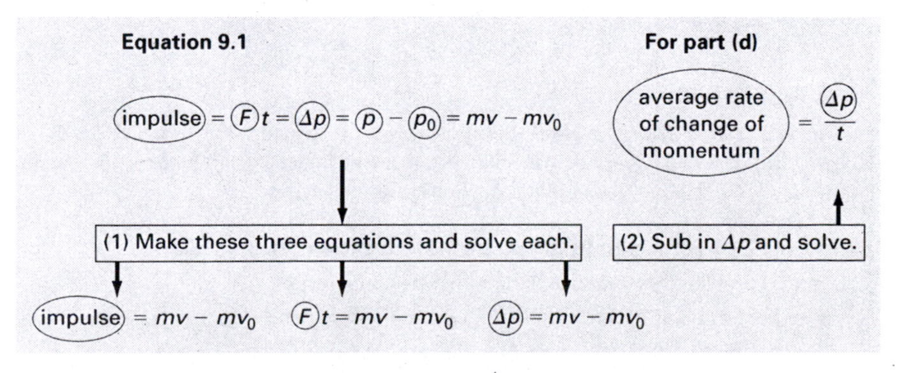

Answers $\Delta p = -40\ \mathrm{kg}\,\frac{\mathrm{m}}{\mathrm{s}}$, impulse $= -40\ \mathrm{kg}\,\frac{\mathrm{m}}{\mathrm{s}}$, $F = -8.0 \times 10^3$ N, $\frac{\Delta p}{t} = -8.0 \times 10^3$ N ■

Be careful with signs! All of these quantities are *negative,* meaning the directions are *to the left,* or in the *opposite direction of the ball's original motion.*

IMPULSE = CHANGE OF MOMENTUM, AND FORCE = RATE OF CHANGE OF MOMENTUM!

Did you notice that impulse = Δp and $F = \Delta p/t$? Both of these are always true!

9.3 2D IMPULSE AND MOMENTUM

EXERCISE 9.2

A hose sprays 25 kg of water per minute at a wall. The water hits the wall at 12 m/s at an angle of 30 degrees and then bounces off at the same speed and angle. Determine the magnitude and direction of the force that the wall exerts on the water.

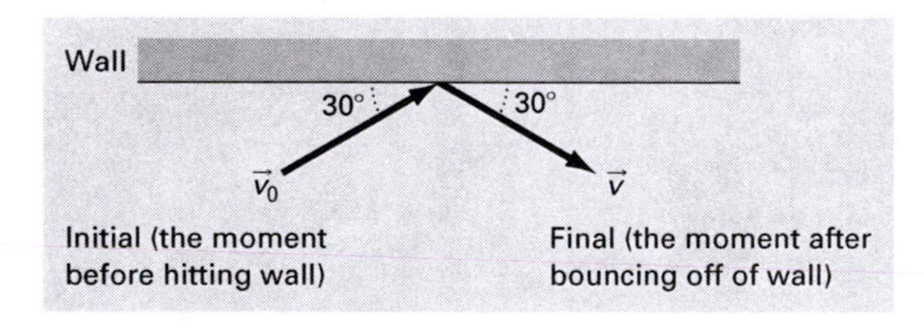

Solution

HOW DO WE DEAL WITH "kg PER MINUTE"?

We have "25 kg of water per minute," so we make t = 1 minute = 60 s, and then m = 25 kg for that amount of elapsed time.

(1) Type of problem

This is a *motion-changing event* (think of it as a 60-second event in which 25 kg of water bounce off the wall). Since the velocity vectors are at angles, the problem is 2D: Use Equation 9.1a and/or 9.1b, along with right triangle trig for the vectors.

(2) Sort by vector and interval, and (3) Equations & unknowns

For a 2D problem, we combine these steps into three parts:

- **Vectors:** Sort by vector and use trig equations to determine components.
- ***x*-direction**: Sort by interval for Equation 9.1a.
- ***y*-direction:** Sort by interval for Equation 9.1b.

This is very similar to the setup for projectile motion in Chapter 4. As with projectile motion, we see the 2D interval here as two separate 1D intervals:

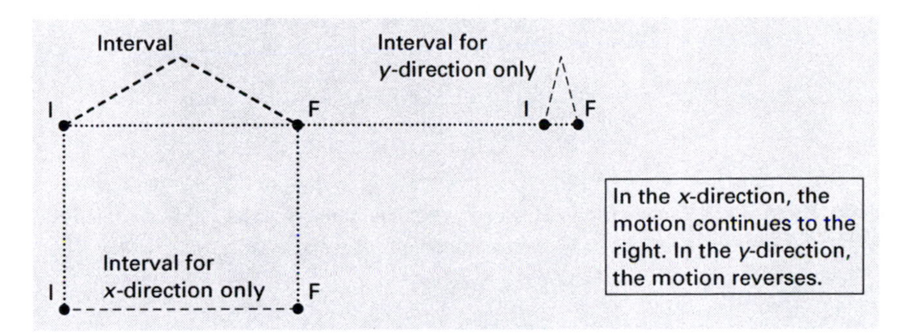

Vectors
We have three vectors to deal with: the known initial and final velocity vectors, $\vec{v}_0$ and $\vec{v}$, and the unknown force vector, $\vec{F}$.

BACK TO VECTOR BASICS!

Even though we are trying to get the "magnitude and direction," F and θ, of the force vector, $\vec{F}$, we first work toward finding the x- and y-components, F_x and F_y. Then we use these, and a right triangle if necessary, to get F and θ. Review vectors in Chapter 3 if you need a refresher.

We start with $\vec{v}_0$:

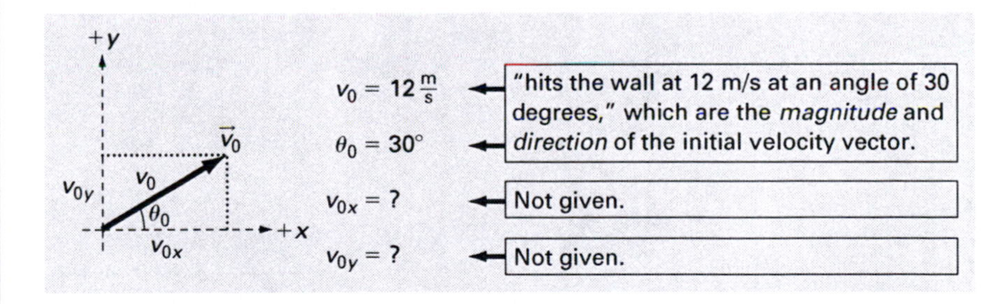

The right triangle and trig equations for $\vec{v}_0$:

v_0 , θ_0 , v_{0y} , v_{0x}

$$\frac{v_{0x}}{v_0} = \cos\theta_0 \qquad \frac{v_{0y}}{v_0} = \sin\theta_0 \qquad \frac{v_{0y}}{v_{0x}} = \tan\theta_0 \qquad v_0^2 = v_{0x}^2 + v_{0y}^2$$

Solve both of these.

We go ahead and calculate the values:

$v_{0x} \cong +10.4\,\frac{m}{s}$ ← Make this *positive,* since it is in the $+x$-direction.

$v_{0y} = +6.0\,\frac{m}{s}$ ← Make this *positive,* since it is in the $+y$-direction.

We could do the same for $\vec{v}$. However, it is the same size as $\vec{v}_0$, but angled away from the wall instead of toward it. So the x-components of $\vec{v}$ and $\vec{v}_0$ are equal, but the y-components have opposite signs:

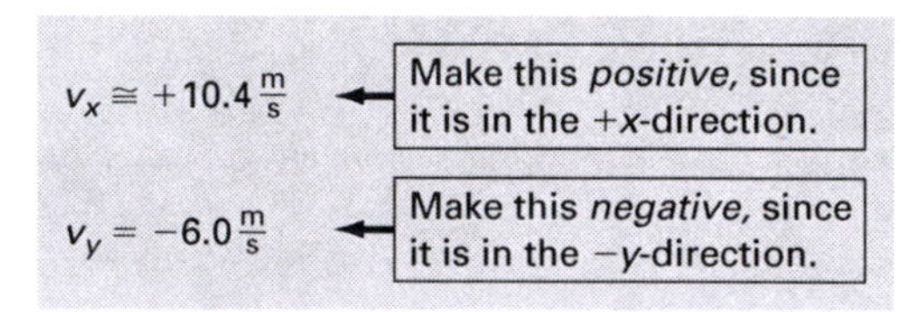

Finally, all quantities for $\vec{F}$ are unknown (we can draw a right triangle and write trig equations later if needed):

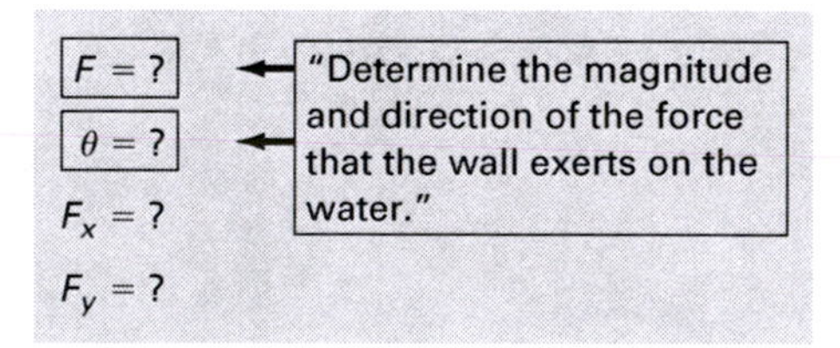

x-direction

The interval for the x-components:

I — Interval — F $+x$ →

$p_{0x} = ?$
$v_{0x} \cong +10.4\,\frac{m}{s}$

$\text{impulse}_x = ?$
$F_x = ?$
$\Delta p_x = ?$
$t = 60$ s

$p_x = ?$
$v_x \cong +10.4\,\frac{m}{s}$ ← Velocity x-components are from right triangle. No other x-components are given.

$m = 25$ kg ← "25 kg of water per minute" tells us m and t.

This leads to Equation 9.1a:

$$\text{impulse}_x = F_x t = \Delta p_x = p_x - p_{0x} = mv_x - mv_{0x}$$

y-direction

The interval for the y-components:

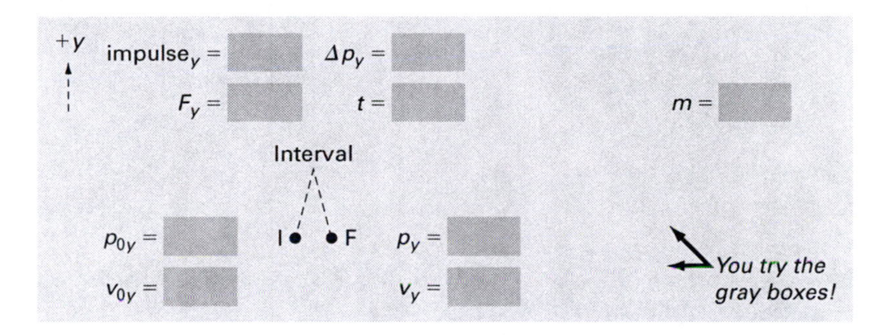

And Equation 9.1b:

$$\text{impulse}_y = F_y t = \Delta p_y = p_y - p_{0y} = mv_y - mv_{0y}$$ ← *You circle the unknowns!*

These answers are in the next part.

(4) Outline solution

We already solved for the velocity vector components. Putting the other equations together:

***x*-direction**

$$\text{impulse}_x = F_x t = \Delta p_x = p_x - p_{0x} = mv_x - mv_{0x}$$

***y*-direction**

$$\text{impulse}_y = F_y t = \Delta p_y = p_y - p_{0y} = mv_y - mv_{0y}$$

You try an outline to get F_x and F_y!

Check the answers for these components, where we will also give the magnitude and direction of the force vector.

Answers for gray boxes

+y

impulse$_y$ = ? Δp_y = ?

F_y = ? $t = 60$ s $m = 25$ kg

Interval (I → F)

p_{0y} = ? p_y = ?

$v_{0y} = +6.0\,\frac{\text{m}}{\text{s}}$ $v_y = -6.0\,\frac{\text{m}}{\text{s}}$

"25 kg of water per minute" tells us m and t.

Velocity y-components are from right triangle. No other y-components are given.

Answers for (4) Outline solution

This shows how to get the components:

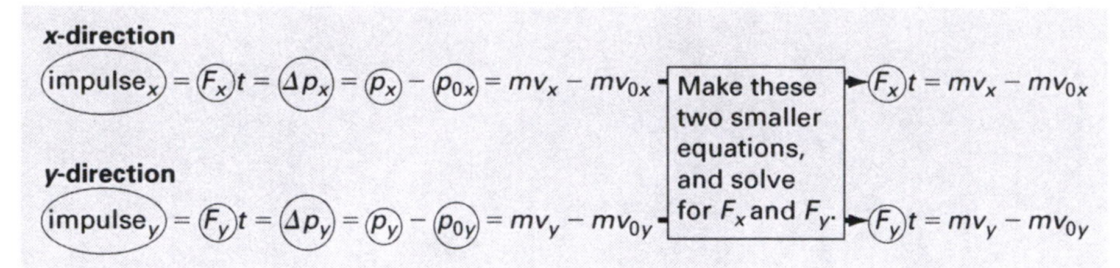

Answers

Force components: $F_x = 0$ and $F_y = -5.0$ N. So $\vec{F}$ is along one axis, and we can determine its magnitude and direction without a right triangle:

- Magnitude: $F = 5.0$ N (magnitude is *always positive*).
- Direction: Along negative y-axis, or perpendicular to and away from the wall.

USE SHORTCUTS IF YOU SEE THEM, BUT ALWAYS USE COMPONENTS FOR 2D!

Before doing any calculations, we might have seen that the wall pushes only in the negative y-direction, and so $F_x = 0$. Nice shortcut! However, *we still need to use y-components* to get F_y before we can get the magnitude, F.

■

9.4 HOW TO SET UP IMPULSE AND MOMENTUM PROBLEMS

Here is a summary of what we did in the previous two exercises:

Impulse and Momentum Problems—Mental and Written Steps

Mental →	**(1) Type of problem** For impulse and/or momentum and a *motion-changing event* (like a ball hitting a wall), use the impulse–momentum theorem: • If 1D: Equation 9.1. • If 2D: Equation 9.1a and/or 9.1b, and right triangle trig. (BUT if it is a collision or explosion involving *two objects of similar size,* like a ball hitting another ball, then use *conservation of momentum,* discussed in the next section.)

Mental and written →	**(2) Sort by (vector and) interval**
	Pay attention to *signs* for all quantities with *directions!*
	• If 1D: Sort quantities in interval for Equation 9.1.
	• If 2D, do three parts:
	  • **Vectors:** Sort by vector and determine components.
	  • **x-direction:** Sort x-components for Equation 9.1a.
	  • **y-direction:** Sort y-components for Equation 9.1b.

Mental and written →	**(3) Equations & unknowns** Write needed equations as described above, and circle the unknowns!
Mental →	**(4) Outline solution**

9.5 CONSERVATION OF MOMENTUM

A collision or explosion, when it involves *two objects of similar size,* changes the motions of *both objects.* There can be more than two objects, but this only occasionally shows up in problems.

COLLISION

Initially, objects 1 and 2 are *apart,* and then come *together.* Two common scenarios:

- Objects 1 and 2 collide and either stick together or bounce apart (what we usually think of as a collision).
- Object 1 drops from above onto object 2.

EXPLOSION

Initially, objects 1 and 2 are *together,* and then move *apart.* Two common scenarios:

- A big object explodes, splitting into objects 1 and 2 that move apart (what we usually think of as an explosion).
- Objects 1 and 2 are initially together but then push apart and move away from each other.

For collisions and explosions, think intervals and instants:

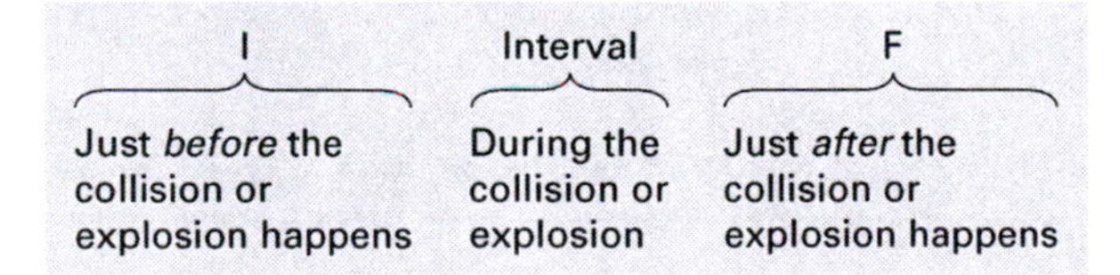

We use *momentum* for collision and explosion problems, and so need *masses* and *velocities* for each object. Here is a summary of the setup and notation we will use:

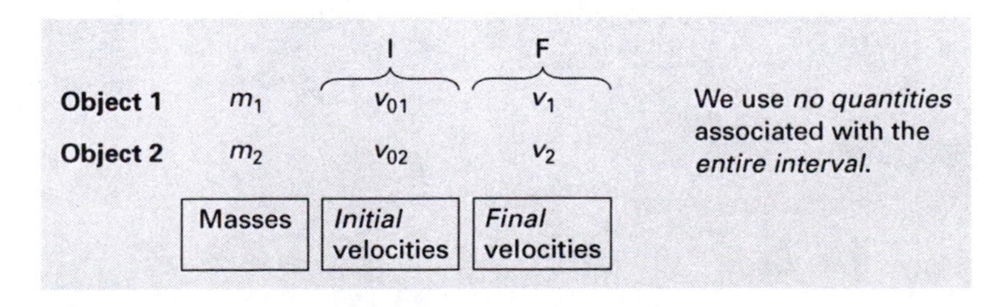

Then the momentum values are:

	Initial momentum	Final momentum
Object 1	m_1v_{01}	m_1v_1
Object 2	m_2v_{02}	m_2v_2
Objects 1 and 2 combined total	$m_1v_{01} + m_2v_{02}$	$m_1v_1 + m_2v_2$

This leads to the following idea, which we use to solve collision or explosion problems:

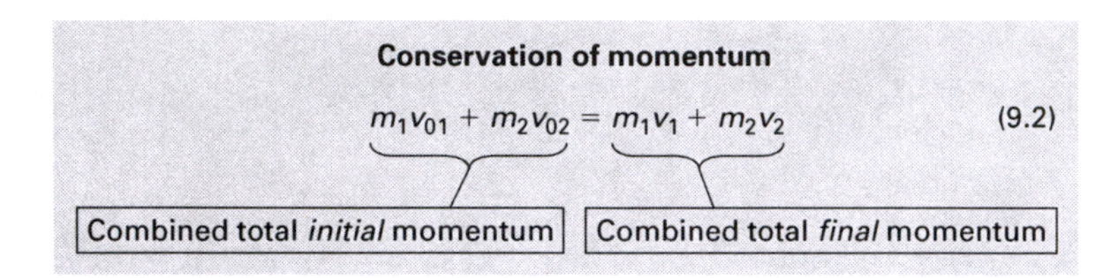

Although written for a 1D situation, Equation 9.2 is really a vector equation. For 2D problems, we use both x- and y-component versions of Equation 9.2:

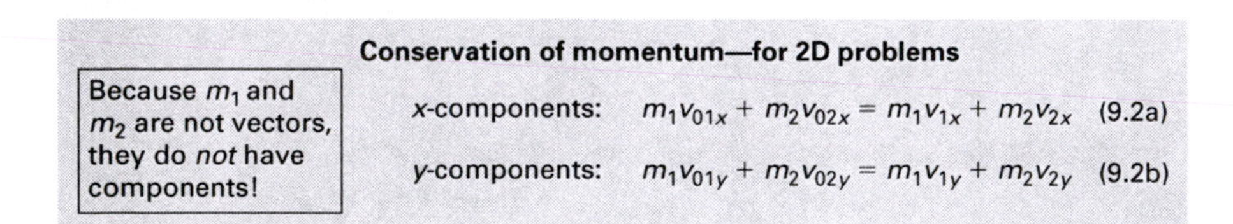

WHY DOES CONSERVATION OF MOMENTUM WORK?

In a collision or explosion, objects 1 and 2 exert *very large forces on each other.* Any *other* forces acting on either object either (i) *cancel out* or (ii) are *very small compared to* the collision/explosion forces. Either way, we IGNORE any external forces on the two-object system, so the *net external force, F, is zero:*

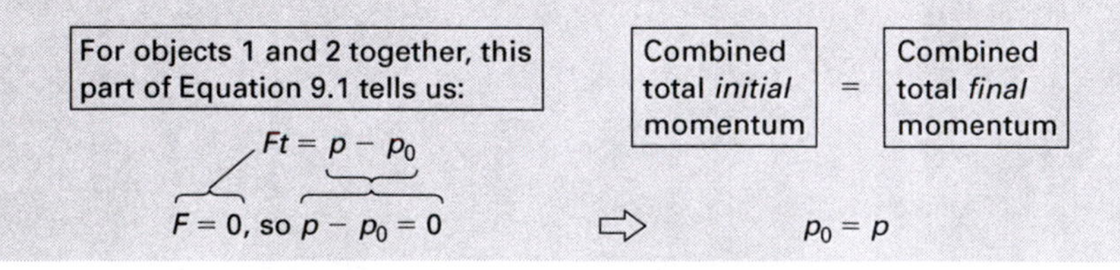

9.6 1D COLLISIONS–OBJECTS COMING TOGETHER

EXERCISE 9.3

A 3.00-kg block travels to the right with an initial speed of 2.00 m/s. It collides inelastically with a 2.00-kg block that initially travels to the left at 1.50 m/s. Just after the collision, the 3.00-kg block travels to the left at 0.500 m/s. (a) What is the velocity of the 2.00-kg block just after the collision? (b) What is the percentage of the initial kinetic energy lost during the collision? (c) What happened to the "lost" kinetic energy?

Solution

(1) Type of problem

Collision, 1D: Use Equation 9.2. We also use the kinetic energy equation ($KE = \frac{1}{2}mv^2$) for part (b).

COLLIDES INELASTICALLY?

An *inelastic* collision has the following two properties:

- The *total momentum is conserved,* so we use Equation 9.2.
- But some *kinetic energy is "lost"* during the collision, so the total *initial* kinetic energy is greater than the total *final* kinetic energy.

(2) Sort by interval and object

For the collision, there is one interval and there are two objects, block 1 (3.00 kg) and block 2 (2.00 kg). We sort the *masses* and *velocities* for Equation 9.2:

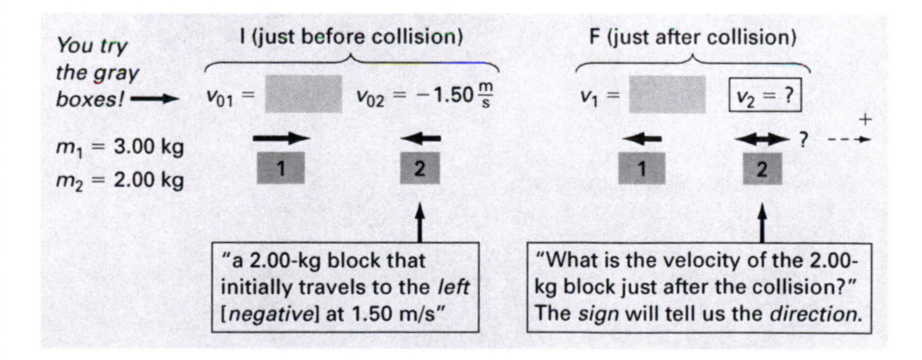

(3) Equations & unknowns

For part (a):

Equation 9.2 $m_1v_{01} + m_2v_{02} = m_1v_1 + m_2 \textcircled{v_2}$

VELOCITY IS NOT CONSERVED! MOMENTUM IS!

Don't make the mistake of writing Equation 9.2 in terms of velocities only:

$$v_{01} + v_{02} = v_1 + v_2 \quad \longleftarrow \text{WRONG!}$$

Include the *masses:* It is conservation of *momentum,* NOT conservation of *velocity!*

For part (b), to get the "percentage of initial kinetic energy lost," we first need the initial and final values of kinetic energy:

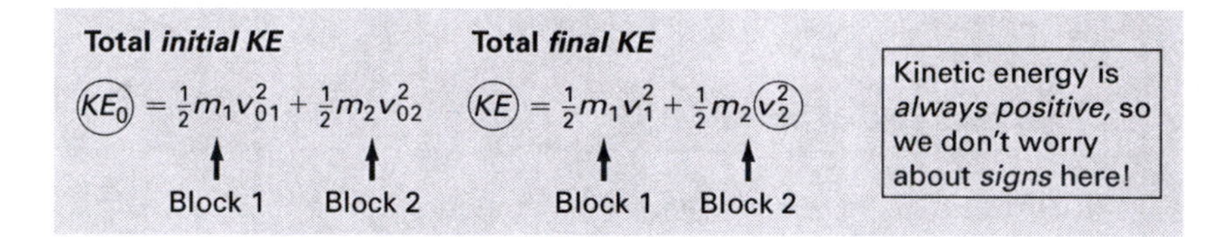

Then, to answer the question:

Percentage of initial *KE* lost

$$\%\text{ of } KE_0 \text{ lost} = \frac{\overbrace{KE_0 - KE}^{\text{Kinetic energy lost}}}{\underbrace{KE_0}_{\text{Initial kinetic energy}}} \times 100\%$$

(4) Outline solution

We write all the equations together:

Equation 9.2 $m_1v_{01} + m_2v_{02} = m_1v_1 + m_2v_2$

Total *initial KE* $KE_0 = \frac{1}{2}m_1v_{01}^2 + \frac{1}{2}m_2v_{02}^2$

Total *final KE* $KE = \frac{1}{2}m_1v_1^2 + \frac{1}{2}m_2v_2^2$

You try an outline!

Percentage of initial *KE* lost $\%\text{ of } KE_0 \text{ lost} = \dfrac{KE_0 - KE}{KE_0} \times 100\%$

Answers for gray boxes

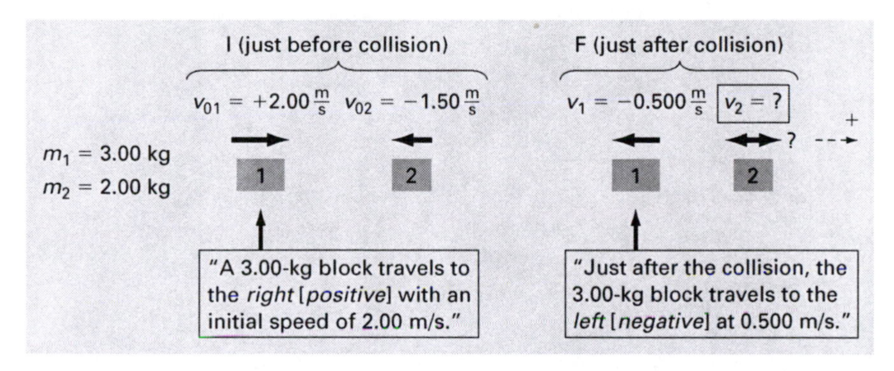

Answers for (4) Outline solution

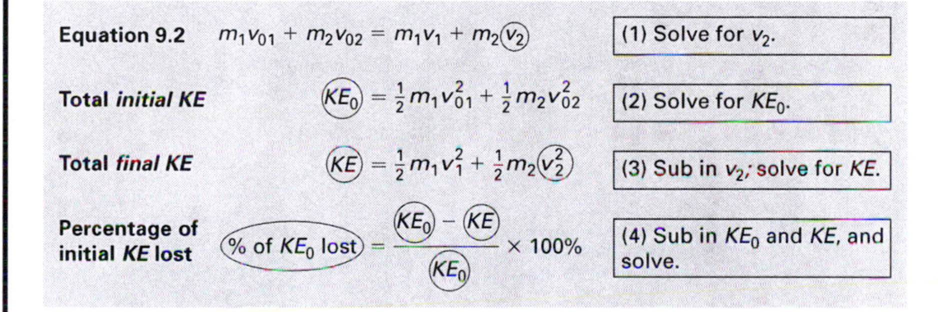

Answers

- $v_2 = +2.25\,\frac{m}{s}$ (*positive* means to the *right*).
- % of KE_0 lost $\cong 34.1\%$.
- During the collision, the "lost" kinetic energy goes into *producing sound, dents, vibration,* and so on. All of these "use up" some of the kinetic energy!

Intermediate answers: $KE_0 = 8.25$ J, $KE \cong 5.438$ J ■

9.7 1D EXPLOSIONS—OBJECTS PUSHING APART

EXERCISE 9.4

A rocket of mass 3000 kg has a front payload of additional mass 500 kg and travels at a speed of 900 m/s. The rocket ejects the payload forward, which causes the rocket to slow to 800 m/s. What is the final speed of the payload?

Solution

(1) Type of problem

Explosion (objects 1 and 2 are initially *together* but then push *apart* and move away from each other), 1D: Use Equation 9.2.

(2) Sort by interval and object

There is one interval and there are two objects. The rocket is object 1 (3000 kg) and the payload is object 2 (500 kg), and we set up for Equation 9.2:

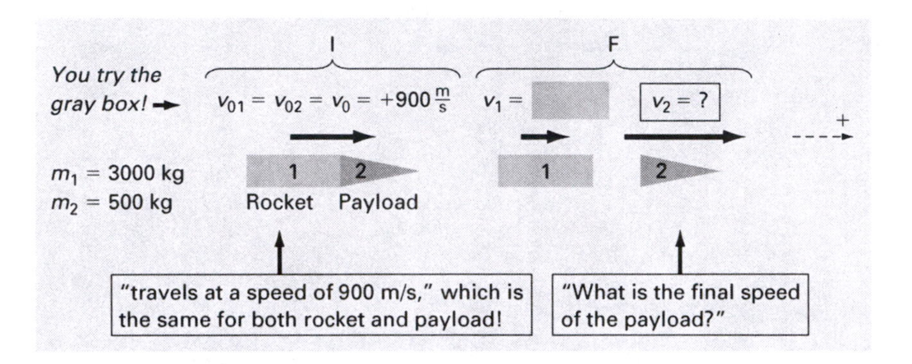

All velocities are in the same direction, so we make them all positive.

(3) Equations & unknowns and (4) Outline solution

We rewrite Equation 9.2 for this situation:

Equation 9.2 $m_1 v_{01} + m_2 v_{02} = m_1 v_1 + m_2 v_2$

$v_{01} = v_{02} = v_0$

$(m_1 + m_2)v_0 = m_1 v_1 + m_2 v_2$ Solve.

Answers for gray boxes

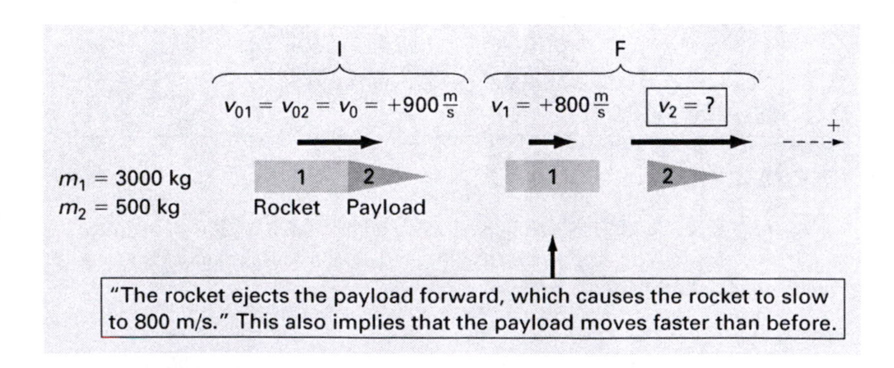

Answers $v_2 = +1500\frac{m}{s}$ (*positive* means to the *right*). ■

We did not calculate kinetic energy in the previous exercise, but if we had, we would have seen that kinetic energy was *gained* (instead of lost, as in an inelastic collision). Why? The ejection or explosion process gives kinetic energy to both objects. *Kinetic energy is increased* even though *momentum is conserved!*

THE BEST WAY TO LEARN: TRY TO SOLVE A PROBLEM!

If you don't feel *thoroughly* confident with conservation of momentum, that is okay! Just use what you *do* know, and try to work through a problem. Even if you make lots of mistakes, you will *learn in the process* and understand more than when you started!

9.8 1D ELASTIC COLLISIONS

A 1D collision is called *elastic* if it has two more properties in addition to the fact that the *total momentum is conserved:*

Properties of 1D *elastic* collisions	Equations to use	
Total *momentum* is conserved.	$m_1v_{01} + m_2v_{02} = m_1v_1 + m_2v_2$	(9.2)
Total *kinetic energy* is conserved.	$\frac{1}{2}m_1v_{01}^2 + \frac{1}{2}m_2v_{02}^2 = \frac{1}{2}m_1v_1^2 + \frac{1}{2}m_2v_2^2$	(9.3)
Relative velocity is reversed.	$v_{01} - v_{02} = v_2 - v_1$	(9.4)

To solve any 1D elastic collision problem, we need *any two* of these three equations. We will use *only Equations 9.2 and 9.4* because the math is *easier* than if we use Equation 9.3! Key word: EASIER! Although Equation 9.4 is rarely given in textbooks, it can be derived (with difficulty) from Equations 9.2 and 9.3.

EXERCISE 9.5

Block 1 (1.00 kg) travels to the right with an initial speed of 6.00 m/s on a horizontal frictionless surface. It collides *elastically* with block 2 (3.00 kg), initially at rest. (a) What is the velocity of each block immediately after the collision? (b) How high does block 2 go up the hill?

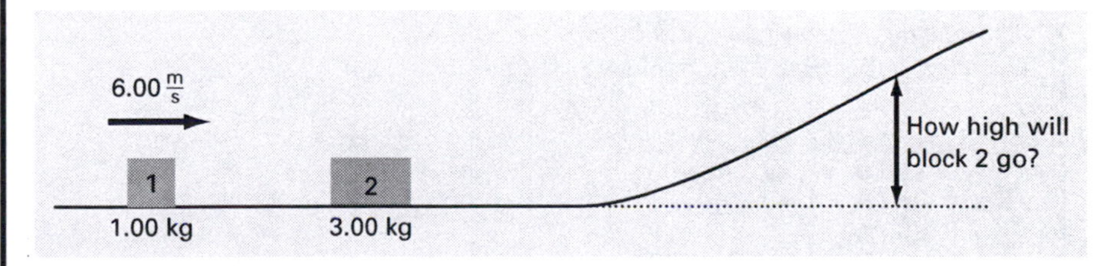

Solution

(1) Type of problem

We set this up as two separate problems:

- **Elastic collision for blocks 1 and 2:** For part (a) we have a 1D *elastic* collision and so use Equations 9.2 and 9.4.

- **Conservation of energy for block 2:** For part (b), after the collision is over, there are NO nonconservative forces that do work on block 2, so we use conservation of energy, Equation 8.4b, for block 2.

(2) Sort by interval and object, and (3) Equations & unknowns

Elastic collision for blocks 1 and 2

The setup is identical to that for other 1D collision problems:

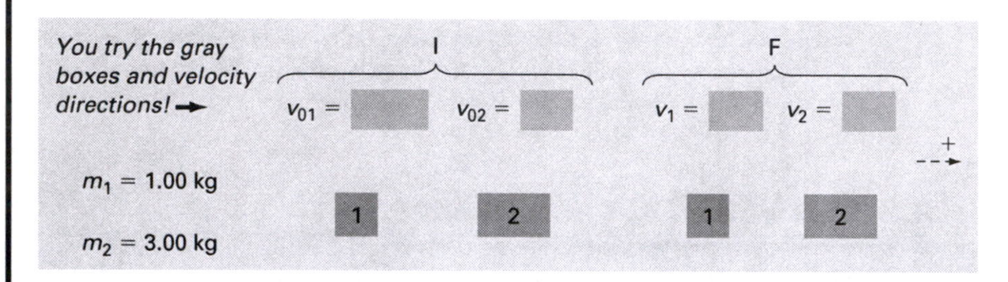

This leads to:

Equation 9.2 $m_1v_{01} + m_2v_{02} = m_1v_1 + m_2v_2$ →

Equation 9.4 $v_{01} - v_{02} = v_2 - v_1$ →

You simplify (sub in zero values) and then circle the unknowns!

OR USE THE "CANNED" EQUATIONS!

If your instructor wants you to use them, textbooks have special "canned" equations for v_1 and v_2 for 1D elastic collisions. However, you still need to be careful to correctly *set up* the problem as we did here!

Conservation of energy for block 2

The setup for Equation 8.4b, for block 2 after the collision:

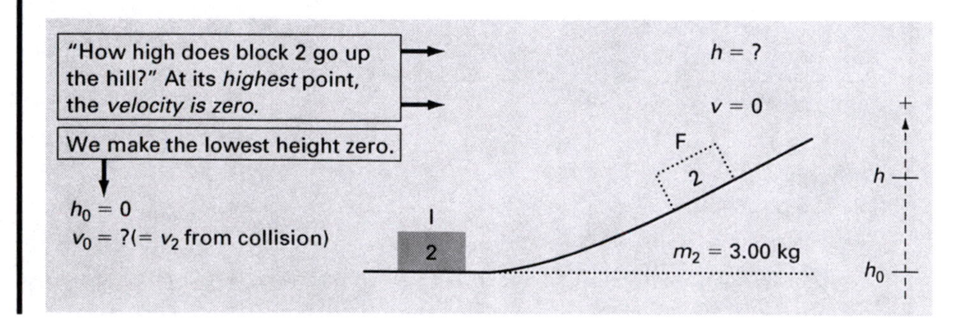

Equation 8.4b for this interval:

$$\tfrac{1}{2}m_2 v_0^2 + m_2 g h_0 = \tfrac{1}{2}m_2 v^2 + m_2 g h$$

Divide by m_2, sub in v_2 for v_0, and sub in zero for h_0 and v!

$$\tfrac{1}{2}v_2^2 = gh$$

(4) Outline solution

Combining all the equations:

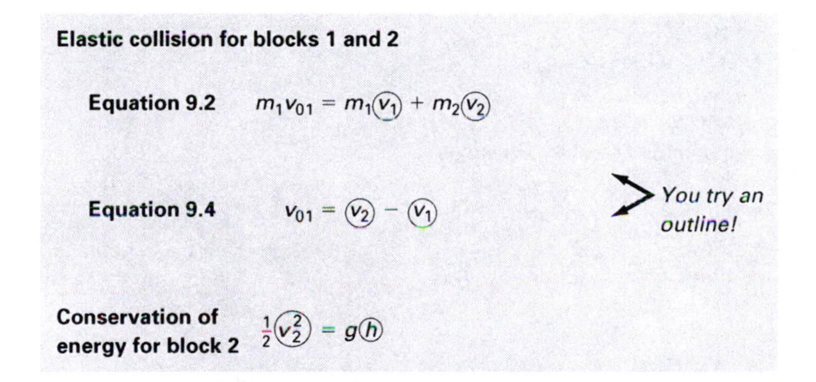

We will also do the math in the answers.

Answers for elastic collision for blocks 1 and 2

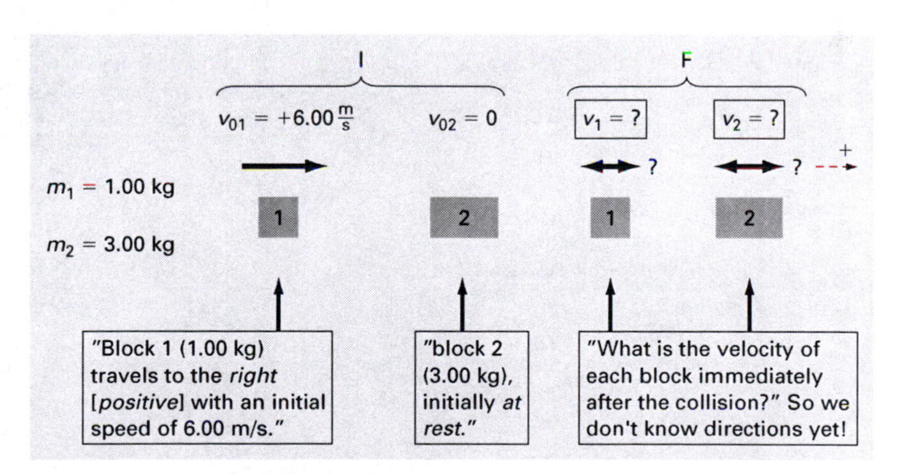

Equation 9.2 $m_1v_{01} + m_2v_{02} = m_1 v_1 + m_2 v_2$

Equation 9.4 $v_{01} - v_{02} = v_2 - v_1$

Sub in *zero* for v_{02}.

$$m_1v_{01} = m_1 v_1 + m_2 v_2$$

$$v_{01} = v_2 - v_1$$

Answers for (4) Outline solution

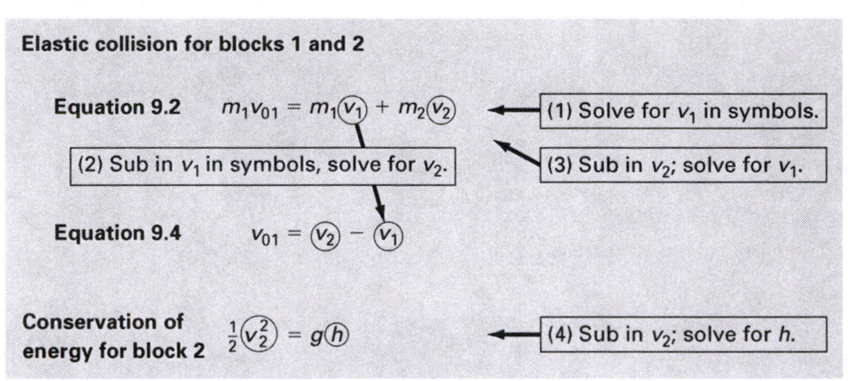

Answers for the math

First, we simplify two of the equations:

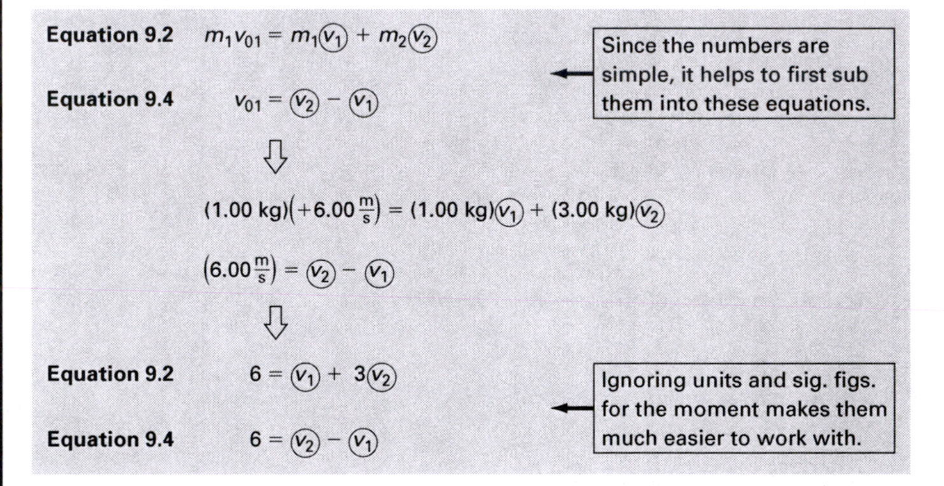

Then we solve:

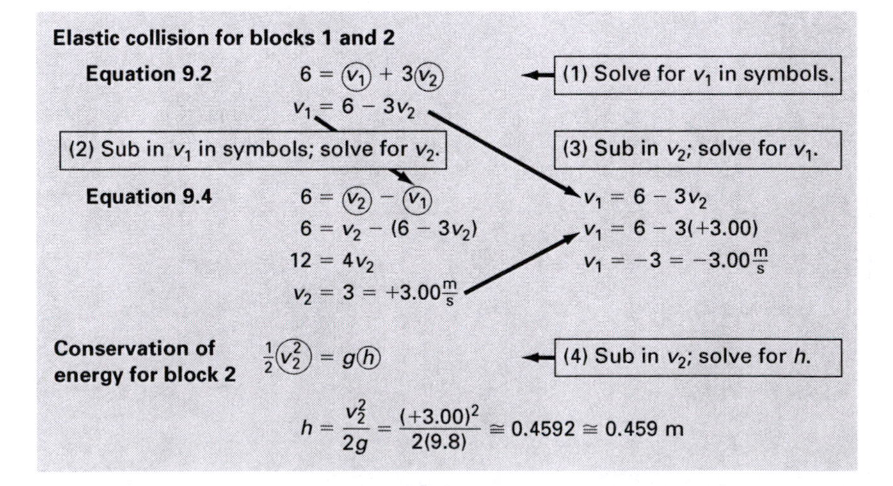

Answers

- $v_1 = -3.00\,\frac{\text{m}}{\text{s}}$ (*negative* means to the *left*).
- $v_2 = +3.00\,\frac{\text{m}}{\text{s}}$ (*positive* means to the *right*).
- $h \cong 0.459$ m.

■

9.9 2D COLLISIONS

The following exercise involves a 2D *completely inelastic* collision.

COMPLETELY (OR PERFECTLY) INELASTIC COLLISION

If the objects are *STUCK together* in the end, it is a *completely* inelastic collision!

EXERCISE 9.6

Initially, a 7.00-kg block moves horizontally at 2.00 m/s on a frictionless surface. From above the block, a 4.00-kg clay ball is tossed with velocity 5.00 m/s at 60.0 degrees below the horizontal. The clay lands on the block, and they move off horizontally together. What is their final velocity?

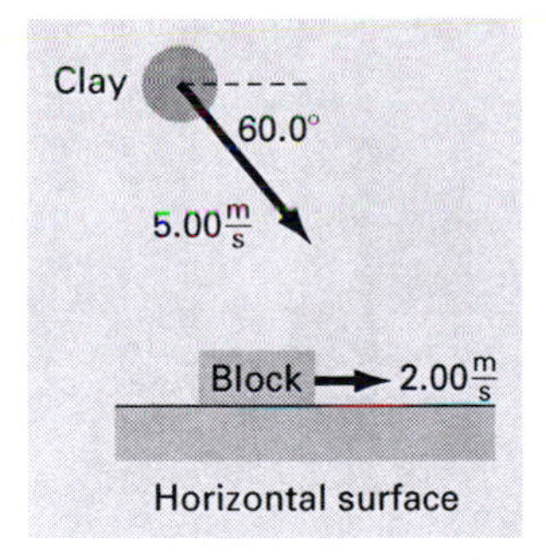

Solution

(1) Type of problem

Collision with velocity vectors at angles, so it is 2D: Use Equation 9.2a and/or 9.2b.

(2) Sort by velocity vector, interval and object, and (3) Equations & unknowns

For 2D, we combine these steps in three parts:

- **Velocity vectors:** Sort *initial* and *final* velocity vectors for *each object,* and get all x- and/or y-components.
- **Momentum x-components:** Sort x-components by interval and object for Equation 9.2a, and/or . . .
- **Momentum y-components:** Sort y-components by interval and object for Equation 9.2b.

HERE, MOMENTUM IS CONSERVED IN THE x-DIRECTION, BUT NOT IN THE y-DIRECTION!

Upon impact, the *surface under the block* causes a large *upward normal force* that takes away any *downward momentum* of the clay/block. So the total momentum *changes* in the y-direction. However, there are *no external forces* on the clay/block in the *horizontal* direction, and so momentum *is conserved* in the x-direction.

So, for this problem, we need *only x-components* and Equation 9.2a.

Velocity vectors
The clay is object 1 and the block is object 2.

The *clay* "is tossed with velocity 5.00 m/s at 60.0 degrees below the horizontal." That means we use a right triangle:

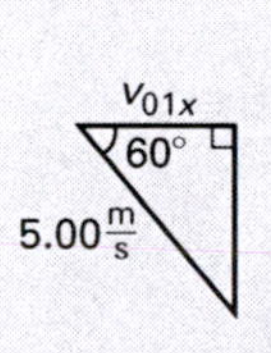

You try the equation, and solve for v_{01x}!

From Chapter 4, the *x-component* of a projectile's velocity is *constant.* We use the given velocity to get the *constant x-component,* even though the *y-component changes* as the ball falls.

WHAT IF THE CLAY WERE DROPPED VERTICALLY ONTO THE BLOCK?

Then the clay would have no horizontal motion, and so would have $v_{01x} = 0$, regardless of the y-component of velocity. The rest of the solution would be exactly the same as for this exercise.

The *block* initially "moves horizontally at 2.00 m/s" *to the right,* so its x-component is *positive:* $v_{02x} = +2.00\,\frac{\text{m}}{\text{s}}$.

The *clay and block* "move off horizontally together," so the final velocity is to the right: $v_x = +v$.

Momentum x-components

We use only x-components of the velocities to set up for Equation 9.2a:

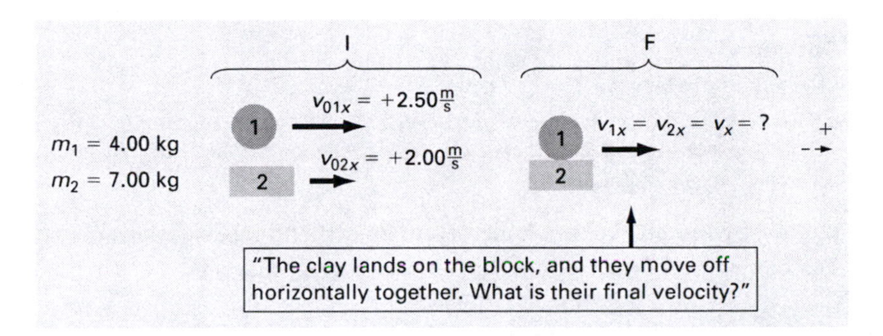

This leads to:

Equation 9.2a $m_1 v_{01x} + m_2 v_{02x} = m_1 v_{1x} + m_2 v_{2x}$

$v_{1x} = v_{2x} = v_x$

$$m_1 v_{01x} + m_2 v_{02x} = (m_1 + m_2)v_x$$

Solve.

(4) Outline solution

Shown on the previous figure!

Answers for velocity vector

v_{01x}, 60°, $5.00\frac{\text{m}}{\text{s}}$

$$\frac{v_{01x}}{\left(5.00\frac{\text{m}}{\text{s}}\right)} = \cos 60.0^\circ$$

$$v_{01x} = +2.50\frac{\text{m}}{\text{s}}$$

Make *positive,* since it is *to the right.*

Answer $v_x \cong +2.18\frac{\text{m}}{\text{s}}$ or . . .

- Magnitude: $v \cong 2.18\frac{\text{m}}{\text{s}}$.
- Direction: to the right.

■

EXERCISE 9.7

A truck of mass 2000 kg traveling east collides with a car of mass 1200 kg traveling north. The two vehicles stick together as a result of the collision. Just after the collision, they slide together at 10.0 m/s, at an angle of 30.0 degrees north of east. The road is slick, so friction between the cars and the road can be ignored. (a) What was the speed of each vehicle before the collision? (b) How much kinetic energy was lost as a result of the collision?

Solution

(1) Type of problem

Collision in 2D, and there are *no external forces* to change the momentum along either the x- (east/west) or y- (north/south) axis: Use *both* Equations 9.2a and 9.2b. We also use the kinetic energy equation for part (b).

(2) Sort by velocity vector, interval and object, and (3) Equations & unknowns

The truck is object 1 and the car is object 2. Here is an initial setup:

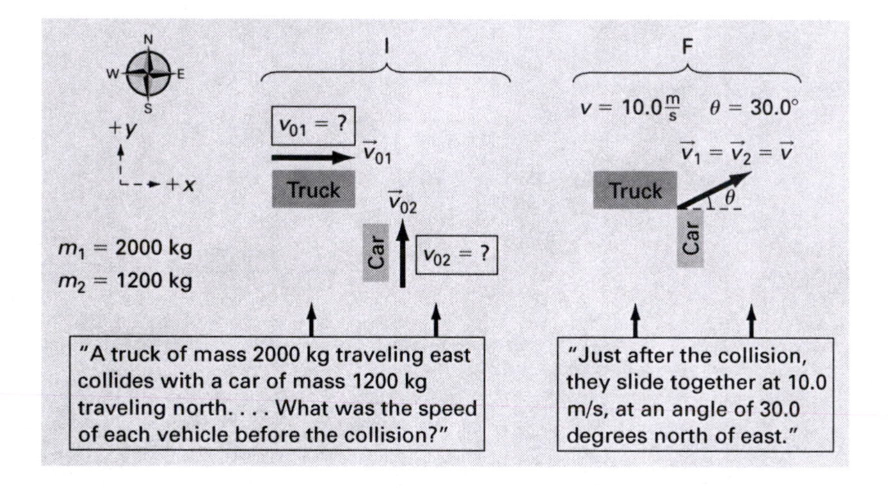

Velocity vectors

For 2D, we need all velocity vector components, *initial* and *final,* for both the *car* and *truck.* The *initial* velocities are each along an axis, so we get all of these components just by looking at the previous figure:

	Truck	Car
Initial velocity x-component	$v_{01x} = +v_{01} = ?$	$v_{02x} = 0$
Initial velocity y-component	$v_{01y} = 0$	$v_{02y} = +v_{02} = ?$

There is just one *final* velocity vector, which is at an angle, and so we use a right triangle (enlarged):

Momentum* x-*components

We set up the interval for only the x-components:

$m_1 = 2000$ kg, $m_2 = 1200$ kg

I: $v_{01x} = +v_{01} = ?$, $v_{02x} = 0$

F: $v_{1x} = v_{2x} = v_x \cong +8.660\,\frac{\text{m}}{\text{s}}$

Then we use Equation 9.2a:

$$m_1 v_{01x} + m_2 v_{02x} = m_1 v_{1x} + m_2 v_{2x} \xrightarrow{v_{02x} = 0,\; v_{1x} = v_{2x} = v_x} m_1 \boxed{v_{01}} = (m_1 + m_2) v_x$$

Momentum* y-*components

Now the interval for only the y-components:

$m_1 = 2000$ kg, $m_2 = 1200$ kg

I: $v_{01y} =$ ▢, $v_{02y} =$ ▢

F: $v_{1y} = v_{2y} = v_y =$ ▢ ← *You try the gray boxes!*

This leads to Equation 9.2b:

$$m_1 v_{01y} + m_2 v_{02y} = m_1 v_{1y} + m_2 v_{2y} \rightarrow$$

You simplify and then circle the unknowns!

Kinetic energy

For this problem, we also need equations for the total *initial* and *final* kinetic energies:

I: $\boxed{KE_0} = \frac{1}{2} m_1 \boxed{v_{01}^2} + \frac{1}{2} m_2 \boxed{v_{02}^2}$ (Truck, Car)

F: $\boxed{KE} = \frac{1}{2} m_1 v_1^2 + \frac{1}{2} m_2 v_2^2 = \frac{1}{2}(m_1 + m_2) v^2$ (Truck, Car, $v_1 = v_2 = v$)

Kinetic energy is a *scalar,* not a vector. So, for kinetic energy calculations, use *only* VELOCITY MAGNITUDES, *not* velocity components.

Then, to answer the question:

$$\text{kinetic energy lost} = KE_0 - KE$$

(4) Outline solution

We have already done the velocity vector calculations. The rest of the equations:

Momentum x-components $m_1 v_{01} = (m_1 + m_2)v_x$

Momentum y-components $m_2 v_{02} = (m_1 + m_2)v_y$

Kinetic energy $KE_0 = \frac{1}{2}m_1 v_{01}^2 + \frac{1}{2}m_2 v_{02}^2$

$KE = \frac{1}{2}(m_1 + m_2)v^2$

$\text{kinetic energy lost} = KE_0 - KE$

You try an outline!

Answers for velocity vector

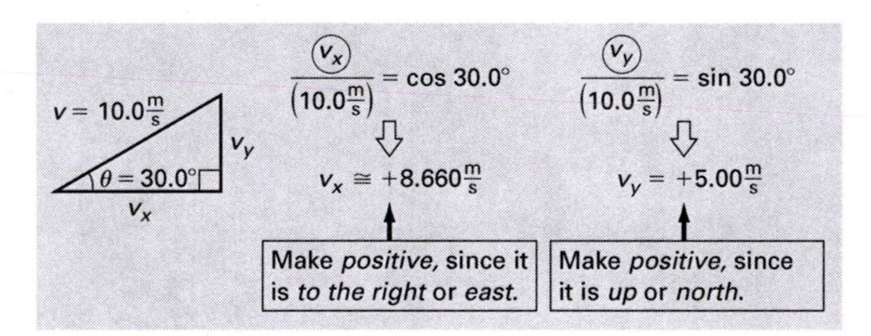

Answers for momentum* y*-components

$m_1 = 2000\text{ kg}$

$m_2 = 1200\text{ kg}$

I: $v_{01y} = 0$, $v_{02y} = +v_{02} = ?$

F: $v_{1y} = v_{2y} = v_y = +5.00\frac{\text{m}}{\text{s}}$

From velocity vector calculations

$$m_1 v_{01y} + m_2 v_{02y} = m_1 v_{1y} + m_2 v_{2y} \xrightarrow{v_{01y} = 0,\; v_{1y} = v_{2y} = v_y} m_2 v_{02} = (m_1 + m_2)v_y$$

Answers for (4) Outline solution

Momentum x-components	$m_1 (v_{01}) = (m_1 + m_2)v_x$	(1) Solve for v_{01}.
Momentum y-components	$m_2 (v_{02}) = (m_1 + m_2)v_y$	(2) Solve for v_{02}.
Kinetic energy	$(KE_0) = \frac{1}{2}m_1(v_{01}^2) + \frac{1}{2}m_2(v_{02}^2)$	(3) Sub in v_{01} and v_{02}; solve for KE_0.
	$(KE) = \frac{1}{2}(m_1 + m_2)v^2$	(4) Solve for KE.
	(kinetic energy lost) = $(KE_0) - (KE)$	(5) Sub in KE_0 and KE; solve.

Answers $v_{01} \cong 13.86 \cong 13.9\frac{\text{m}}{\text{s}}$, $v_{02} \cong 13.33 \cong 13.3\frac{\text{m}}{\text{s}}$, kinetic energy lost $\cong 1.39 \times 10^5$ J

Intermediate answers: $KE_0 \cong 2.988 \times 10^5$ J, $KE = 1.60 \times 10^5$ J ■

There is a shortcut vector diagram that can be used specifically in the previous exercise to get the initial velocities. If you know it, use it! If you don't know it, stick to the basics to get you through *any collision or explosion problem.* For 2D explosions, the setup is just like that for 2D collisions.

9.10 HOW TO SET UP CONSERVATION OF MOMENTUM PROBLEMS

Here is a summary of what we have done in the previous several exercises:

Conservation of Momentum Problems—Mental and Written Steps

Mental → **(1) Type of problem**
For a collision or explosion, use conservation of momentum:
- If 1D: Equation 9.2.
- If 2D: Equation 9.2a and/or 9.2b, and right triangle trig.

Mental and written → **(2) Sort by (velocity vector and) interval and object**
Pay attention to *signs* for all *velocities!*
- If 1D:
  - Sort by interval and object for Equation 9.2.
  - If *elastic collision,* ALSO use Equation 9.4.
- If 2D:
  - **Velocity vectors:** Sort *initial* and *final* velocity vectors for *each object,* and get all *x*- and/or *y*-components.
  - **Momentum *x*-components:** Sort *x*-components only for Equation 9.2a, and/or . . .
  - **Momentum *y*-components:** Sort *y*-components only for Equation 9.2b.

Also, use the *kinetic energy* equation if needed.

Mental and written → **(3) Equations & unknowns**
Write needed equations as described above, and circle the unknowns!

Mental → **(4) Outline solution**

9.11 CENTER OF MASS

The center of mass of an object is the point that can act as if all of the object's mass is concentrated there. A system of more than one object also has a center of mass, the position of which can be calculated with the following equations:

Center of mass

x-component: $$x_{cm} = \frac{m_1x_1 + m_2x_2 + m_3x_3}{m_1 + m_2 + m_3} \quad (9.5a)$$

y-component: $$y_{cm} = \frac{m_1y_1 + m_2y_2 + m_3y_3}{m_1 + m_2 + m_3} \quad (9.5b)$$

Each term here will be explained in detail in the next two exercises. These equations are written for a system of three objects, but we can change them to fit the number of objects in the problem.

For a 1D problem, we use either Equation 9.5a or 9.5b. For a 2D problem, we use both.

9.12 1D CENTER OF MASS

EXERCISE 9.8

A girl and a boy stand at opposite ends of a 3.0-m-long uniform board with mass 50 kg. The girl's mass is 30 kg and the boy's is 40 kg. Where is the center of mass of the girl/board/boy system?

Solution

(1) Type of problem

Center of mass, 1D, three objects: Use Equation 9.5a for three objects.

(2) Sort by object

We assign a number (1, 2, 3) to each object with mass (girl, board, boy). This figure illustrates the meaning of each term in Equation 9.5a:

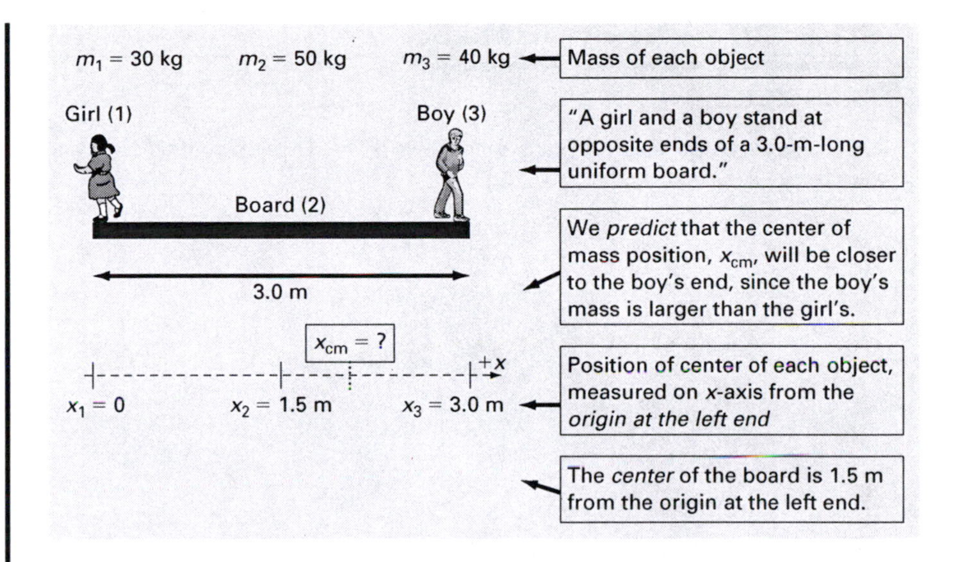

PUT THE ORIGIN AT THE POSITION OF ONE OF THE OBJECTS!

This makes x equal *zero* for that object, the girl in this case. When something is *zero*, the calculations are *easier.* Key word: EASIER!

(3) Equations & unknowns and (4) Outline solution

Being careful in the setup makes this part easy! Equation 9.5a:

$$x_{cm} = \frac{m_1 x_1 + m_2 x_2 + m_3 x_3}{m_1 + m_2 + m_3} \quad \boxed{\text{Solve.}}$$

Answer $x_{cm} \cong 1.63 \cong 1.6$ m

This is measured *from the origin* at the left end of the board, so the center of mass is closer to the boy's end as predicted. If we had chosen a *different origin,* the *value* of x_{cm} would be *different,* measured from *that origin,* but the *physical location* of the center of mass would be *exactly the same.* ■

If x_{cm} is given, and an object's position or mass is unknown, do the setup in *exactly the same way,* and just do some algebra in the end to solve for the unknown.

9.13 2D CENTER OF MASS

EXERCISE 9.9

Determine the center of mass of the system shown below. The masses of objects 1 and 3 are each twice the mass of object 2.

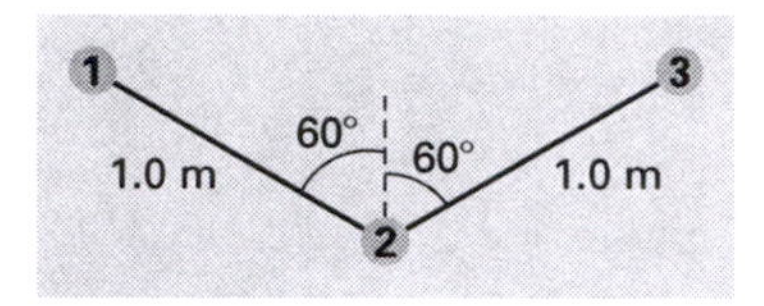

Solution

(1) Type of problem

Center of mass, 2D, three objects: Use Equations 9.5a and 9.5b for three objects.

(2) Sort by object

For Equations 9.5a and 9.5b, we need *masses* and x- and *y-coordinates* for *each object,* as well as the unknown x- and y-coordinates for the center of mass:

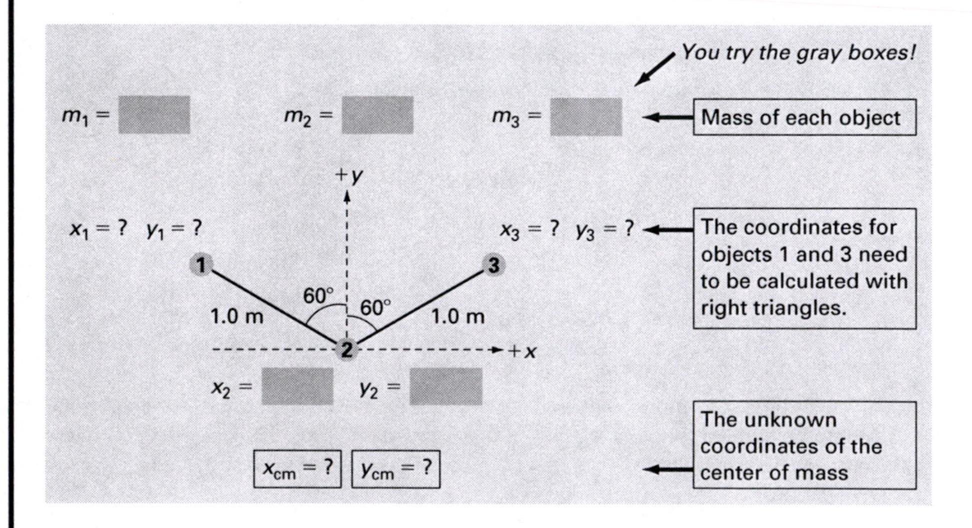

The answers for the gray boxes will be given when we redraw this figure, after we calculate the unknown x- and y-coordinates for each object.

We redraw the coordinate system for object 1 to determine its coordinates:

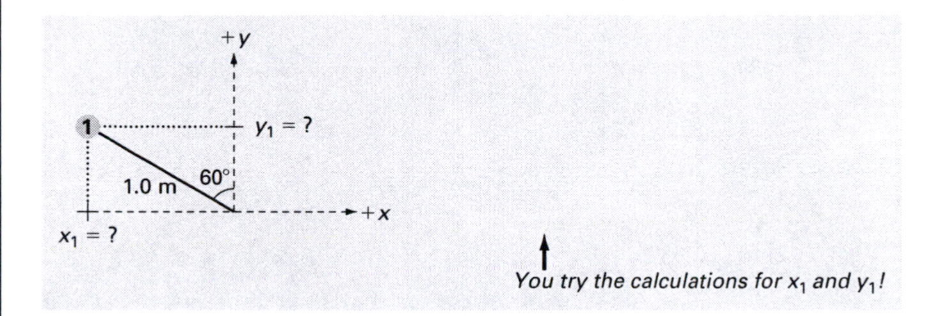

For object 3, the coordinates are the same, except *both are positive.* Now we redraw the setup figure with all the known coordinates and answers for the gray boxes:

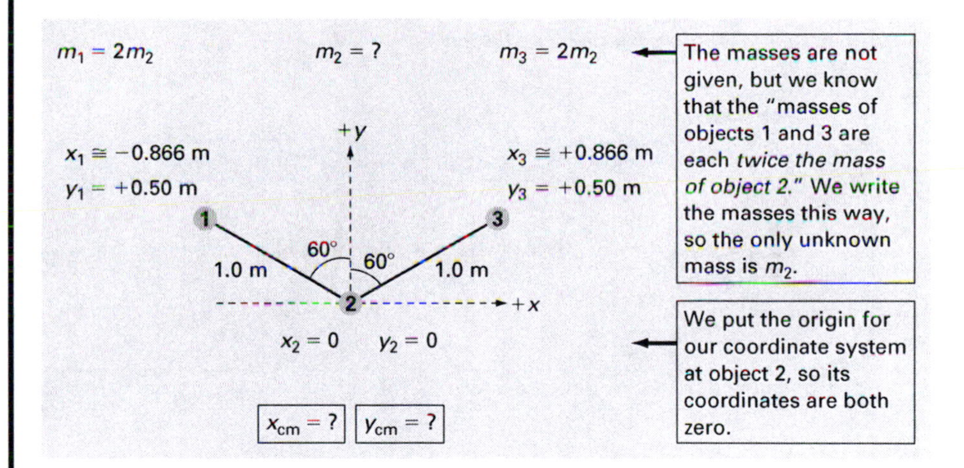

(3) Equations & unknowns and (4) Outline solution

Again, if we are careful in the setup, this part is easy! We write Equations 9.5a and 9.5b and go ahead and do the math:

$$x_{cm} = \frac{m_1 x_1 + m_2 x_2 + m_3 x_3}{m_1 + m_2 + m_3}$$

$$y_{cm} = \frac{m_1 y_1 + m_2 y_2 + m_3 y_3}{m_1 + m_2 + m_3}$$

You sub in the values and symbols, and solve!

Answers for . . .

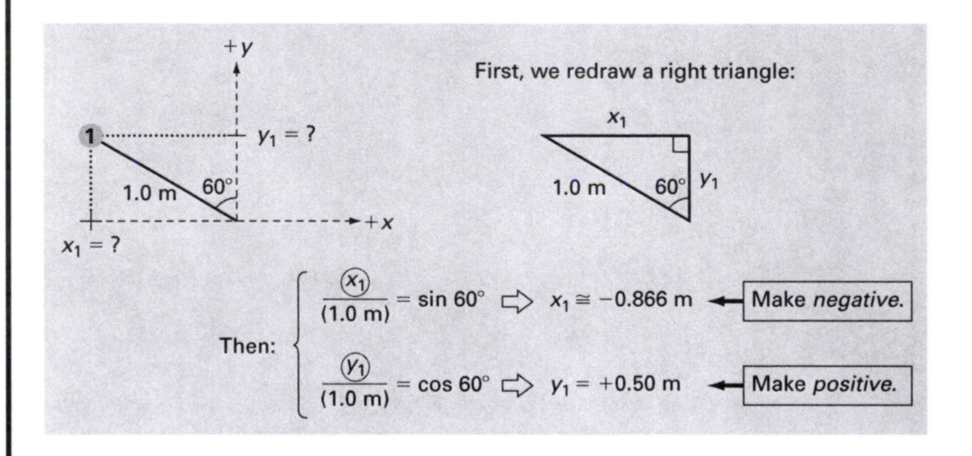

Answers for (4) Outline solution, and the math

$$x_{cm} = \frac{m_1x_1 + m_2x_2 + m_3x_3}{m_1 + m_2 + m_3} = \frac{(2m_2)(-0.866) + m_2(0) + (2m_2)(+0.866)}{(2m_2) + m_2 + (2m_2)} = 0$$

$$y_{cm} = \frac{m_1y_1 + m_2y_2 + m_3y_3}{m_1 + m_2 + m_3} = \frac{(2m_2)(+0.50) + m_2(0) + (2m_2)(+0.50)}{(2m_2) + m_2 + (2m_2)} = \frac{2m_2}{5m_2} = +0.40 \text{ m}$$

Leave out m (for meters) until the end.

The unknown mass, m_2, cancels out.

Answers $x_{cm} = 0$, $y_{cm} = +0.40$ m

This means the center of mass is 0.40 m directly above object 2.

USE SYMMETRY TO AVOID CALCULATIONS!

Because objects 1 and 3 have identical masses, there is symmetry. We can just *see* that $x_{cm} = 0$ without calculating it. If we see this, we can skip all the *x*-coordinate calculations and only do the *y*-coordinates.

■

9.14 HOW TO SET UP CENTER OF MASS PROBLEMS

Here is a summary of what we did in the previous two exercises:

Center of Mass Problems—Mental and Written Steps

Mental →	**(1) Type of problem** Center of mass: Use Equation 9.5a and/or 9.5b.
Mental and written →	**(2) Sort by object** • Number each object: 1, 2, and so on. • Pick an origin at the position of one object so that its coordinate(s) are zero. • Sort masses and *x*- and/or *y*-coordinates for each object, for Equation 9.5a and/or 9.5b.
Mental and written →	**(3) Equations & unknowns** Equation 9.5a and/or 9.5b, and circle the unknowns.
Mental →	**(4) Outline solution**

CHAPTER 10

ANGULAR VELOCITY AND ACCELERATION

ANGULAR, OR ROTATIONAL, motion problems do not usually involve complicated math. The main difficulty is in the large number of new quantities and equations. For this reason, we pay special attention to definitions and units, along with our usual focus on how to set up several types of problems.

10.1 HOW TO RELATE ANGULAR AND TANGENTIAL OR LINEAR QUANTITIES

First, compare the quantities for a wheel in either of these two similar motions:

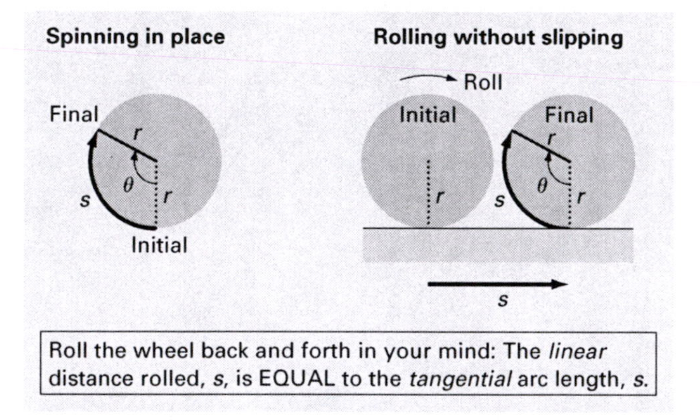

For either the tangential motion or the linear motion, we use the same equation to relate these quantities:

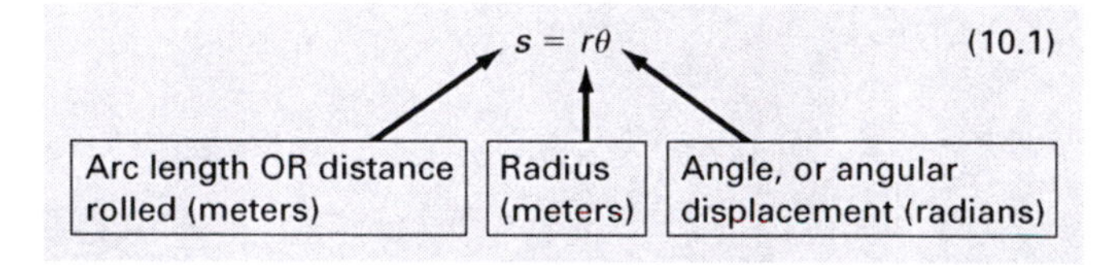

Watch the units! For calculations with Equation 10.1, always convert θ to radians.

REVOLUTIONS, DEGREES, RADIANS: ALL UNITS FOR θ!

When you see any of these units given in a problem, call that quantity θ in your setup. Textbooks do not always use the words *angle* or *angular displacement* but often mention only the units. Try to recognize angular quantities by their units as well as their names!

Now look at tangential or linear velocity (v) and angular velocity (ω) magnitudes for the two motions:

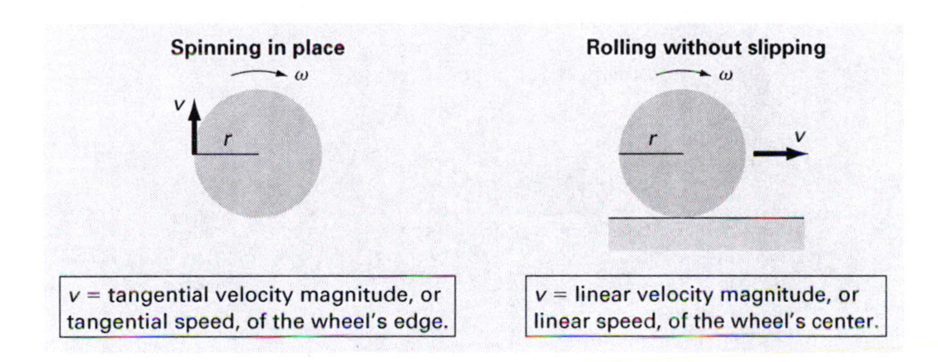

Again, one equation relates the quantities for either situation:

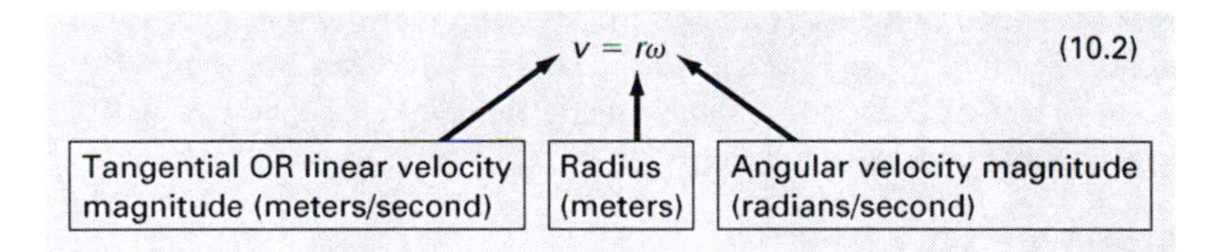

Always convert ω to radians/second for calculations with Equation 10.2.

REVOLUTIONS/MINUTE (RPM), REVOLUTIONS/SECOND, RADIANS/SECOND: ALL UNITS FOR ω!

When you see any of these units, call that quantity ω in your setup. Textbook problems may only mention units, like *rpm,* and not actually say *angular velocity.*

Finally, tangential or linear acceleration (a) and angular acceleration (α) magnitudes for the two motions:

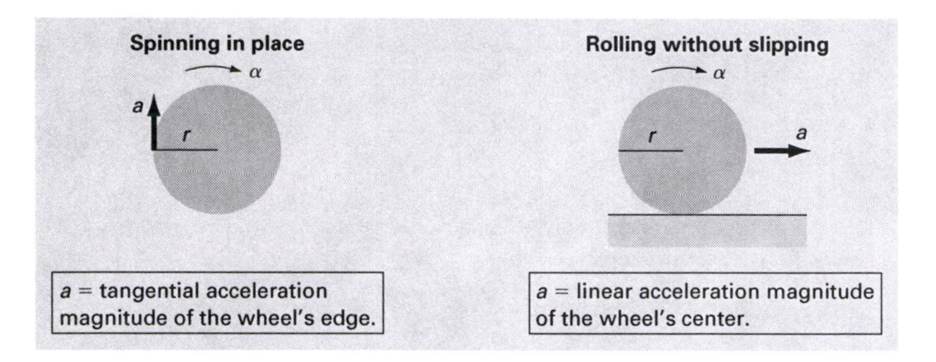

Here, too, one equation works for either situation:

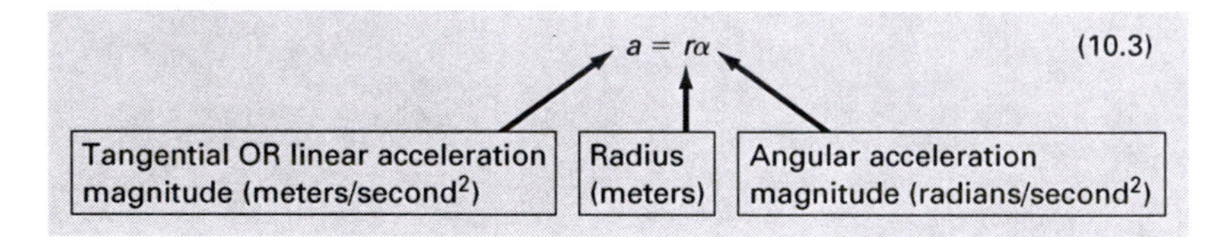

Again, be careful of the units!

10.2 TWO-OBJECT, TWO-CIRCLE PROBLEMS

For a problem in which two objects are somehow coupled in circular motion, we first ask: Do the two objects have *equal tangential/linear motions* (s, v, and a) OR *equal angular motions* (θ, ω, and α)? The next two exercises show how to answer this question. Remember that we show many mental steps that you will not normally write down in your own solutions.

EXERCISE 10.1

A large pulley of radius 0.190 m is connected by a belt to a small pulley of radius 0.0400 m, as shown in the figure. The large pulley rotates through 625 revolutions. Through how many revolutions does the small pulley rotate?

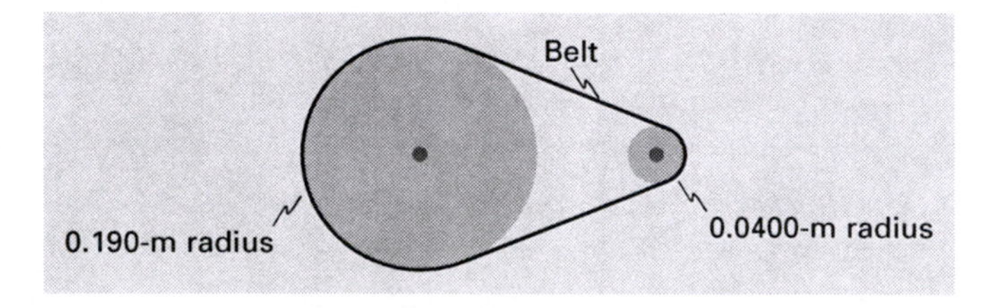

Solution

(1) Type of problem

We have two objects coupled in circular motion: Use Equation 10.1 for each because the question is about the angle (revolutions).

(2) Sort by object/circle

First, do the two objects have *equal tangential motions* OR *equal angular motions?*

OBJECTS COUPLED AT THE OUTER EDGES HAVE EQUAL TANGENTIAL/LINEAR MOTIONS!

As they each spin in place, the pulleys are coupled by the belt *at the outer edges,* so they have *equal tangential motions:*

- Arc lengths ($s = s_1 = s_2$)
- Tangential velocity magnitudes ($v = v_1 = v_2$)
- Tangential acceleration magnitudes ($a = a_1 = a_2$)

Two objects that *roll without slipping* would have equal *linear* motions (also s, v, and a) if they both roll together along the same surface. For example, two bicycle wheels are *coupled at the outer edges* because they both roll together on the road.

The large pulley is object 1 and the small pulley is object 2. We sort quantities in Equation 10.1 for each object/circle:

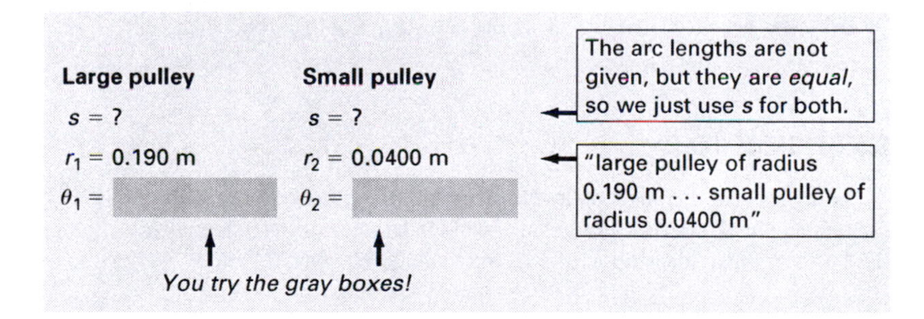

(3) Equations & unknowns and (4) Outline solution

Equation 10.1 for each:

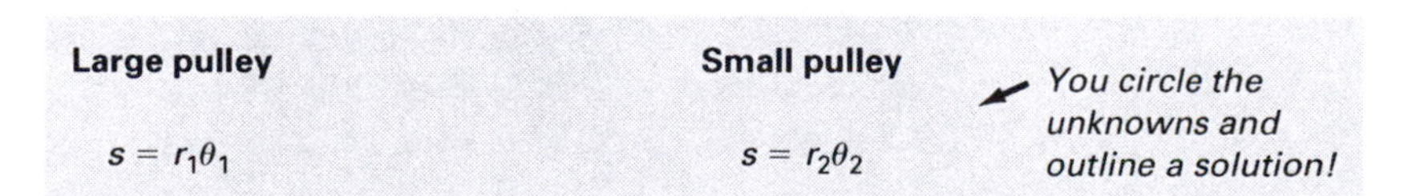

Answers for gray boxes

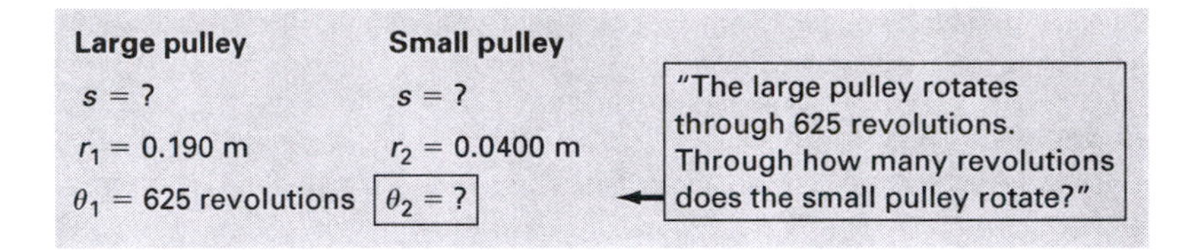

Answers for (3) Equations & unknowns and (4) Outline solution

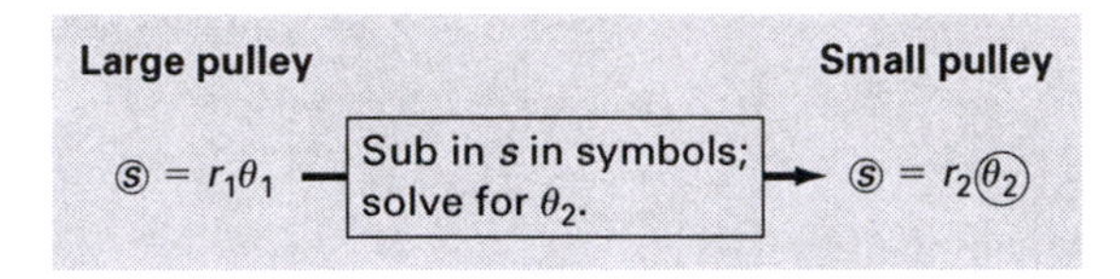

Answer $\theta_2 \cong 2969 \cong 2970$ revolutions

BUT DON'T WE HAVE TO CONVERT TO RADIANS FOR EQUATION 10.1? YES AND NO!

- YES, if we had used $s = r_1\theta_1$ to calculate a *numerical value* for *s*. Then we would need to first convert θ_1 to radians to get *s* in meters.
- NO, since we *never actually did* calculate a *numerical value* for *s*. After we substitute *in symbols,* we solve for $\theta_2 = \theta_1(r_1/r_2)$. The units cancel for r_1 and r_2, so the units of θ_2 and θ_1 are the same (revolutions). If you are not sure, however, convert to radians. You can always convert back to revolutions later.

■

EXERCISE 10.2

A fishing reel has a crank handle with length 0.035 m. A cylinder with radius 0.012 m holds the fishing line. Your hand moves the end of the handle with a tangential velocity of magnitude 0.60 m/s. At what linear velocity does the fishing line come in?

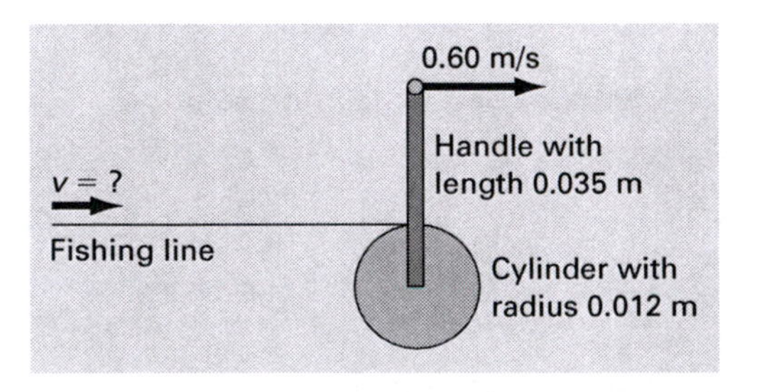

Solution

(1) Type of problem

Two objects (handle and cylinder) are coupled in circular motion. There is also the *fishing line,* but *its velocity* is the same as the *cylinder's tangential velocity* at its top edge, which is what we will actually solve for. We use Equation 10.2 because the question is about velocity.

(2) Sort by object/circle

First, do the two objects have *equal tangential motions* OR *equal angular motions?*

OBJECTS ROTATING TOGETHER ABOUT THE SAME AXIS HAVE EQUAL ANGULAR MOTIONS!

This is the case for the handle and cylinder, so they have *equal:*

- Angles, or angular displacements ($\theta = \theta_1 = \theta_2$)
- Angular velocities ($\omega = \omega_1 = \omega_2$)
- Angular accelerations ($\alpha = \alpha_1 = \alpha_2$)

The quantities in Equation 10.2, sorted for the handle (object 1) and the cylinder (object 2):

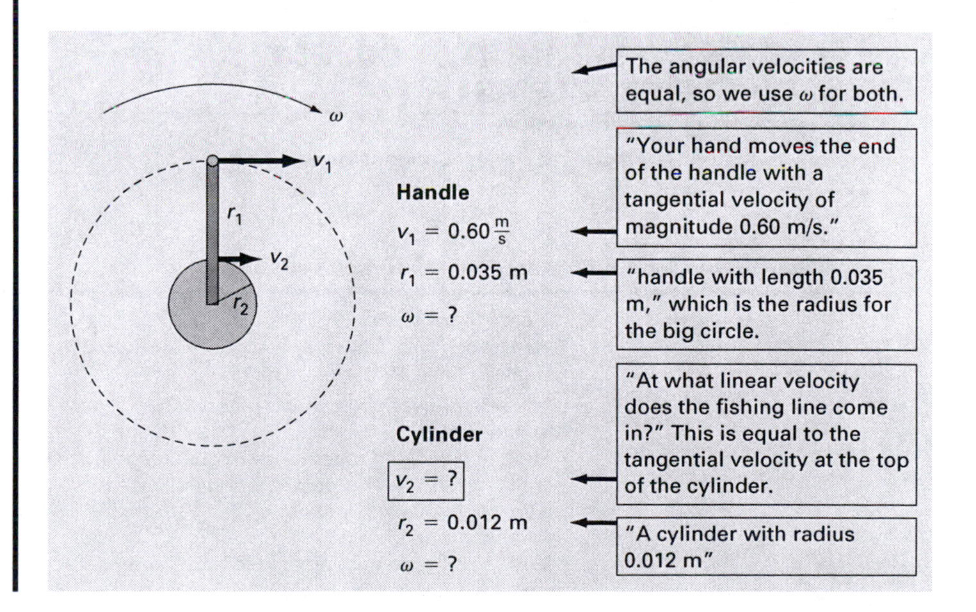

(3) Equations & unknowns and (4) Outline solution

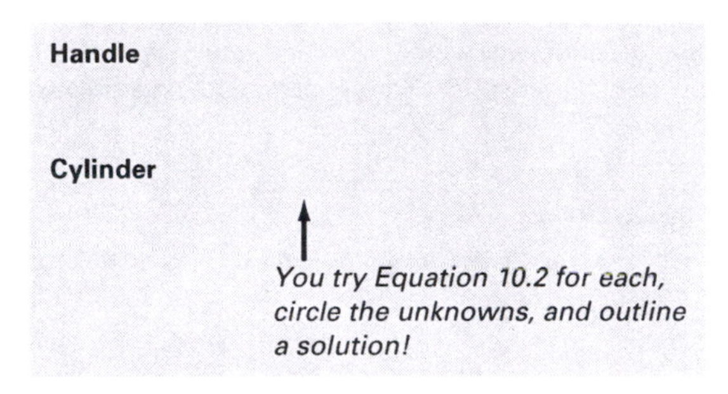

Answers for (3) Equations & unknowns and (4) Outline solution

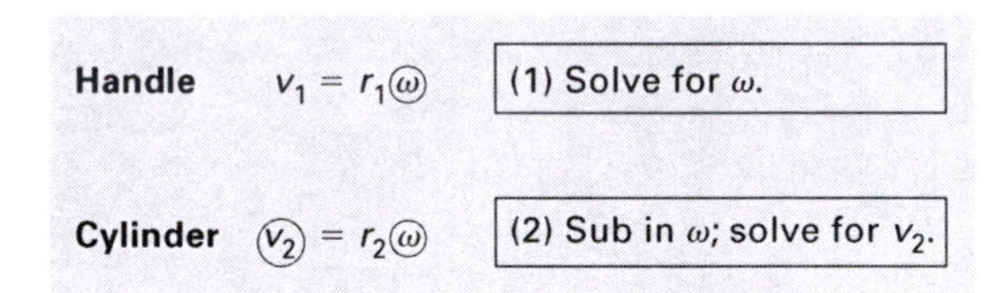

Answer $v_2 \cong 0.206 \cong 0.21 \frac{\text{m}}{\text{s}}$ ■

10.3 HOW TO SET UP TWO-OBJECT, TWO-CIRCLE PROBLEMS

Here is a summary of what we did in the previous two exercises:

Two-Object, Two-Circle Problems—Mental and Written Steps

Mental →	**(1) Type of problem** Two objects/circles coupled in circular motion: Use Equations 10.1, 10.2, and/or 10.3.
Mental and written →	**(2) Sort by object/circle and (3) Equations & unknowns** • First, decide if the objects have *equal tangential/linear motions* OR *equal angular motions.* • Sort quantities for Equations 10.1, 10.2, and/or 10.3 for each object/circle. • Write out equation(s) and circle unknowns.
Mental →	**(4) Outline solution**

10.4 CONSTANT/AVERAGE ANGULAR VELOCITY

As with any *motion,* think in terms of *intervals!* First, we have:

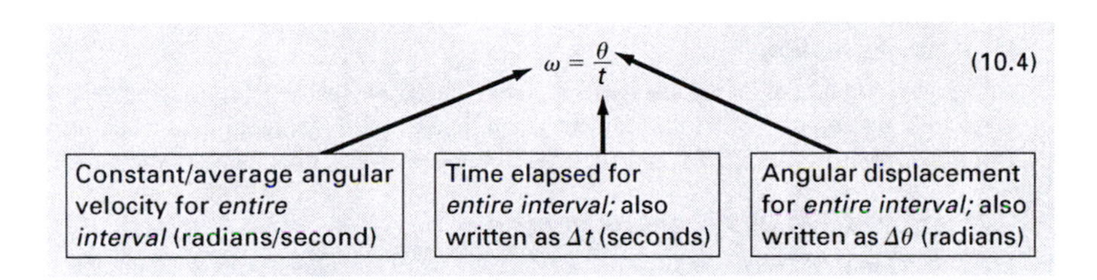

Be consistent with units! There are actually a few sets of units that we can use with Equation 10.4:

Quantity	ω	θ	t
Most common set of units (Always use these if *also* using Equations 10.1 and/or 10.2.)	$\frac{\text{radians}}{\text{second}}$	radians	seconds
Less common set of units	$\frac{\text{revolutions}}{\text{second}}$	revolutions	seconds
Another less common set of units	$\frac{\text{revolutions}}{\text{minute}} = \text{rpm}$	revolutions	minutes

For either *spinning in place* OR *rolling without slipping,* we also have this equation:

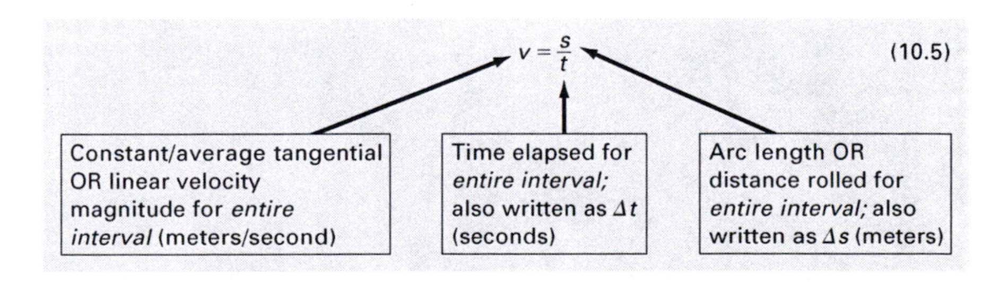

This is just Equation 2.2, with s instead of x.

We often use Equations 10.1 and 10.2 together with Equations 10.4 and 10.5, as we show in the next exercise.

EXERCISE 10.3

A spool of thread has an average diameter of 2.6 cm. The spool releases thread as it spins at 35 rpm for 8.0 seconds. (a) What is the tangential velocity magnitude at a point on the rim of the spool? (b) Through what angle does the spool turn? (c) What length of thread is released?

Solution

(1) Type of problem

Arc length (s, length of thread released), angle (θ), angular velocity (ω, rpm), tangential velocity magnitude (v), radius (r, half the diameter), and time elapsed (t) are all mentioned or implied: Use Equations 10.1, 10.2, 10.4, and 10.5, because together they contain all of these quantities.

(2) Sort by object, circle, and/or interval

Preliminary unit conversions:

One-half of "diameter of 2.6 cm," which we convert to meters.	$r = \frac{1}{2}(2.6\text{ cm}) = 1.3\text{ cm} \times \left(\frac{1\text{ m}}{100\text{ cm}}\right) = 0.013\text{ m}$
"spins at 35 rpm," which we convert to rad/s, so we can use Equation 10.2 to calculate values.	$\omega = 35\text{ rpm} = 35\frac{\text{rev}}{\text{min}} \times \left(\frac{2\pi\text{ rad}}{1\text{ rev}}\right) \times \left(\frac{1\text{ min}}{60\text{ s}}\right) \cong 3.67\frac{\text{rad}}{\text{s}}$

There is one object (the spool) and one interval. We sort the quantities by separating *angular quantities, tangential quantities,* and *radius:*

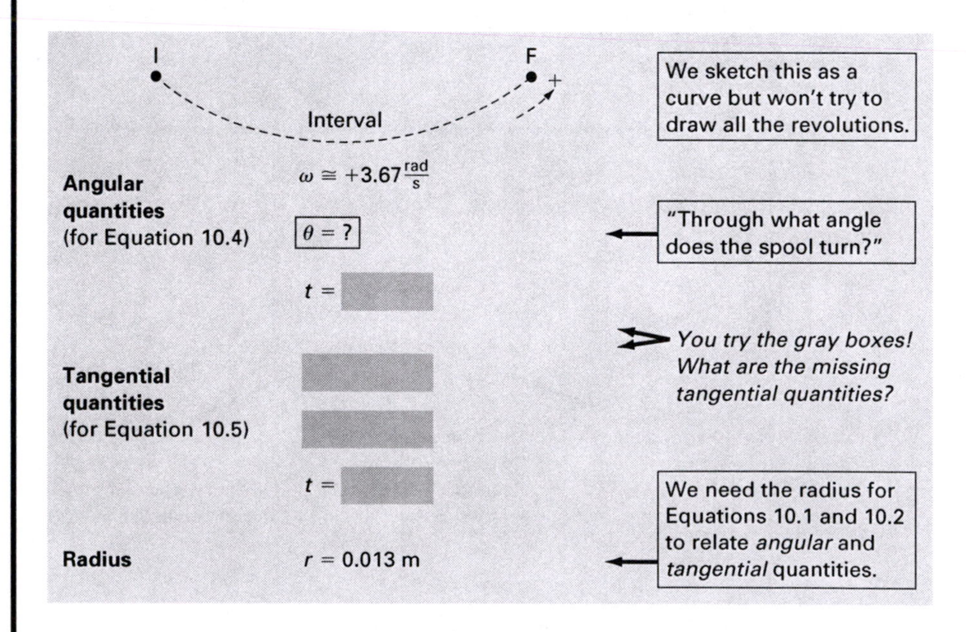

The rotation in this problem is all in *one direction,* so we make it *counterclockwise,* which is usually *positive* for angular quantities.

(3) Equations & unknowns and (4) Outline solution

The equations:

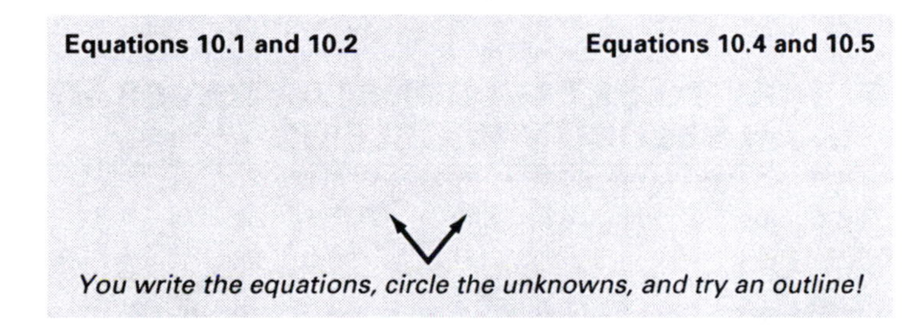

Answers for gray boxes

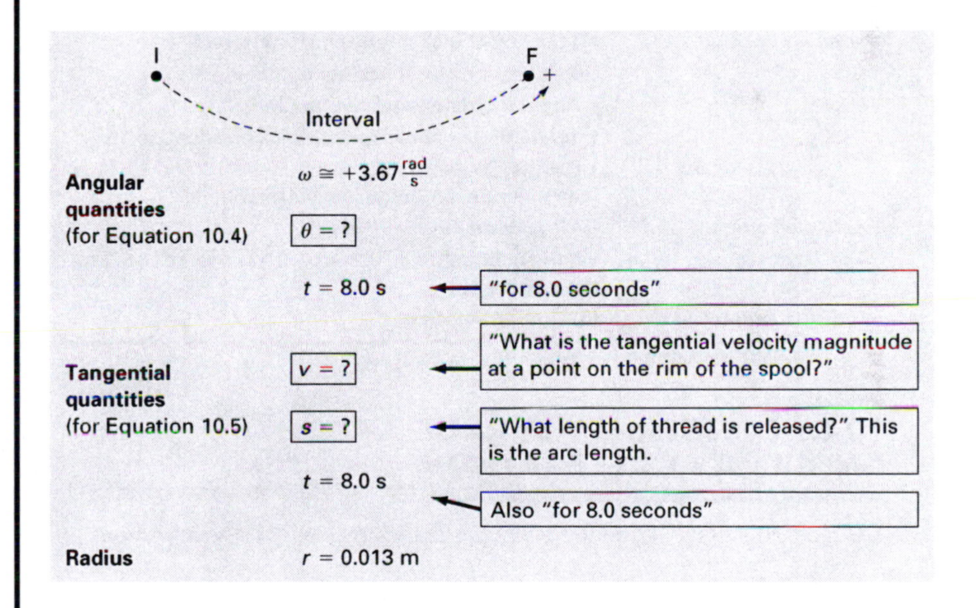

Answers for (3) Equations & unknowns and (4) Outline solution

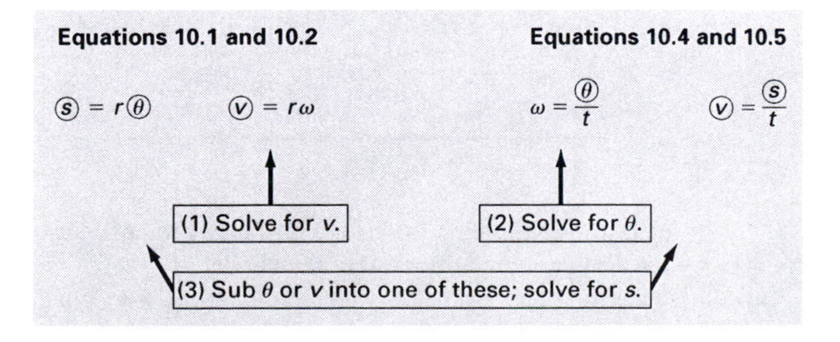

We did not actually need all four equations, but it helps to know all four, in case we get stuck, or to use the extra equation to check our answers.

Answers $v \cong +0.0476 \cong +0.048\frac{\text{m}}{\text{s}}$, $\theta \cong +29.3 \cong +29$ radians, $s \cong +0.381 \cong +0.38$ m ■

10.5 HOW TO SET UP CONSTANT/AVERAGE ANGULAR VELOCITY PROBLEMS

Here is a summary of what we just did:

Constant/Average Angular Velocity Problems—Mental and Written Steps

Mental →	**(1) Type of problem**
	Constant/average angular velocity: Use Equations 10.1, 10.2, 10.4, and/or 10.5.
Mental and written →	**(2) Sort by object, circle, and/or interval**
	Sort quantities for these equations in groups:
	• **Angular quantities:** For Equation 10.4.
	• **Tangential or linear quantities:** For Equation 10.5.
	• **Radius:** So we can use Equations 10.1 and 10.2 to relate angular and tangential/linear quantities.
Mental and written →	**(3) Equations & unknowns**
	Write out Equations 10.1, 10.2, 10.4, and/or 10.5, and circle the unknowns!
Mental →	**(4) Outline solution**

10.6 CONSTANT/AVERAGE ANGULAR ACCELERATION

The main equations for motion with constant/average angular acceleration:

Constant/Average Angular Acceleration

$\theta = \omega_0 t + \frac{1}{2}\alpha t^2$	(10.6a)
$\theta = \left(\frac{\omega_0 + \omega}{2}\right)t$	(10.6b)
$\omega = \omega_0 + \alpha t$	(10.6c)
$\omega^2 = \omega_0^2 + 2\alpha\theta$	(10.6d)

These look a lot like Equations 2.3a–d for constant/average linear acceleration, except here we have angular quantities rather than linear.

Be consistent with units! Some possible combinations for Equations 10.6a–d:

Quantities	θ	t	ω_0 and ω	α
Most common set of units (Always use these if *also* using any of Equations 10.1–3.)	radians	seconds	$\frac{\text{radians}}{\text{second}}$	$\frac{\text{radians}}{\text{second}^2}$
Less common set of units	revolutions	seconds	$\frac{\text{revolutions}}{\text{second}}$	$\frac{\text{revolutions}}{\text{second}^2}$

Each quantity fits in an interval like so:

Part of interval	Initial instant (I)	Entire interval	Final instant (F)
Quantities that correspond	Initial angular velocity, ω_0	• Angular displacement, θ • Angular acceleration, α • Time elapsed, t	Final angular velocity, ω

For motion with constant/average angular acceleration, we set up with these quantities sorted by interval, *regardless of the question(s) being asked.*

EXERCISE 10.4

A motor slows at a constant rate from 400 rpm to 250 rpm while turning through 12 revolutions. (a) What is the angular acceleration of the motor? (b) How much time does this process take?

Solution

(1) Type of problem

Constant angular acceleration because it *slows at a constant rate:* Use Equations 10.6a–d.

(2) Sort by object, circle, and/or interval

We first convert everything to radians, seconds, and so on, which are the most common units for Equations 10.6a–d:

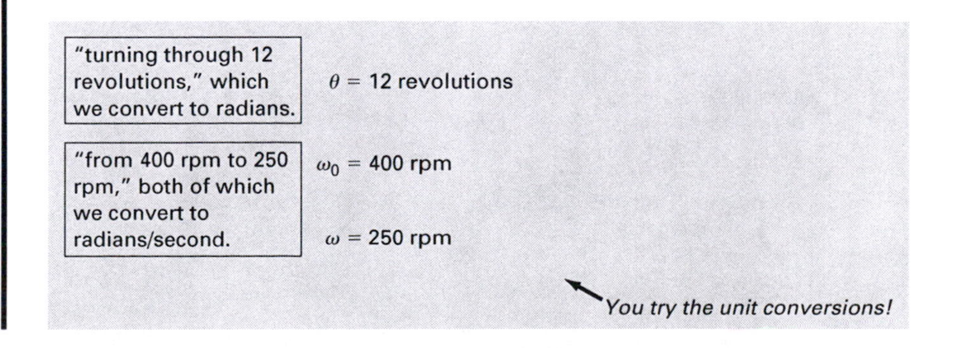

There is one object/circle and there is one interval. Also, there are only angular quantities, which we sort for Equations 10.6a–d:

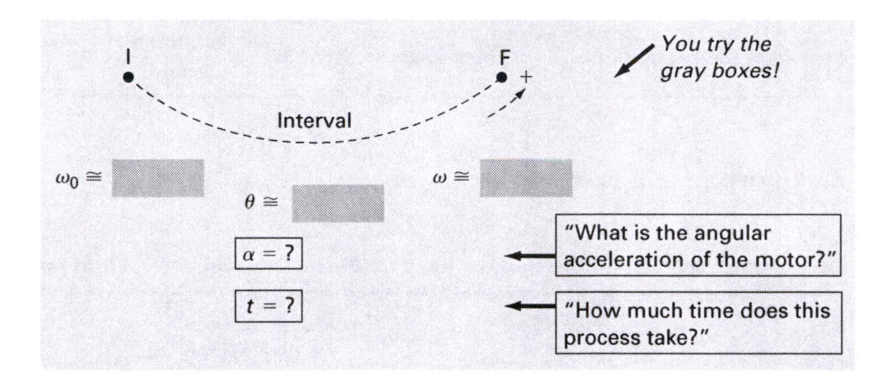

(3) Equations & unknowns and (4) Outline solution

Equations 10.6a–d:

$$\theta = \omega_0 t + \tfrac{1}{2}\alpha t^2$$

$$\theta = \left(\frac{\omega_0 + \omega}{2}\right) t$$

$$\omega = \omega_0 + \alpha t$$

$$\omega^2 = \omega_0^2 + 2\alpha\theta$$

You circle the unknowns and try an outline!

Answers for unit conversions

"turning through 12 revolutions," which we convert to radians.

$$\theta = 12 \text{ rev} \times \left(\frac{2\pi \text{ rad}}{1 \text{ rev}}\right) \cong 75.4 \text{ rad}$$

"from 400 rpm to 250 rpm," both of which we convert to radians/second.

$$\omega_0 = 400 \text{ rpm} = 400\frac{\text{rev}}{\text{min}} \times \left(\frac{2\pi \text{ rad}}{1 \text{ rev}}\right) \times \left(\frac{1 \text{ min}}{60 \text{ s}}\right) \cong 41.9\frac{\text{rad}}{\text{s}}$$

$$\omega = 250 \text{ rpm} = 250\frac{\text{rev}}{\text{min}} \times \left(\frac{2\pi \text{ rad}}{1 \text{ rev}}\right) \times \left(\frac{1 \text{ min}}{60 \text{ s}}\right) \cong 26.2\frac{\text{rad}}{\text{s}}$$

Answers for gray boxes

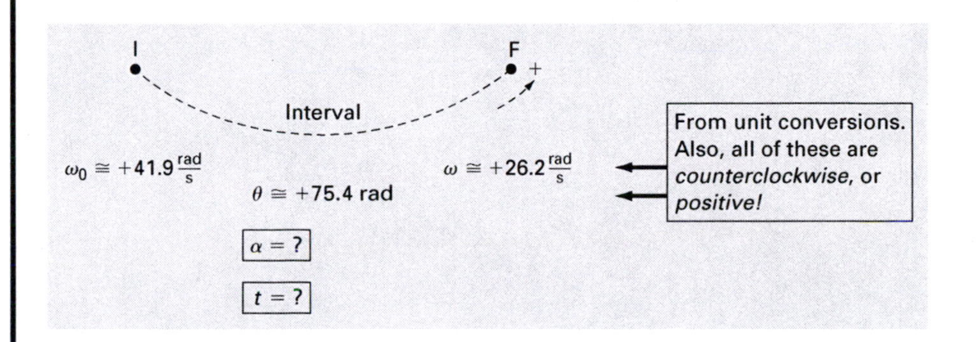

Answers for (3) Equations & unknowns and (4) Outline solution

$$\theta = \omega_0 t + \frac{1}{2}\alpha t^2$$

$$\theta = \left(\frac{\omega_0 + \omega}{2}\right) t \quad \leftarrow \text{(1) Solve for } t.$$

$$\omega = \omega_0 + \alpha t$$

$$\omega^2 = \omega_0^2 + 2\alpha\theta \quad \leftarrow \text{(2) Solve for } \alpha.$$

We can check our answers with the extra equations!

Answers

- $\alpha \cong -7.09 \cong -7.1\,\frac{\text{rad}}{\text{s}^2}$ (*negative* means *clockwise,* or the *opposite* direction of the *angular velocity,* so it is *slowing,* which makes sense).
- $t \cong 2.21 \cong 2.2$ s (elapsed time should always come out *positive*). ■

10.7 CONSTANT/AVERAGE ANGULAR ACCELERATION—MULTIPLE INTERVALS

EXERCISE 10.5

A wheel slows at a constant rate while rotating through 31 radians and finally comes to a stop. At one moment during the rotation, its angular velocity is 4.0 rad/s, and from this moment it rotates through the final 19 radians. (a) What was the angular velocity of the wheel at the beginning of this entire motion? (b) How much total time elapsed?

Solution

(1) Type of problem

There are two overlapping intervals:

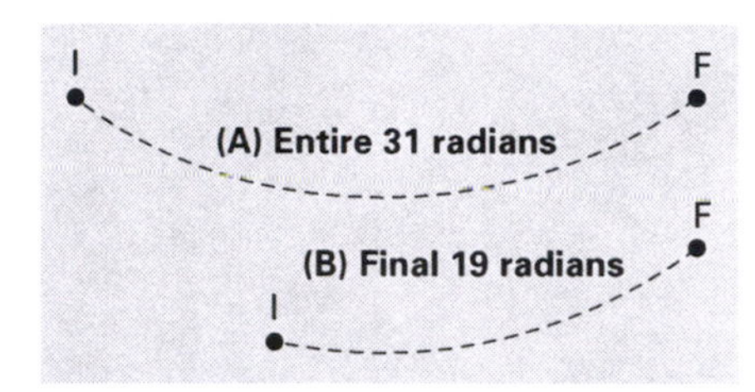

Both have constant angular acceleration: Use Equations 10.6a–d for each.

(2) Sort by object, circle, and/or interval

We sort quantities for each interval separately.

WHAT QUANTITIES ARE THE SAME FOR THE TWO INTERVALS?

The angular acceleration, α, although unknown, is the *same* for both intervals! This is one key to solving this problem!

We start with interval (A):

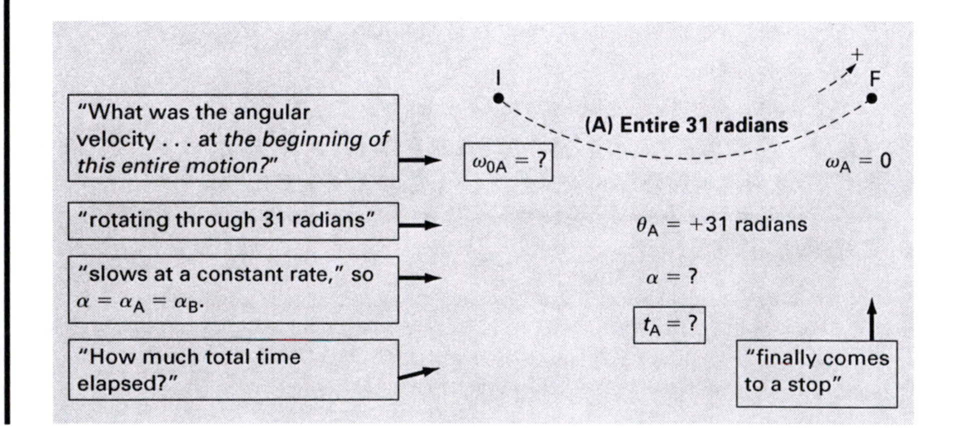

Then interval (B):

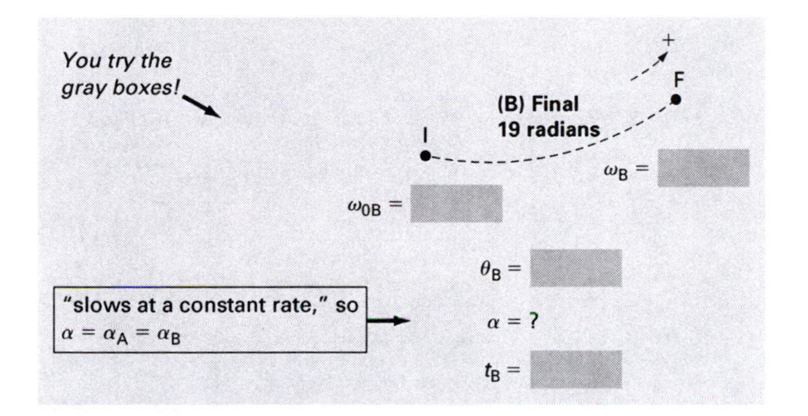

(3) Equations & unknowns and (4) Outline solution

Equations 10.6a–d for each interval:

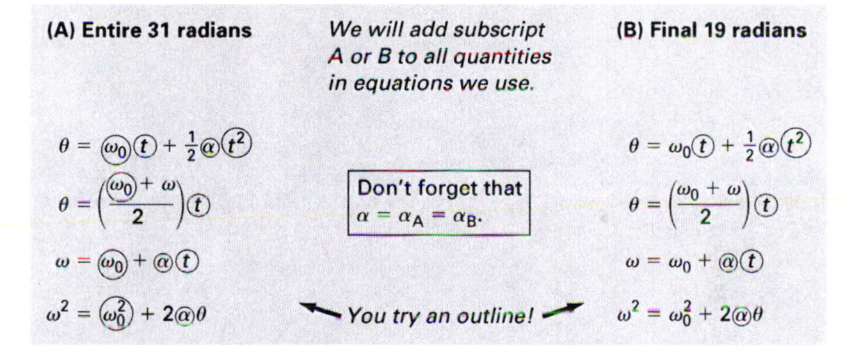

Answers for gray boxes

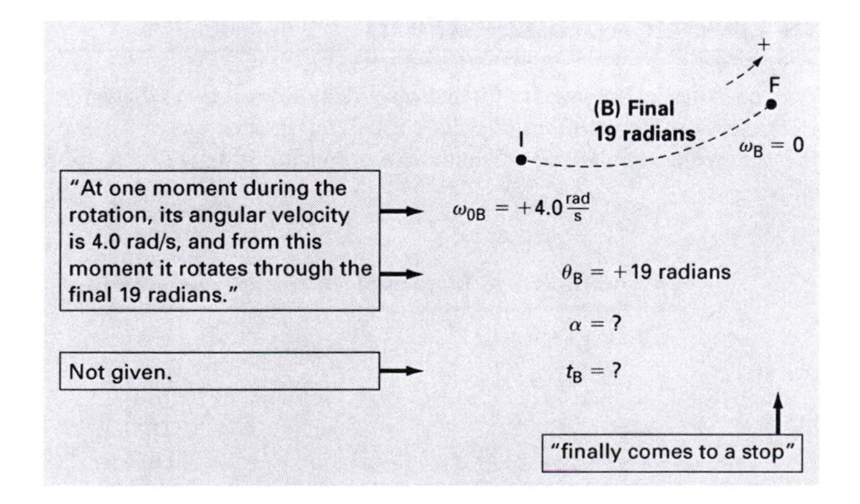

Answers for (4) Outline solution

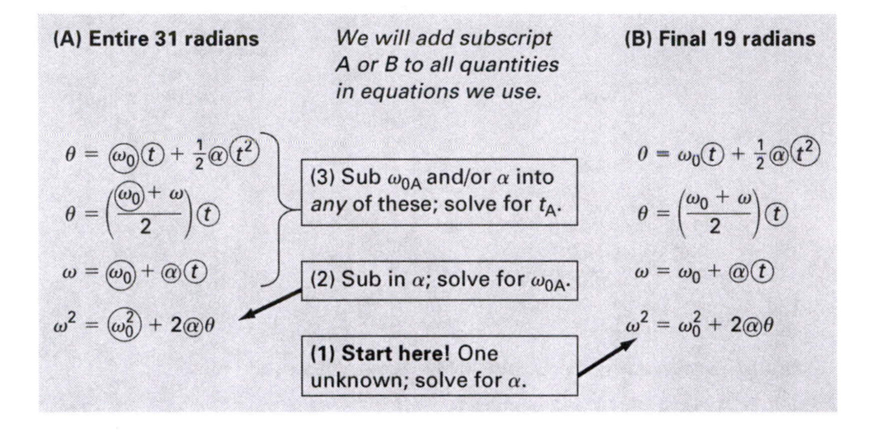

Answers $\omega_{0A} \cong +5.11 \cong +5.1\frac{\text{rad}}{\text{s}}$, $t_A \cong 12.1 \cong 12$ s

Intermediate answer: $\alpha \cong -0.421\frac{\text{rad}}{\text{s}^2}$ (*negative* means *clockwise,* or opposite the motion, so it makes sense that it is *slowing*). ■

10.8 CONSTANT/AVERAGE ANGULAR ACCELERATION—WITH TANGENTIAL OR LINEAR ACCELERATION

Some questions about angular motion also include *tangential* (if *spinning in place*) or *linear* (if *rolling without slipping*) acceleration. For these, we use Equations 10.1–3 to relate angular and tangential/linear quantities. In addition, we can use the following:

Constant/Average Tangential OR Linear Acceleration

$s = v_0 t + \frac{1}{2}at^2$	(10.7a)
$s = \left(\frac{v_0 + v}{2}\right)t$	(10.7b)
$v = v_0 + at$	(10.7c)
$v^2 = v_0^2 + 2as$	(10.7d)

These are just Equations 2.3a–d with s instead of x, so the units here are the same as for Equations 2.3a–d. Each of the five quantities fits in an interval like this:

Part of interval	Initial instant (I)	Entire interval	Final instant (F)
Quantities that correspond	Initial tangential OR linear velocity, v_0	• Arc length OR distance rolled, s • Tangential OR linear acceleration, a • Time elapsed, t	Final tangential OR linear velocity, v

EXERCISE 10.6

A bicycle has wheels with radius 0.300 m and travels on a straight, level road. The bicycle has an initial velocity of 5.00 m/s and speeds up at a rate of 0.250 m/s^2 for 4.00 s. (a) Through how many revolutions do each of the wheels go during this time? (b) What is the final angular velocity of each of the wheels?

Solution

(1) Type of problem

The bicycle and wheels have constant *linear acceleration,* and the wheels roll (assume: without slipping) with constant *angular acceleration,* so we use:

- Equations 10.1–3 to relate the *angular* and *linear* quantities, and
- EITHER (i) Equations 10.6a–d for constant *angular* acceleration
- OR (ii) Equations 10.7a–d for constant *linear* acceleration

(2) Sort by object, circle, and/or interval

First:

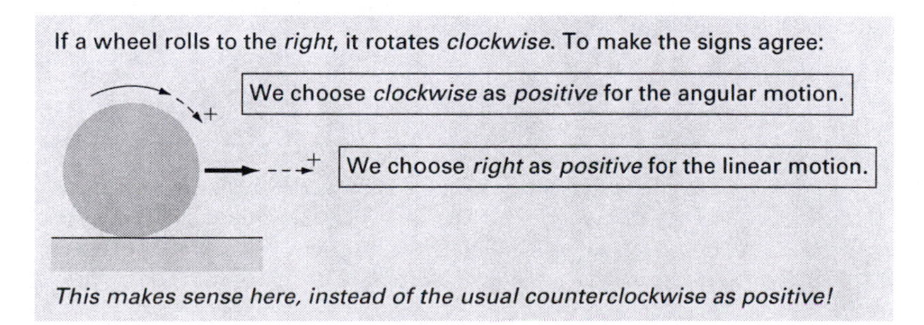

There is one object/circle (one of the wheels) and there is one interval. We separate the *angular quantities* and *linear quantities,* and we also need the *radius.* First, the angular quantities sorted for Equations 10.6a–d:

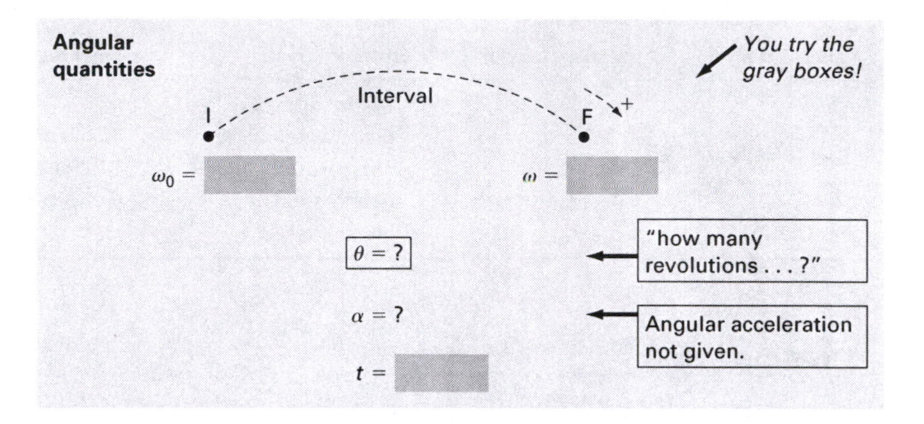

And the linear quantities sorted for Equations 10.7a–d:

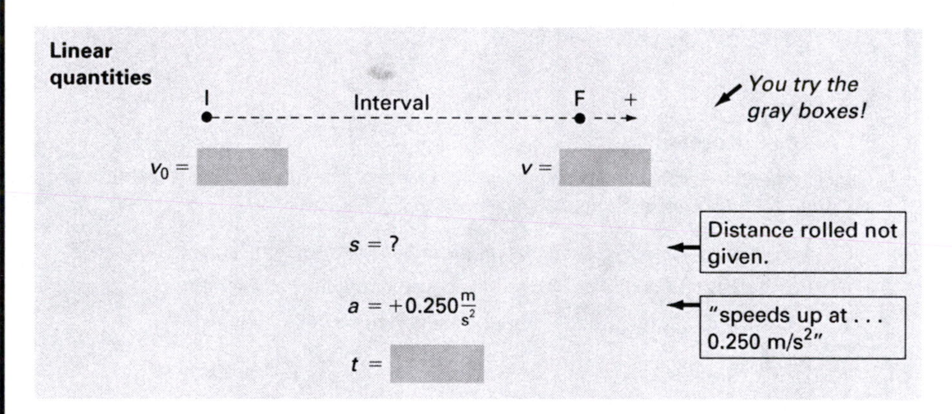

And:

Radius $r = 0.300$ m ← "radius 0.300 m"

SET IT UP FIRST!

Don't worry about the order of the questions or what the unknowns are! Focus on trying to recognize the *type of problem!* Then you know how to *set it up.* For each type of problem, set up the problem according to the equations you plan to use, regardless of which quantities are unknown and regardless of the question order.

(3) Equations & unknowns and (4) Outline solution

As explained earlier, we have two options here: Equations 10.1–3, and EITHER (i) Equations 10.6a–d OR (ii) Equations 10.7a–d. We show both possibilities:

Option (i)	**Equations 10.1–3**	**and**	**Equations 10.6a–d**
	$s = r\theta$		$\theta = \omega_0 t + \frac{1}{2}\alpha t^2$
You circle → the unknowns and try an outline!	$v_0 = r\omega_0$, $v = r\omega$	Equation 10.2 for both I & F!	$\theta = \left(\frac{\omega_0 + \omega}{2}\right)t$
			$\omega = \omega_0 + \alpha t$
	$a = r\alpha$		$\omega^2 = \omega_0^2 + 2\alpha\theta$

Or:

Option (ii)	**Equations 10.1–3**	**and**	**Equations 10.7a–d**
	$s = r\theta$		$s = v_0 t + \frac{1}{2}at^2$
You circle → the unknowns and try an outline!	$v_0 = r\omega_0$, $v = r\omega$	Equation 10.2 for both I and F!	$s = \left(\frac{v_0 + v}{2}\right)t$
			$v = v_0 + at$
	$a = r\alpha$		$v^2 = v_0^2 + 2as$

Answers for gray boxes

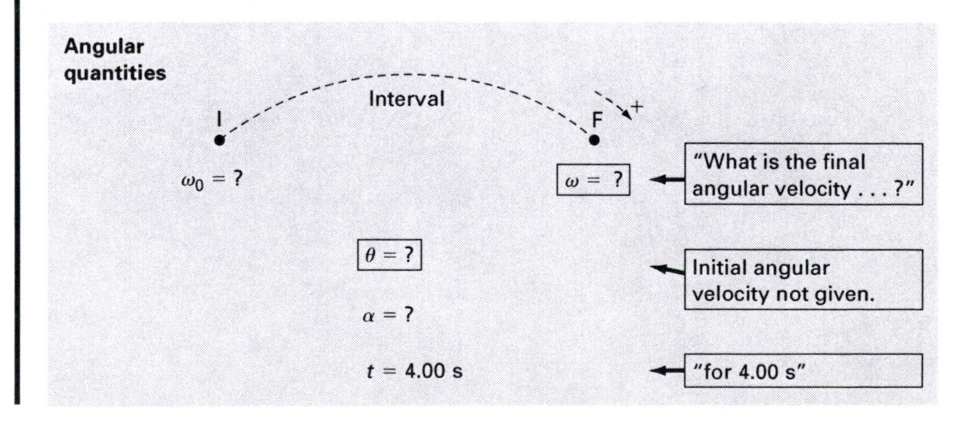

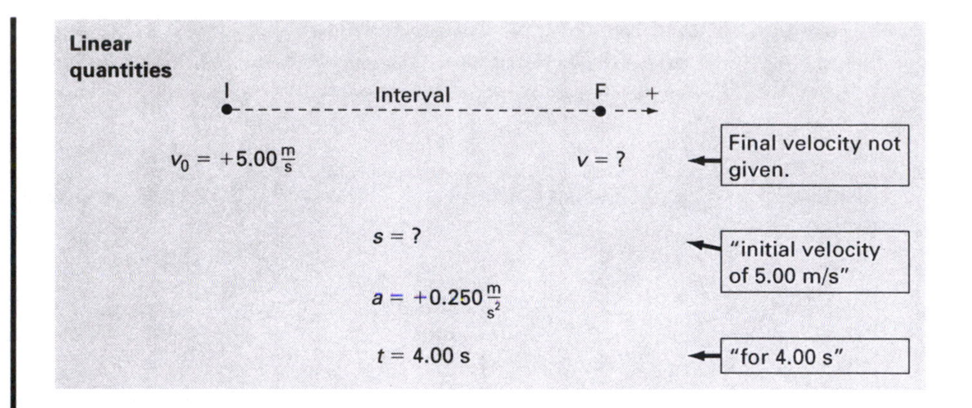

Answers for (4) Outline solution

Here are two possible outlines:

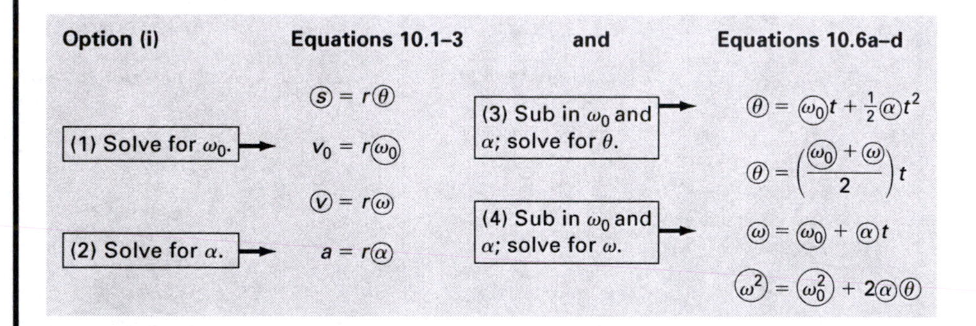

Option (ii) Equations 10.1–3 and Equations 10.7a–d

(3) Sub in s; solve for θ. → $s = r\theta$

$v_0 = r\omega_0$

(4) Sub in v; solve for ω. → $v = r\omega$

$a = r\alpha$

(1) Solve for s. → $s = v_0 t + \frac{1}{2}at^2$

$s = \left(\frac{v_0 + v}{2}\right)t$

(2) Solve for v. → $v = v_0 + at$

$v^2 = v_0^2 + 2as$

Answers $\theta \cong +73.33$ radians $\times \left(\frac{1\text{ rev}}{2\pi\text{ rad}}\right) \cong +11.7$ revolutions, $\omega = +20.0\frac{\text{rad}}{\text{s}}$

Intermediate answers: $\omega_0 \cong +16.67\frac{\text{rad}}{\text{s}}, \alpha \cong +0.8333\frac{\text{rad}}{\text{s}^2}, s = +22.0\text{ m}, v = +6.00\frac{\text{m}}{\text{s}}$ ■

10.9 CONSTANT/AVERAGE ANGULAR ACCELERATION—WITH CENTRIPETAL ACCELERATION

Remember from Chapter 6 that centripetal acceleration happens when velocity changes *direction,* even if the speed is constant. Here we combine Equations 6.2 and 10.2 to get a new equation for centripetal acceleration:

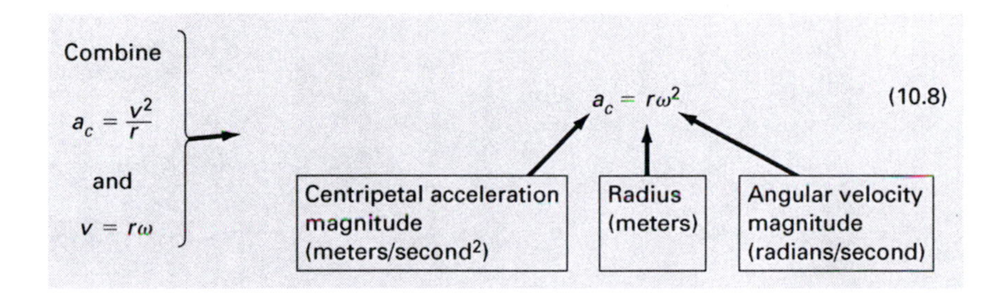

For an object spinning in place, there can be *three types of acceleration:*

Type	Symbol	Units	Magnitude equation	Direction of each
Angular acceleration	α	$\frac{\text{rad}}{\text{s}^2}$	$a_t = r\alpha$ (10.3)	
Tangential acceleration	a_t	$\frac{\text{m}}{\text{s}^2}$	*These two are related!*	
Centripetal acceleration	a_c	$\frac{\text{m}}{\text{s}^2}$	$a_c = \frac{v^2}{r}$ (6.2) Or $a_c = r\omega^2$ (10.8)	

Study this table so you can keep the three types clear in your mind! In this section, we will subscript *tangential* acceleration with t to keep it distinct from *centripetal* acceleration, subscripted with c.

EXERCISE 10.7

A disk of radius 0.12 m begins at rest and angularly accelerates at a constant rate of 1.5 rad/s^2. Determine the tangential acceleration, centripetal acceleration, and the magnitude and direction of the total acceleration at the edge of the disk after 0.60 s has elapsed.

Solution

(1) Type of problem

- Constant angular and tangential acceleration: Use Equation 10.3, and EITHER Equations 10.6a–d OR Equations 10.7a–d.

- Centripetal acceleration: Use Equation 10.8.
- Magnitude and direction of the total acceleration (vector): Use right triangle trig for vector addition of tangential and centripetal acceleration.

(2) Sort by object, circle, and/or interval

There is one object/circle and one interval. First, the *angular quantities:*

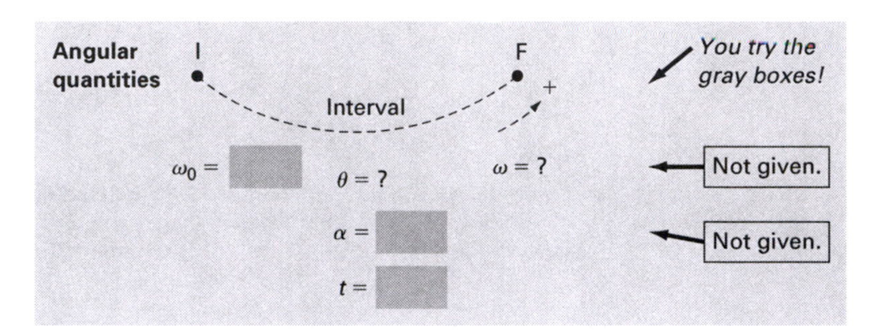

Other quantities we need:

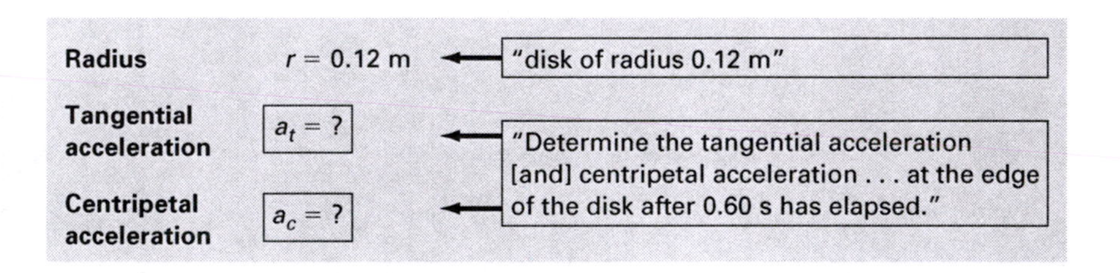

And finally:

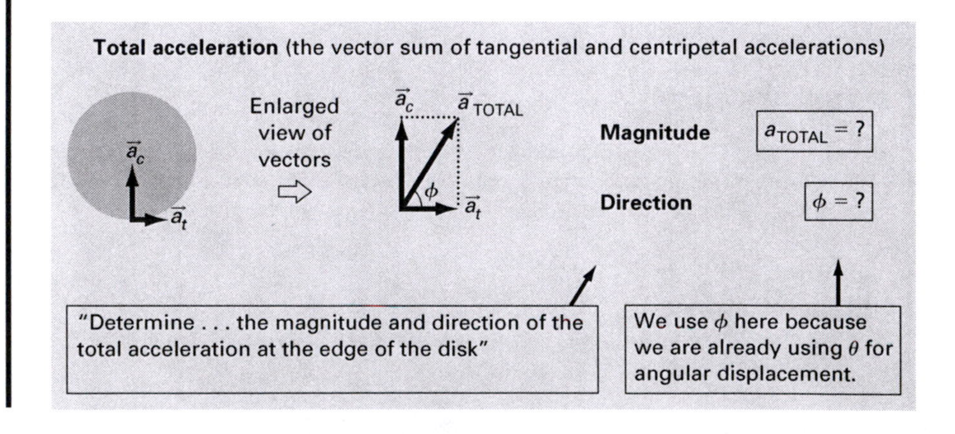

(3) Equations & unknowns and (4) Outline solution

To get the tangential acceleration, we choose Equation 10.3 with Equations 10.6a–d. For the centripetal acceleration, we use Equation 10.8:

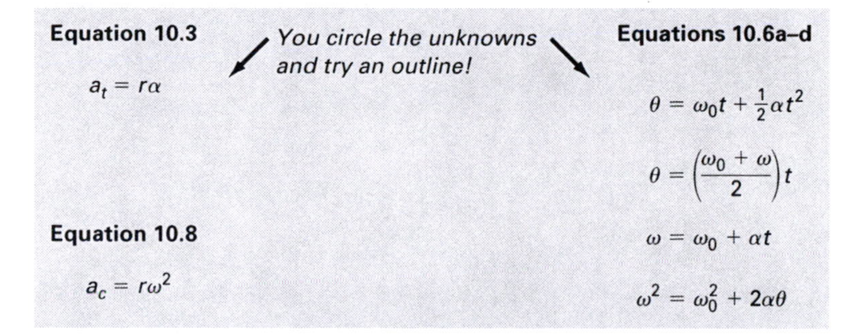

Once we know a_t and a_c, we can get the total acceleration:

Answers for gray boxes

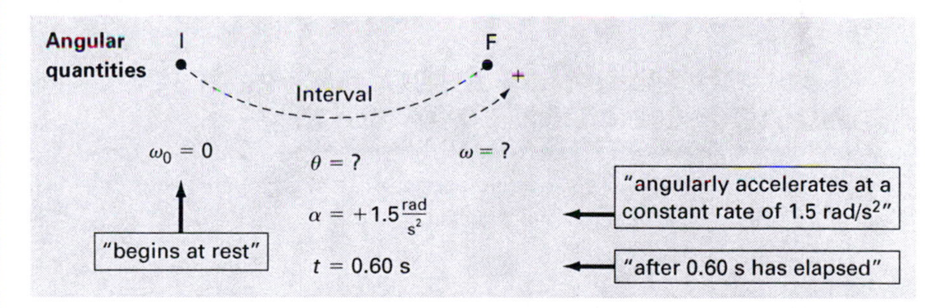

Answers for (3) Equations & unknowns and (4) Outline solution

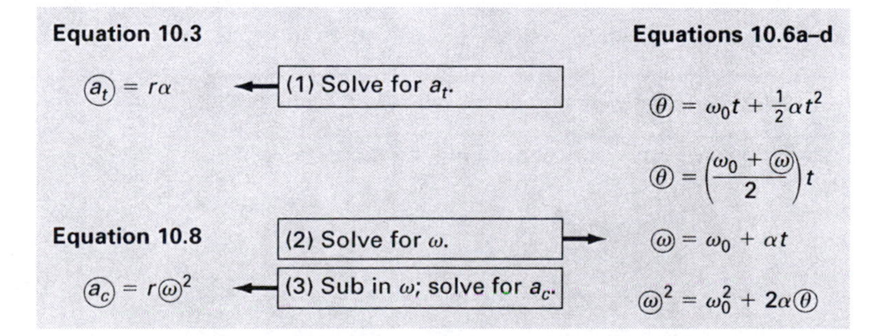

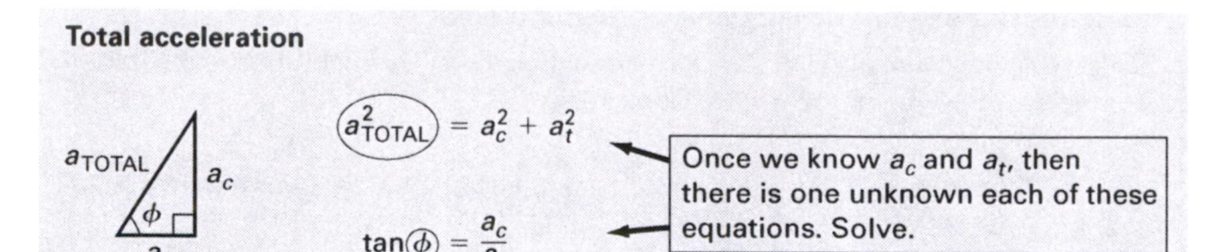

Answers

- $a_t = 0.18\frac{\text{m}}{\text{s}^2}$
- $a_c = 0.0972 \cong 0.097\frac{\text{m}}{\text{s}^2}$
- $a_{\text{TOTAL}} \cong 0.2046 \cong 0.20\frac{\text{m}}{\text{s}^2}$, $\phi \cong 28.4 \cong 28°$ (from the tangential direction)

Intermediate answer: $\omega = +0.90\frac{\text{rad}}{\text{s}}$

WHAT HAPPENS TO THE ACCELERATION VALUES IF THE ELAPSED TIME IS DOUBLED?

If we double 0.60 s to get 1.2 s, we would calculate the *centripetal acceleration* to be *four times larger:* $a_c \cong 0.39\frac{\text{m}}{\text{s}^2}$! However, because the *angular acceleration* is *constant,* the *tangential acceleration* is the *same as before:* $a_t = r\alpha = 0.18\frac{\text{m}}{\text{s}^2}$!

■

10.10 SUMMARY OF ANGULAR VELOCITY AND ACCELERATION EQUATIONS

Because there are so many, we summarize all of the equations we have used for angular velocity and acceleration problems:

Relating angular and tangential or linear quantities (Equations 10.1–3)	$s = r\theta$ $v = r\omega$ $a_t = a = r\alpha$
Constant/average angular velocity (Equation 10.4)	$\omega = \frac{\theta}{t}$
Constant/average tangential or linear velocity (Equation 10.5)	$v = \frac{s}{t}$
Constant/average angular acceleration (Equations 10.6a–d)	$\theta = \omega_0 t + \frac{1}{2}\alpha t^2$ $\theta = \left(\frac{\omega_0 + \omega}{2}\right)t$ $\omega = \omega_0 + \alpha t$ $\omega^2 = \omega_0^2 + 2\alpha\theta$

Constant/average tangential or linear acceleration (Equations 10.7a–d)	$s = v_0t + \frac{1}{2}at^2$ $s = \left(\frac{v_0 + v}{2}\right)t$ $v = v_0 + at$ $v^2 = v_0^2 + 2as$
Centripetal acceleration (Equation 10.8)	$a_c = r\omega^2$

10.11 HOW TO SET UP CONSTANT/AVERAGE ANGULAR ACCELERATION PROBLEMS

Here is a summary of the steps we followed in the previous several exercises:

Constant/Average Angular Acceleration Problems—Mental and Written Steps

Mental → **(1) Type of problem**

- Constant/average angular acceleration: Use Equations 10.6a–d.
- If there is *also* tangential/linear acceleration: Use . . .
  - Equations 10.1–3, and
  - EITHER Equations 10.6a–d
  - OR Equations 10.7a–d
- If there is *also* centripetal acceleration: Use Equation 10.8, and right triangle trig if needed, to get the total acceleration vector.

Mental and written → **(2) Sort by object, circle, and/or interval**

Sort in groups, as needed:

- **Angular quantities**: For Equations 10.6a–d.
- **Tangential/linear quantities:** For Equations 10.7a–d.
- **Radius:** So we can use Equations 10.1–3 to relate angular and tangential/linear quantities.
- **Other:** Such as centripetal acceleration, and so on.

Mental and written → **(3) Equations & unknowns**

Write the equations as described above, and circle the unknowns.

Mental → **(4) Outline solution**

CHAPTER 11

TORQUE AND EQUILIBRIUM

11.1 TORQUE

We will use two different equations for torque, which is a measure of a force's ability to cause angular acceleration. Both equations work for any torque calculation, but each one has its own advantages, depending on the problem. The first torque equation:

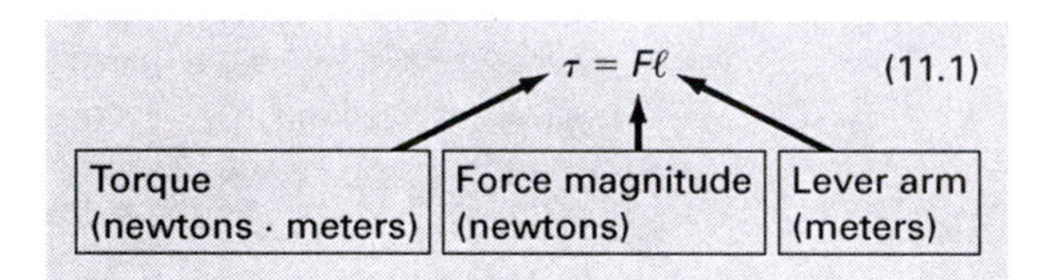

As usual, in the following exercises we explain both the written and mental steps, and so our solutions are much longer than your handwritten solutions will be.

EXERCISE 11.1

A 30-N force is applied to a 0.20-m-long beam at its right end, at an angle of 65° to the beam. Determine the torque caused about an axis perpendicular to the beam and through its left end.

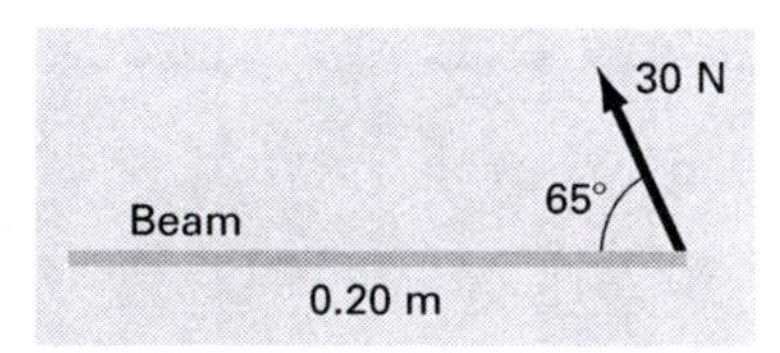

Solution

(1) Type of problem

Torque: Use Equation 11.1.

(2) Sort by force vector, (3) Equations & unknowns, and (4) Outline solution

We have one force vector for which we need the *magnitude* (given, $F = 30$ N) and the *lever arm*, ℓ, in Equation 11.1. This is how to get the lever arm:

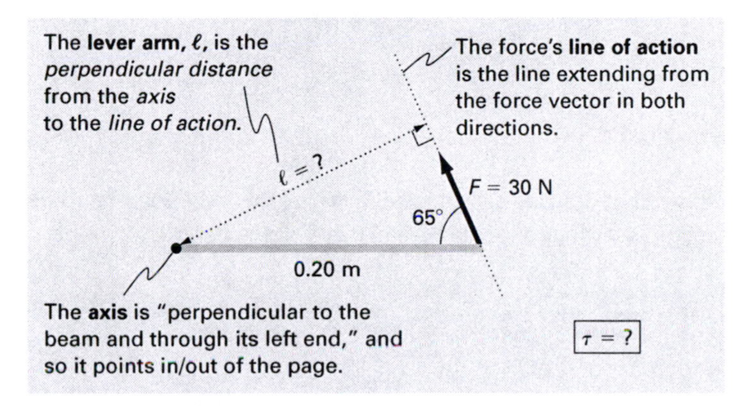

We use a right triangle to determine ℓ:

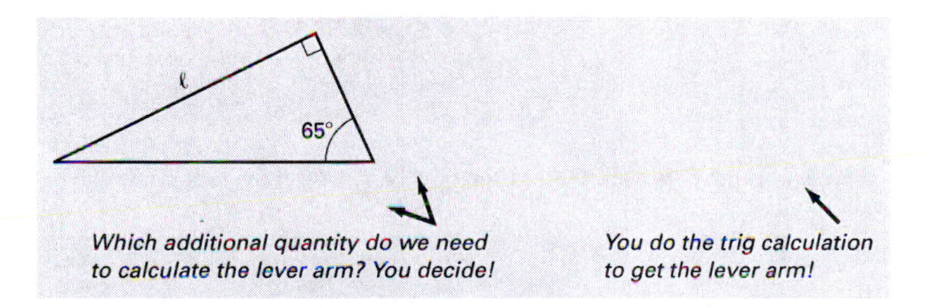

Now that we know ℓ, we use Equation 11.1:

$$\tau = F\ell \qquad \text{Solve.}$$

Answers for right triangle calculations

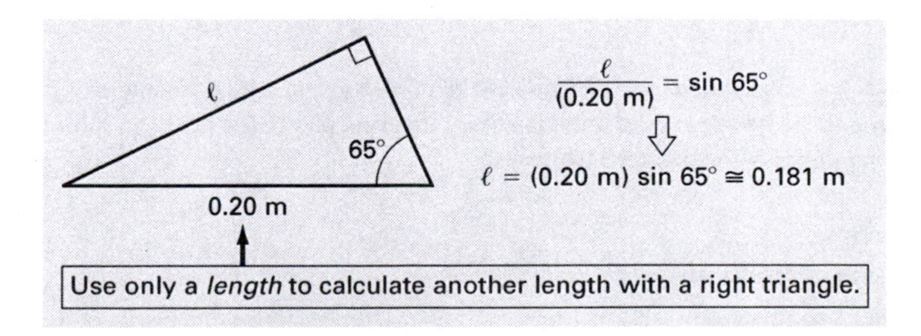

DON'T MIX LENGTHS AND FORCES IN RIGHT TRIANGLE CALCULATIONS!

Here we use *only lengths.* For some right triangles, we use *only forces.* Don't mix them when doing right triangle calculations!

Answer $\tau \cong 5.4\ \text{N} \cdot \text{m}$ ■

Instead of drawing a right triangle to get the lever arm, the following equation *always* works, as long as we use r and θ as defined here:

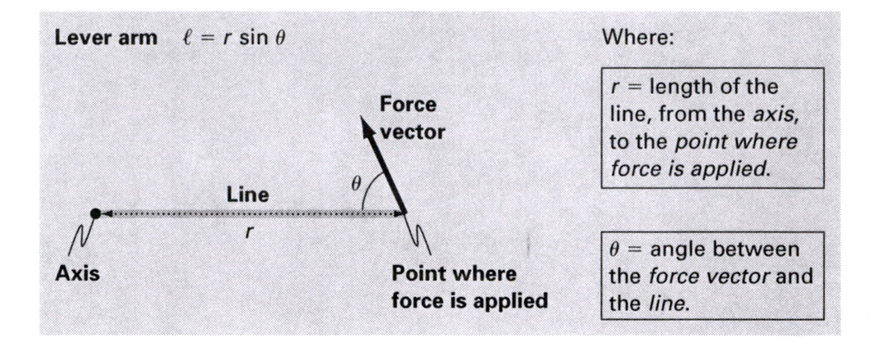

If we substitute $\ell = r \sin \theta$ into Equation 11.1, we get the second torque equation:

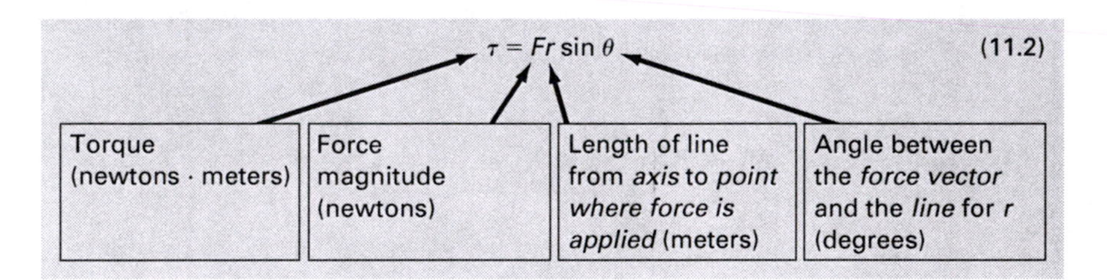

With this equation, we don't need to calculate the lever arm separately. Here we repeat the previous exercise, but now with Equation 11.2.

EXERCISE 11.2

A 30-N force is applied to a 0.20-m-long beam at its right end, at an angle of 65° to the beam. Determine the torque caused about an axis perpendicular to the beam and through its left end. (See the figure in the previous exercise.)

Solution

(1) Type of problem

Torque: Use Equation 11.2 this time.

(2) Sort by force vector

We sort the quantities in Equation 11.2:

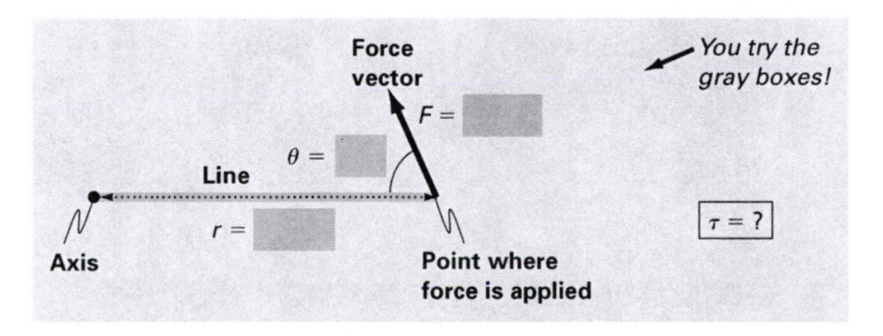

(3) Equations & unknowns and (4) Outline solution

We substitute these quantities into Equation 11.2:

$$\textcircled{\tau} = Fr \sin \theta \qquad \boxed{\text{Solve.}}$$

Answers for (2) Sort by force vector

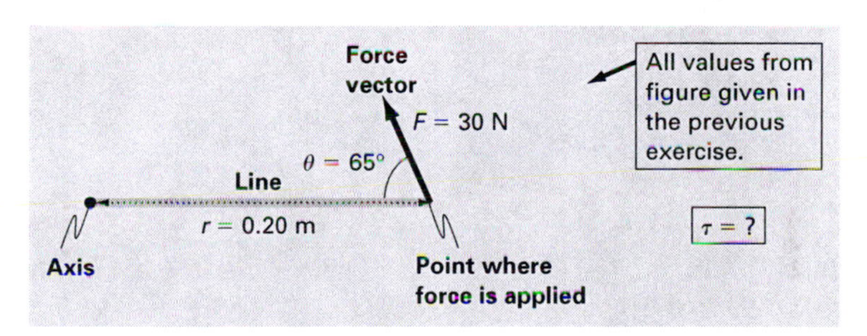

Answer $\tau \cong 5.4$ N · m ■

11.2 HOW TO SET UP TORQUE PROBLEMS

Torque Problems—Mental and Written Steps

Mental →	**(1) Type of problem** Torque: Use Equation 11.1 or 11.2.
Mental and written →	**(2) Sort by force vector** For each force vector that causes torque, sort quantities for Equation 11.1 or 11.2.
Mental and written →	**(3) Equations & unknowns** • Use Equation 11.1 or 11.2 for *each* force vector. • For Equation 11.1, we may also need a right triangle trig equation for the lever arm calculation.
Mental →	**(4) Outline solution**

Which torque equation should we use? Equation 11.2 is easier when there are angles *other than* 90°. When angles are 90°, it is easiest to use Equation 11.1, which is really just Equation 11.2 with $\theta = 90°$:

$\tau = Fr \sin\theta$	⇨	$\tau = F\ell \sin 90°$	⇨	$\tau = F\ell$
Equation 11.2		If $\theta = 90°$, then $r = \ell$.		Equation 11.1

11.3 EQUILIBRIUM FOR "RIGID" BODIES

In Chapter 5, a *point mass* in *equilibrium* had zero linear acceleration, which meant there was zero net force acting on it: $F_{\text{net},x} = 0$ and $F_{\text{net},y} = 0$. Instead, we now look at a *rigid body* like a plank, which can rotate *and* move linearly. For the plank to be in *equilibrium,* it has zero linear *and* angular accelerations. That means:

Equilibrium for a *rigid body* has both of these conditions.

Zero net *torque* $\{ \tau_{\text{net}} = 0$ (11.3)

Zero net *force*
$$F_{\text{net},x} = 0 \quad (11.4a)$$
$$F_{\text{net},y} = 0 \quad (11.4b)$$

WHAT IS THE DIFFERENCE BETWEEN A POINT MASS AND A RIGID BODY?

Up to now, we have treated each object as a *point mass* when drawing its FBD. In other words, we would draw the FBD as if all forces acted at the *same point.* There were no *lever arms,* and so we completely ignored *torques* caused by any forces.

A *rigid body,* on the other hand, is a *mass that has width, height, and other* dimensions that we do NOT ignore. Now when we draw the FBD, we care about *where* on the object each force acts. This lets us determine *lever arms* and *torques* caused by each force.

11.4 EQUILIBRIUM—WITH ONLY 90° ANGLES

If all angles in the problem are 90°, then we use *lever arms* and *Equation 11.1* to get the torque caused by each force. If lever arms are *given,* even if there are angles other than 90°, then we would still use Equation 11.1.

In equilibrium problems for rigid bodies, there are *many* mental steps, not usually written down, which we explain in detail.

EXERCISE 11.3

A 6.0-m-long uniform plank weighs 500 N; it is supported by a vertical rope at the left end and another vertical rope 2.0 m from the right end. A person who weighs 800 N stands 1.5 m from the left end of the plank. What is the tension in each rope?

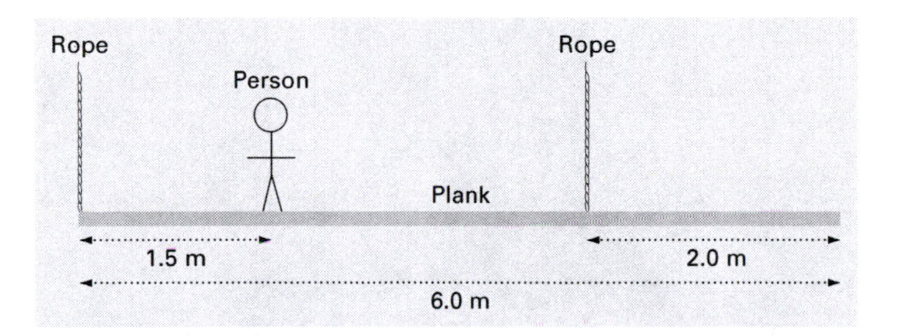

Solution

(1) Type of problem

Equilibrium for a rigid body: Use Equations 11.1 or 11.2, 11.3, 11.4a, and/or 11.4b.

(2) Sort by force vector and (3) Equations & unknowns

For the object in equilibrium (plank/person together), we do these steps in three parts:

- **FBD:** Identify force vectors and draw a FBD of the plank/person (see Chapter 5 for a review of how to draw a FBD).
- **Torque calculations:** Use Equation 11.1 or 11.2 for each force vector that causes torque, and then add and substitute these into Equation 11.3.
- **Force calculations:** use Equations 11.4a and/or 11.4b with force components from the FBD.

FBD

We identify long-range forces (the two weight vectors) and contact forces (the two tension vectors) acting *on the plank/person:*

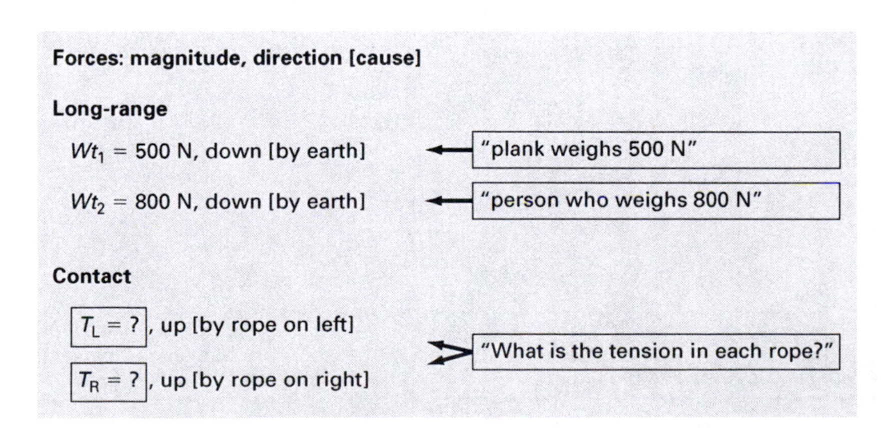

We care *where* the forces act in a FBD for a *rigid body,* so we draw it like this, labeling the distances:

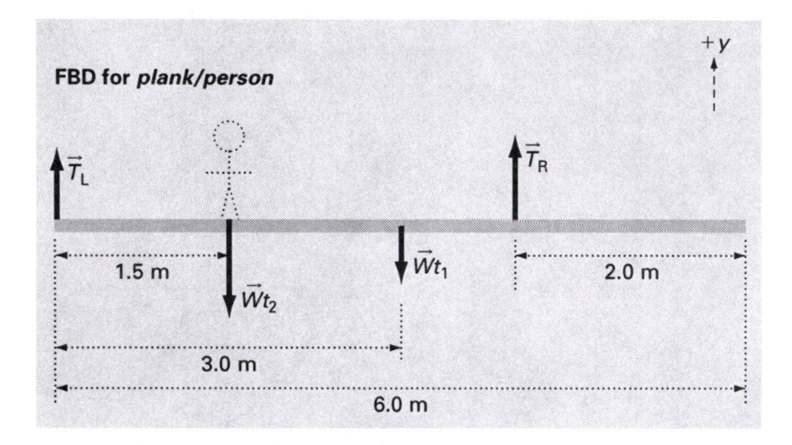

"[U]niform plank" means that $\vec{Wt}_1$ is at its center, or 3.0 m from the end.

Torque calculations

How do we use an *axis* for torque calculations when the plank/person system is *not rotating* (or *not angularly accelerating,* to be exact) about an axis?

Here is the idea: The system is *not rotating* about the left end of the plank. It is also *not rotating* about the right end of the plank, or about the center, or *about any axis anywhere!* So if we *put an imaginary axis* ANYWHERE, then the net torque will be zero, and we can use Equation 11.3.

We will try a horizontal axis *at the left end of the plank,* perpendicular to the plank (the axis points in/out of the page):

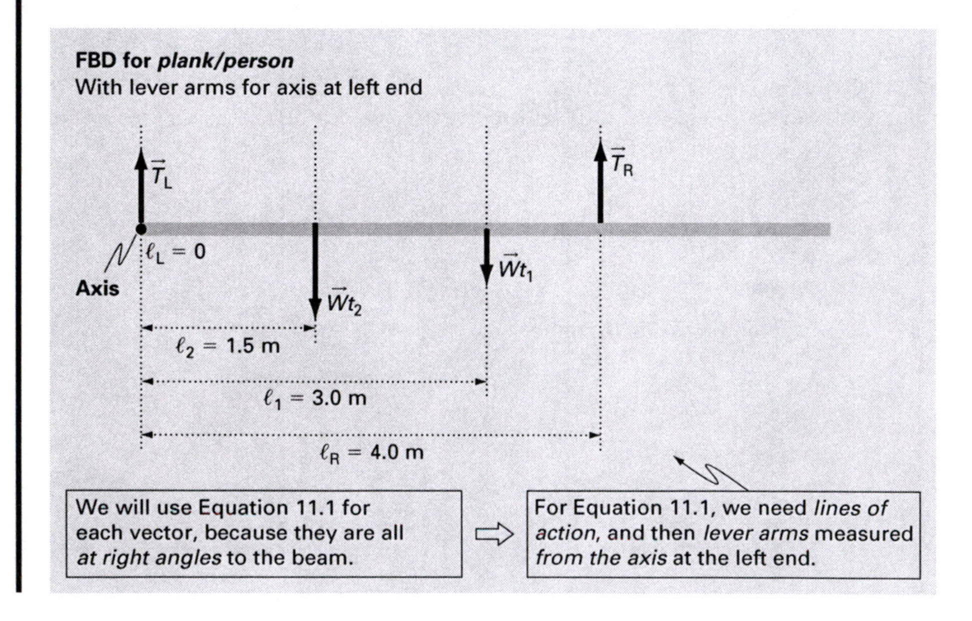

With Equation 11.1, we will calculate torque, including the *sign*, for each force.

HOW DO WE DETERMINE THE SIGN OF THE TORQUE?

In Equations 11.1 and 11.2, we need to *pick* the sign for the torque. Usually, *counterclockwise* is *positive*, and *clockwise* is *negative*. We can have it the other way if we want, as long as we are consistent with all signs.

We will calculate the torques one at a time. First, we *imagine* that $\vec{T}_R$ causes the ONLY torque:

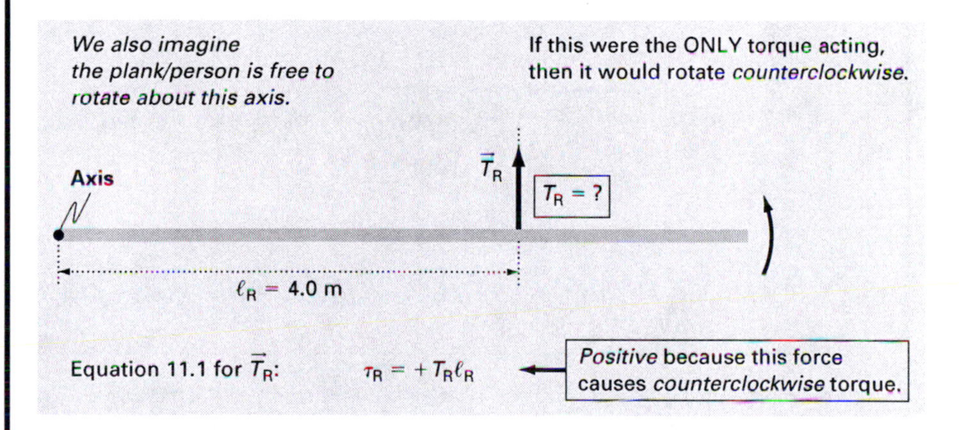

To be picky, we *should* say the *angular acceleration* would be counterclockwise, but we will just say it would *rotate* counterclockwise. Shhh!

We do the same thing for $\vec{Wt}_1$:

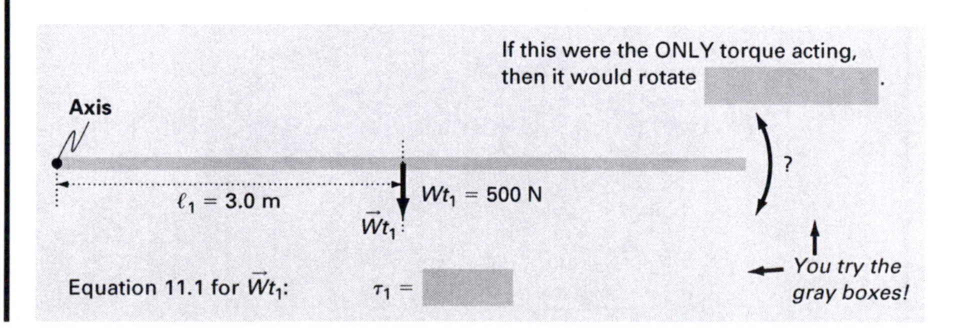

We *mentally* do the same for the other two forces and get:

Equation 11.1 for $\vec{Wt}_2$: $\tau_2 =$ [gray box]

Equation 11.1 for $\vec{T}_L$: $\tau_L =$ [gray box]

You try the gray boxes! Answers in the next figure. Explanations in the answers section.

Finally, we add all of the individual torques and substitute them into Equation 11.3:

Torque for each vector

$\tau_1 = -Wt_1\ell_1$

$\tau_2 = -Wt_2\ell_2$

$\tau_L = T_L\ell_L = 0$

$\tau_R = +T_R\ell_R$

(Add and sub) →

$\tau_{net} = 0$

$(-Wt_1\ell_1) + (-Wt_2\ell_2) + (0) + (+T_R\ell_R) = 0$

$T_R\ell_R = Wt_1\ell_1 + Wt_2\ell_2$

PICK AN AXIS THAT MAKES THE TORQUE ZERO FOR AN UNKNOWN FORCE!

This equation has only one unknown, T_R. That is not by accident. We picked the axis at the left end, *at the point of an unknown force,* $\vec{T}_L$. That made its torque, τ_L, *zero,* and got T_L out of this equation, so it has fewer unknowns. EASIER math!

Force calculations

For *force* calculations, we don't care where the forces act, and so we can redraw the FBD like this, as if for a point mass:

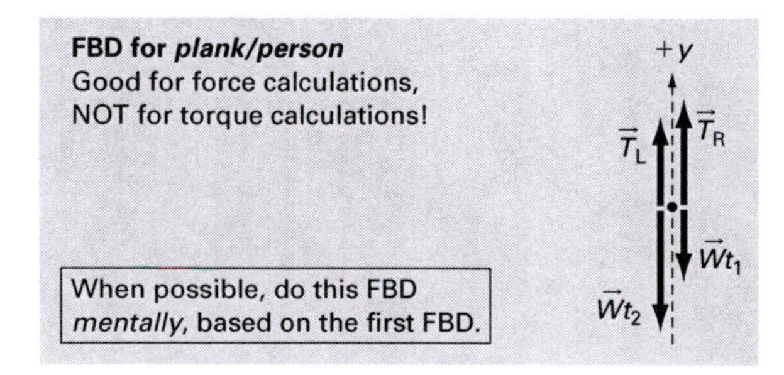

We substitute the y-components from the FBD into Equation 11.4b:

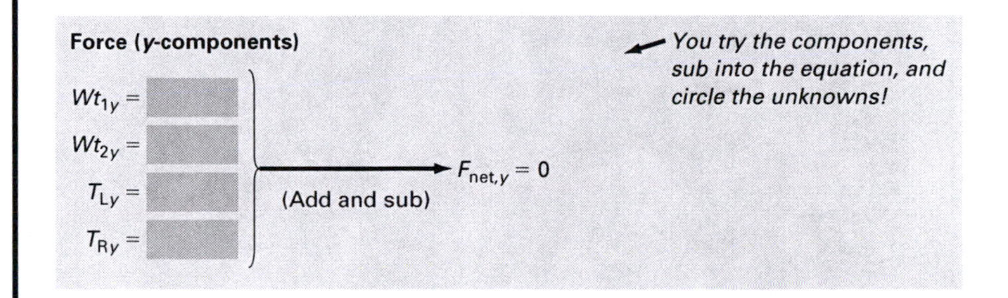

(4) Outline solution

Putting the equations together:

Torque calculations $\boxed{T_R}\ell_R = Wt_1\ell_1 + Wt_2\ell_2$ *You try an outline!*

Force calculations $\boxed{T_L} + \boxed{T_R} = Wt_1 + Wt_2$

Answers for gray boxes

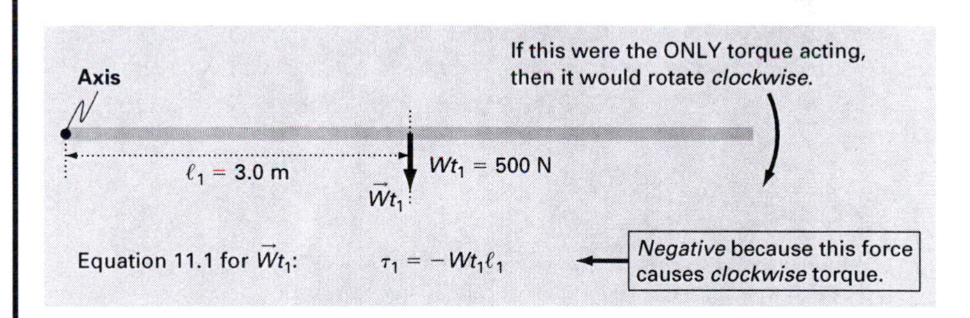

Equation 11.1 for $\vec{Wt}_2$: $\tau_2 = -Wt_2\ell_2$ ← *Negative* because this force causes *clockwise* torque. The diagram is similar to that for Wt_1.

Equation 11.1 for $\vec{T}_L$: $\tau_L = T_L\ell_L = 0$ ← *Zero* because the *lever arm*, ℓ_L, is *zero*.

Answers for . . .

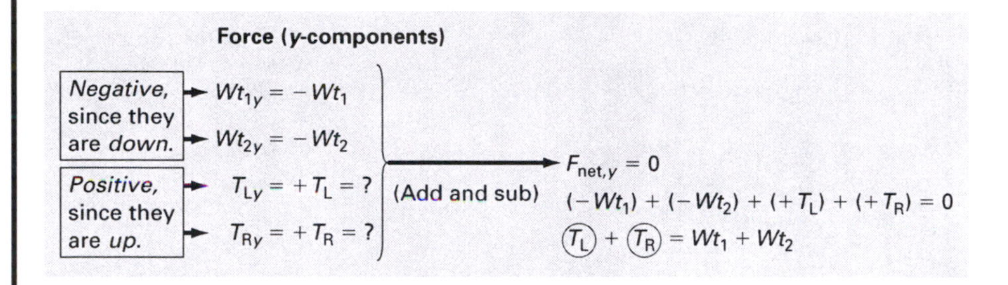

Answers for (4) Outline solution

Torque calculations	$T_R \ell_R = Wt_1 \ell_1 + Wt_2 \ell_2$	(1) Solve for T_R.
Force calculations	$T_L + T_R = Wt_1 + Wt_2$	(2) Sub in T_R; solve for T_L.

Answers $T_R = 675$ N, $T_L = 625$ N ■

In the previous exercise, instead of doing force calculations, we could have done *torque calculations about another axis* to get a second equation. The torque about any axis is zero, as is illustrated in the next exercise.

EXERCISE 11.4

For the situation in the previous exercise, show that the torque is also zero about a horizontal axis 4.0 m from the left end and perpendicular to the plank.

Solution

(1) Type of problem

Net torque: Use Equation 11.1 for each force vector, and then add to get the net torque.

(2) Sort by force vector, (3) Equations & unknowns, and (4) Outline solution

We did most of this in the previous exercise, but now we need to do a new torque calculation for a "horizontal axis 4.0 m from the left end and perpendicular to the plank."

Torque calculations

We have to *redo* all of the *lever arms* because it is a *different axis:*

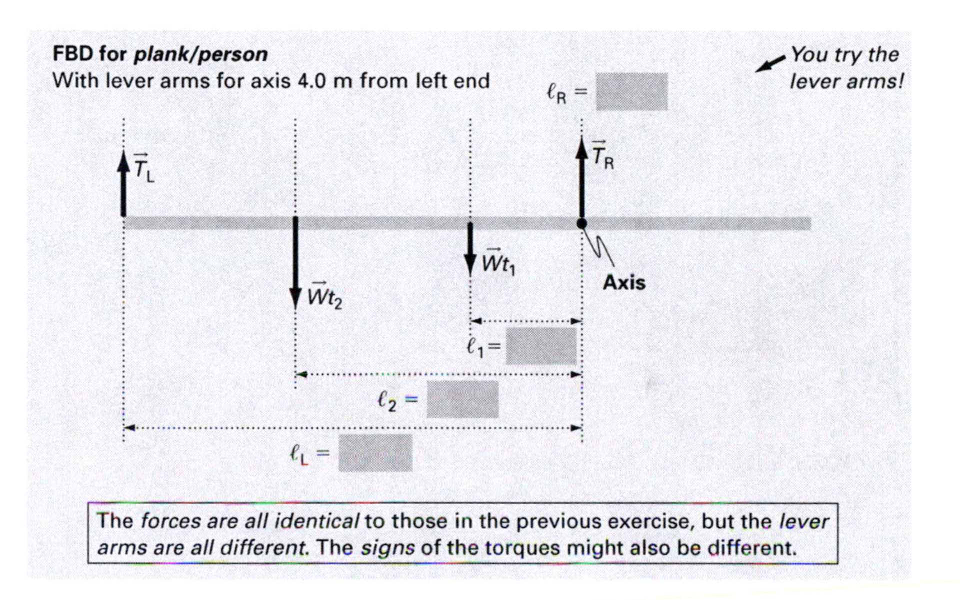

Now we determine the torque for each force, beginning with $\vec{T}_L$:

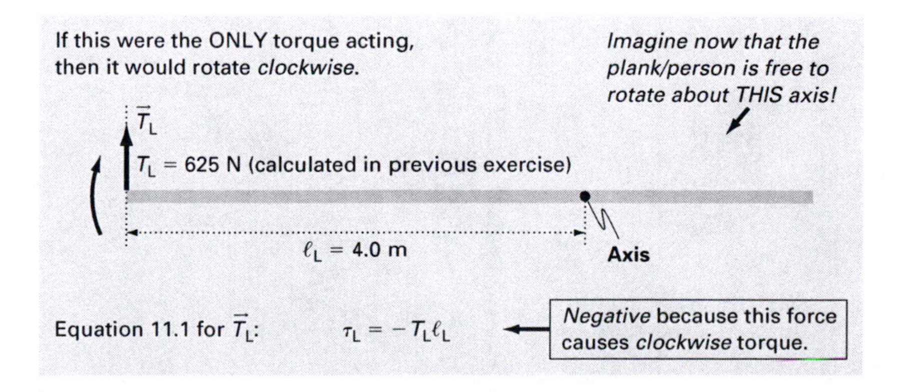

Because of the new axis location, the value of τ_L is *no longer zero* as it was in the previous exercise. Now we do the same thing for $\vec{Wt}_1$:

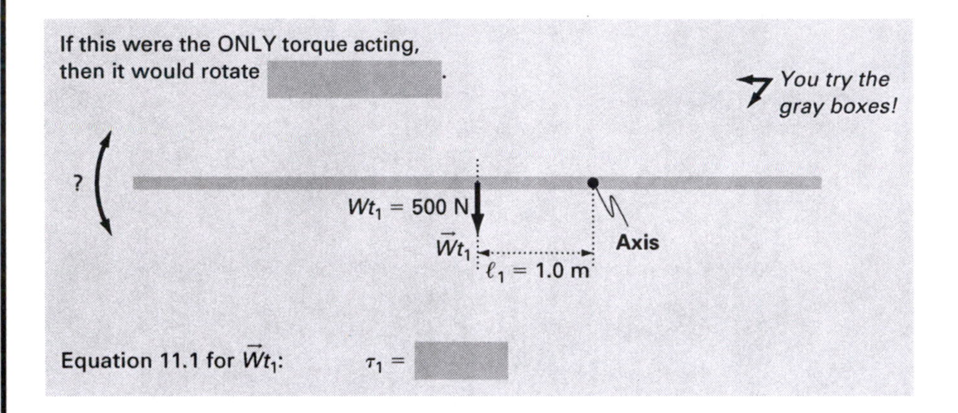

We *mentally* do the same for the other two forces:

Equation 11.1 for $\vec{Wt}_2$: $\tau_2 =$

Equation 11.1 for $\vec{T}_R$: $\tau_R =$

You try the gray boxes! Answers in the next figure. Explanations in the answers section!

Finally, we put the individual torques together:

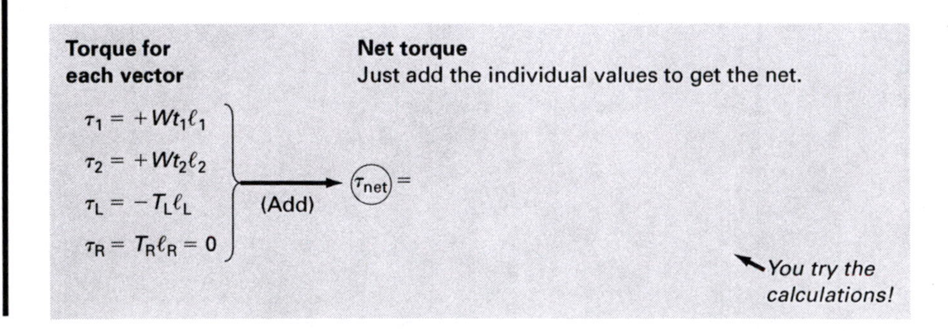

Answers for . . .

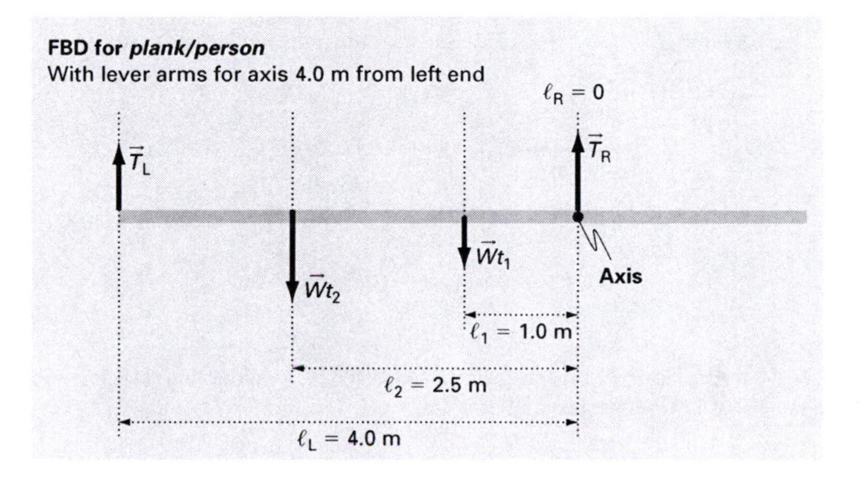

Answers for gray boxes

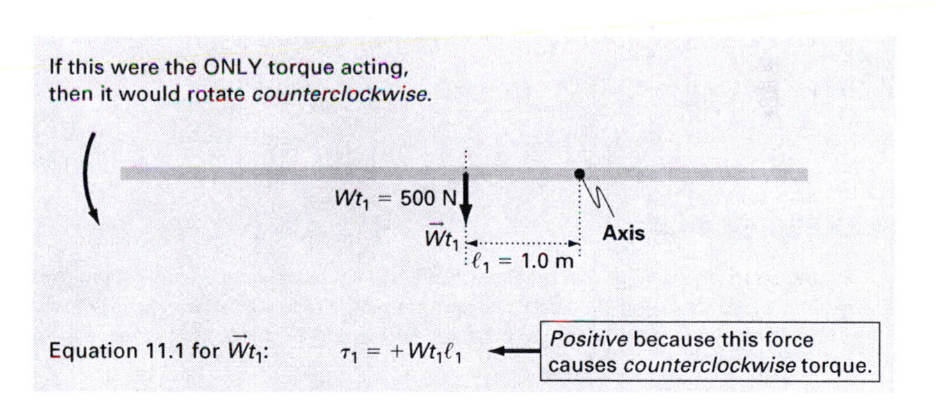

Equation 11.1 for $\vec{Wt}_2$: $\tau_2 = +Wt_2\ell_2$ ← *Positive* because this force causes *counterclockwise* torque. The diagram is similar to that for Wt_1.

Equation 11.1 for $\vec{T}_R$: $\tau_R = T_R\ell_R = 0$ ← *Zero* because the *lever arm*, ℓ_R, is *zero*.

Answers for . . .

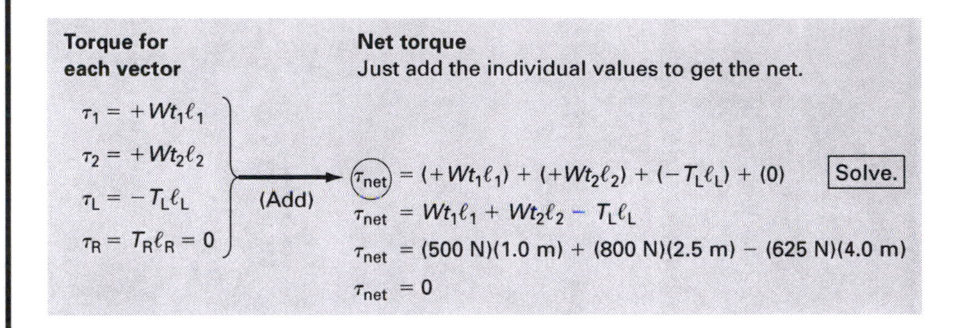

The net torque *has to be zero* since the plank/person is in equilibrium. This is a good check of our value of T_L from the previous exercise.

FOR AN OBJECT IN EQUILIBRIUM, THE NET TORQUE IS ZERO ABOUT ANY AXIS!

This means we can pick *any* axis when using Equation 11.3.

Answer $\tau_{net} = 0$ ■

EXERCISE 11.5

The person in the previous two problems walks slowly past the rope on the right, toward the right end of the plank. (a) How far is the person past the rope when the system is on the verge of tipping clockwise? (b) When the person is in this new position, what is the tension in the rope on the right?

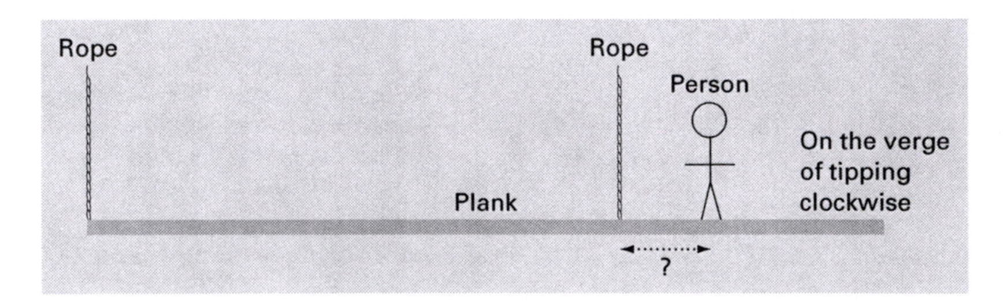

Solution

(1) Type of problem

"[O]n the verge of tipping" means *equilibrium,* but just barely! Equilibrium for a rigid body: Use Equations 11.1 or 11.2, 11.3, 11.4a, and/or 11.4b.

(2) Sort by force vector and (3) Equations & unknowns

FBD and torque calculations

We go ahead and put lever arms on the FBD, which is similar to that in the previous exercise, except now $\vec{T}_L$ is zero and $\vec{W}t_2$ has moved:

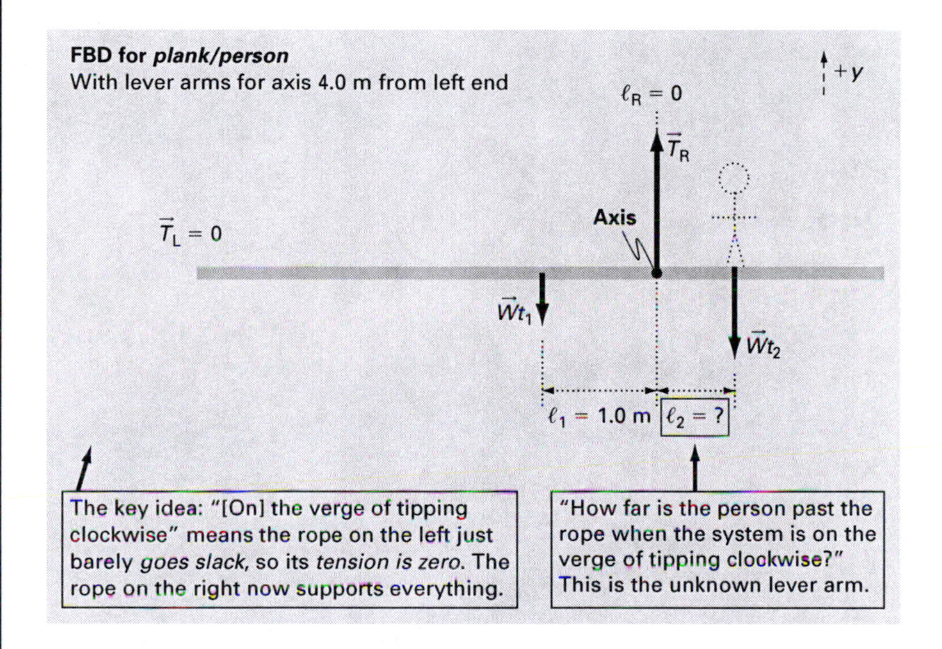

Why did we put the axis at the (unknown) force vector, $\vec{T}_R$? Because this makes its torque, τ_R, *zero* and gets T_R out of the net torque equation.

Now we use Equation 11.1 for each vector, and then we add them with Equation 11.3:

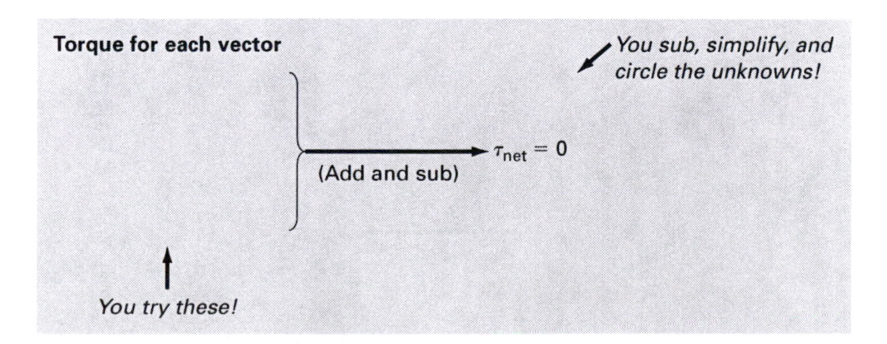

Force calculations

Since all forces are vertical, we can get the y-components without redrawing the FBD, and add them in Equation 11.4b:

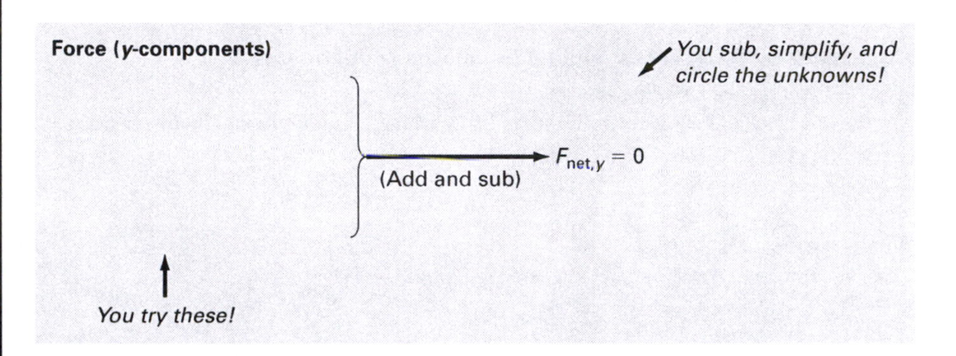

(4) Outline solution

We put the equations together:

Torque calculations $Wt_1\ell_1 = Wt_2\ell_2$ ← (1) Solve.

Force calculations $T_R = Wt_1 + Wt_2$ ← (2) Solve.

Answers for torque calculations

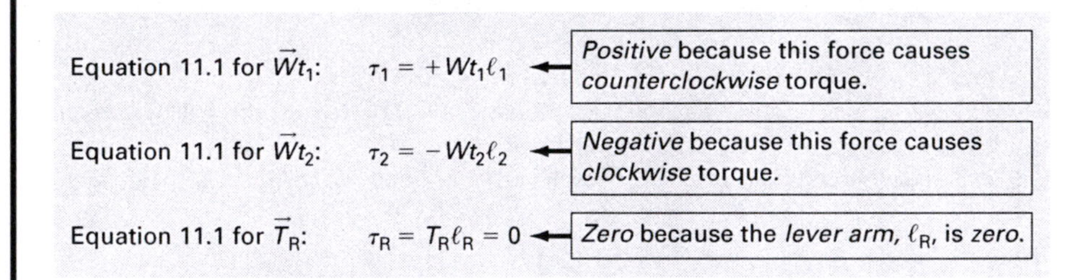

Torque for each vector

$\tau_1 = +Wt_1\ell_1$

$\tau_2 = -Wt_2\ell_2$

$\tau_R = T_R\ell_R = 0$

(Add and sub) →

$\tau_{net} = 0$

$(+Wt_1\ell_1) + (-Wt_2\ell_2) + (0) = 0$

$Wt_1\ell_1 = Wt_2\ell_2$

Answers for force calculations

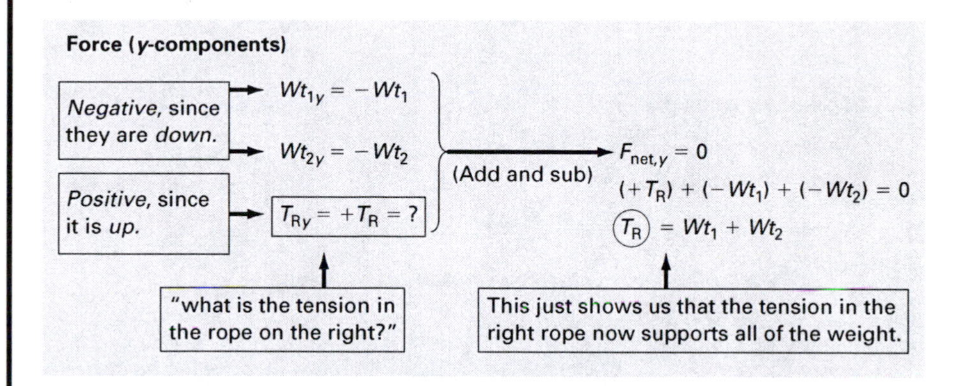

Answers $\ell_2 = 0.625 \cong 0.63$ m, $T_R = 1300$ N ■

11.5 EQUILIBRIUM—WITH NON-90° ANGLES

For situations in which there are angles *other than* 90° and no lever arms are given in the diagram, we use Equation 11.2 to get the torque caused by each force.

EXERCISE 11.6

A person's lower leg is like a beam, attached to the rest of the leg via a joint at the knee. Here, the lower leg is a 0.60-m-long uniform beam that weighs 40 N. A muscle, like a cable, holds the lower leg steady at an angle of 75° from the vertical. The muscle makes an angle of 20° with the lower leg and attaches to the lower leg at a point 0.12 m from the joint. Determine (a) the tension force in the muscle and (b) the horizontal and vertical components of the force exerted on the lower leg by the joint.

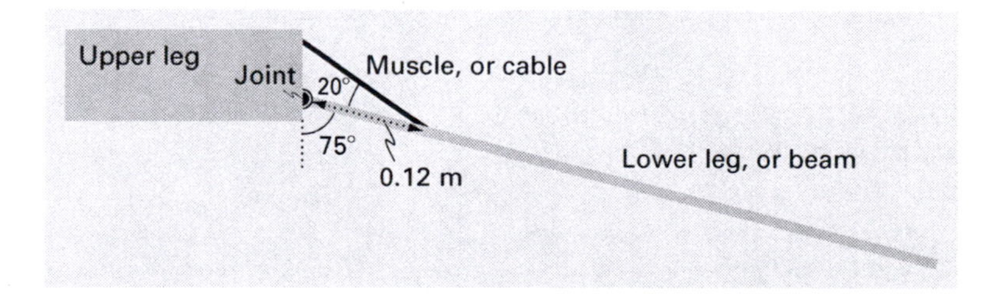

Solution

(1) Type of problem

Equilibrium for a rigid body: Use Equations 11.1 or 11.2, 11.3, 11.4a, and/or 11.4b.

(2) Sort by force vector and (3) Equations & unknowns

FBD

First, we mentally identify forces *acting on the lower leg:*

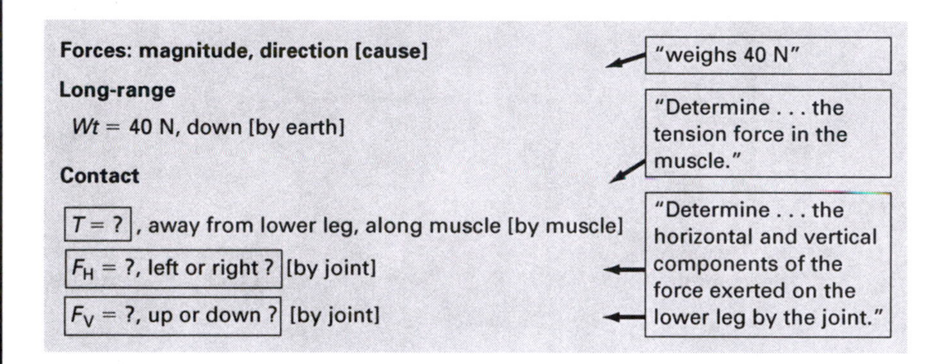

This leads to:

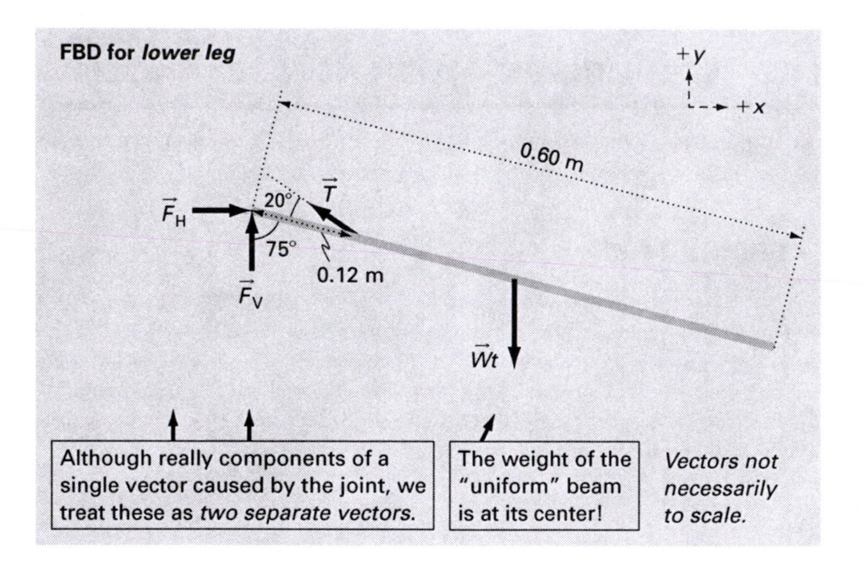

To make our FBD easier to read, we can draw a vector ANYWHERE along its *line of action.* These are *equivalent:*

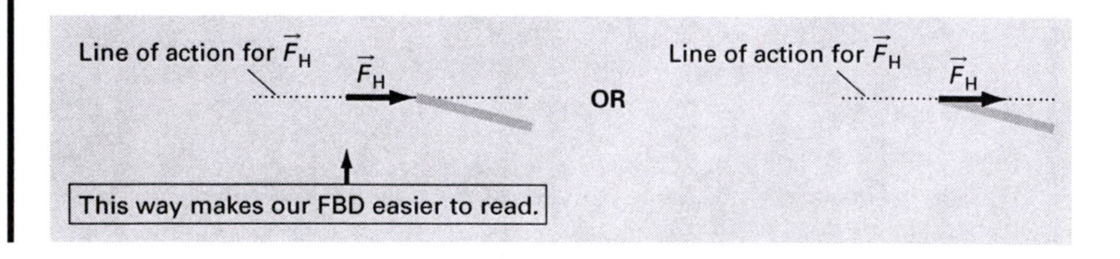

HOW TO DEAL WITH FORCES WITH UNKNOWN DIRECTION: GUESS!

We *guess* that $\vec{F}_H$ is to the *right*, to counteract the *x*-component of the tension force—this makes sense. And we *guess* that $\vec{F}_V$ is *up*—this one is harder to know beforehand.

If we guess *incorrectly* for a vector, its *magnitude* (which should be positive) will come out *negative*. We do NOT redo the problem, but just note the correct direction.

Torque calculations

First, we use Equation 11.2 for each force vector.

We put an axis at the upper left end because two unknown forces, $\vec{F}_H$ and $\vec{F}_V$, are there, making them both have *zero torque* for this axis. EASIER math! Hooray!

For each of the other two forces, we need r and θ for Equation 11.2. First, $\vec{Wt}$:

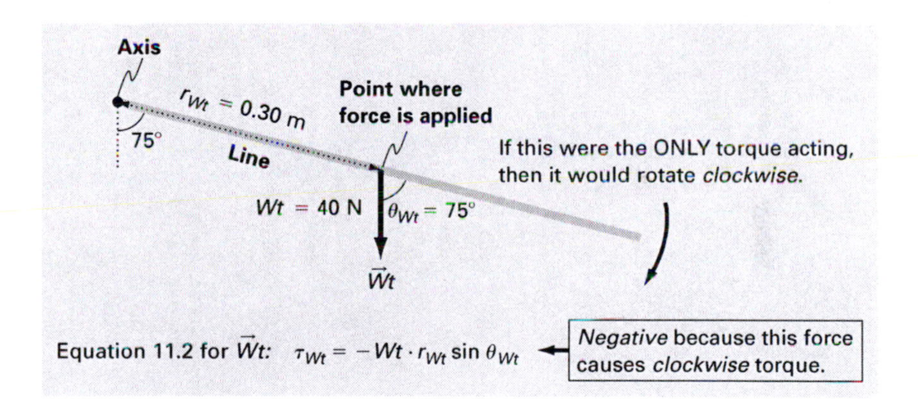

We can actually draw the force above or below the leg, and we can use ANY of the following angles as θ_{Wt} in Equation 11.2:

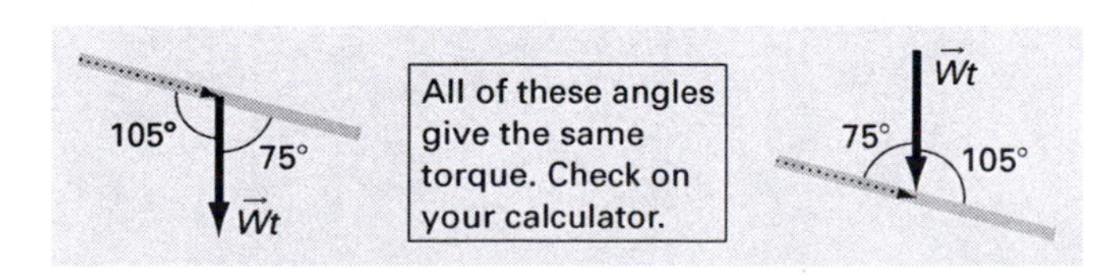

Now the torque caused by $\vec{T}$:

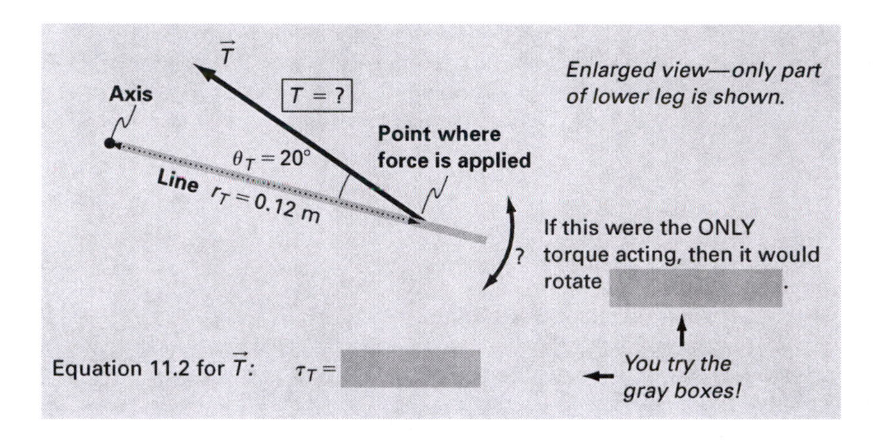

Finally, we substitute the torques into Equation 11.3:

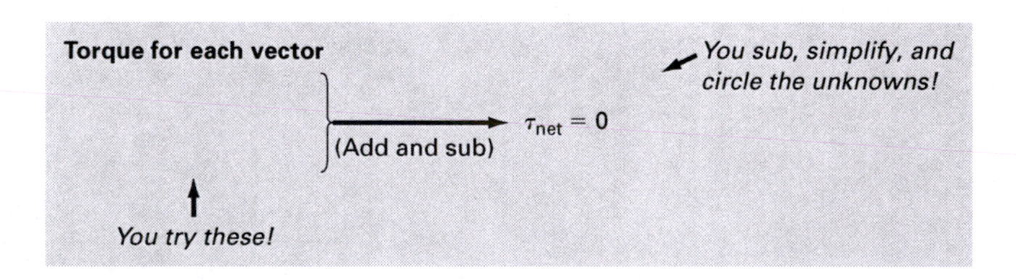

Force calculations

Because of the angle, the FBD for force calculations is tougher to do mentally:

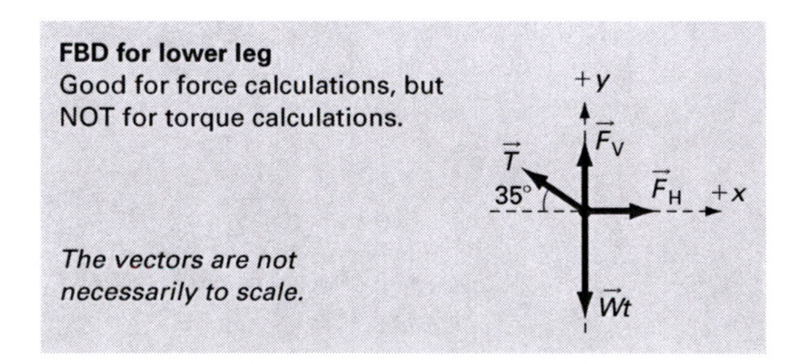

How did we get the 35° angle? Here is the idea:

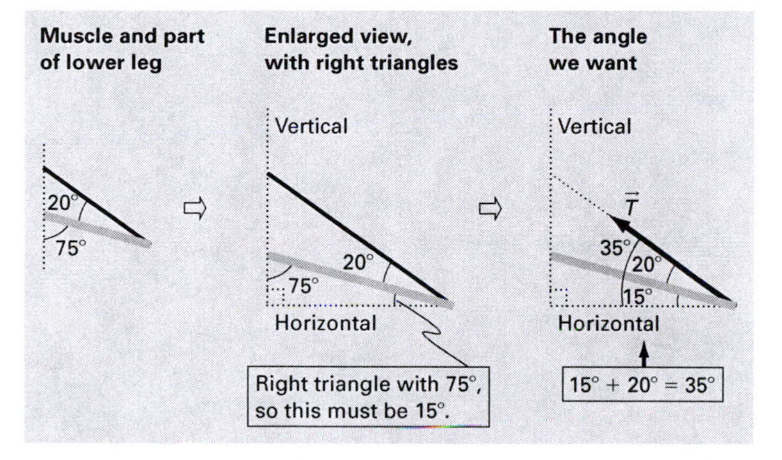

We find the components of $\vec{T}$ from a right triangle:

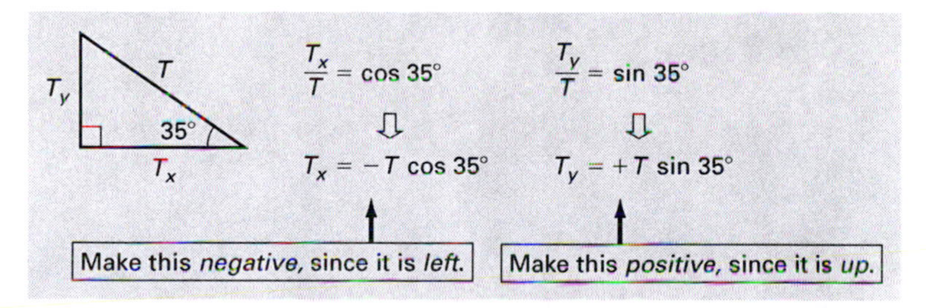

We substitute all the x-components into Equation 11.4a:

Force (x-components)

$$Wt_x = 0$$
$$T_x = -T \cos 35°$$
$$F_{Hx} = +F_H$$
$$F_{Vx} = 0$$

$$\rightarrow F_{\text{net},x} = 0$$

$$(0) + (-T \cos 35°) + (+F_H) + (0) = 0$$

(Add and sub) $F_H = T \cos 35°$

And the y-components into Equation 11.4b:

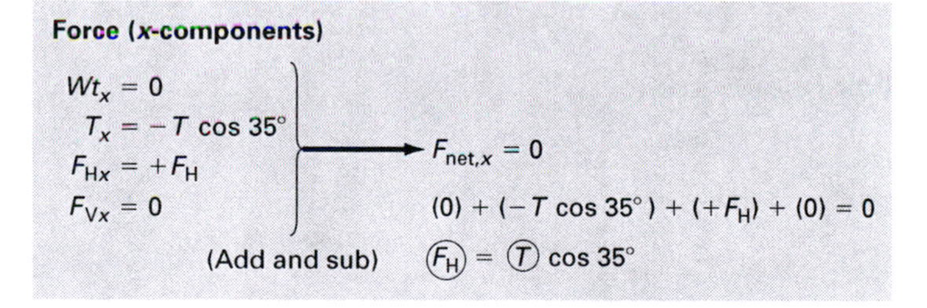

(4) Outline solution

Finally, we put all the equations together:

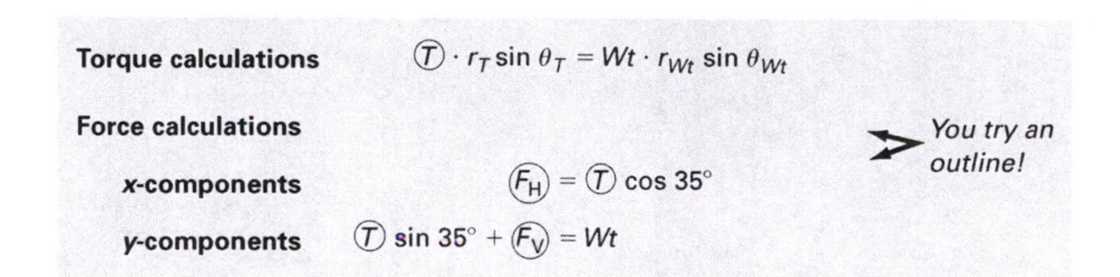

Answers for gray boxes

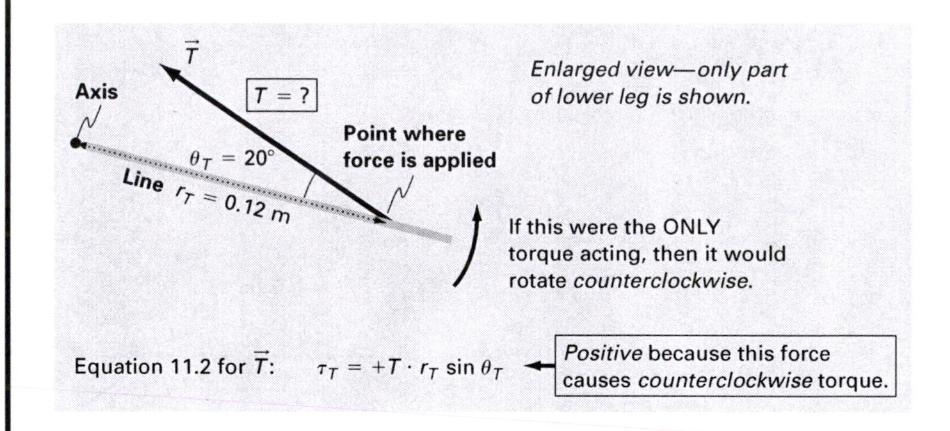

Answers for torque calculations

Torque for each vector

$$\tau_{Wt} = -Wt \cdot r_{Wt} \sin\theta_{Wt}$$

$$\tau_T = +T \cdot r_T \sin\theta_T$$

(Add and sub) →

$$\tau_{net} = 0$$

$$(-Wt \cdot r_{Wt} \sin\theta_{Wt}) + (+T \cdot r_T \sin\theta_T) = 0$$

$$(T) \cdot r_T \sin\theta_T = Wt \cdot r_{Wt} \sin\theta_{Wt}$$

Answers for (4) Outline solution

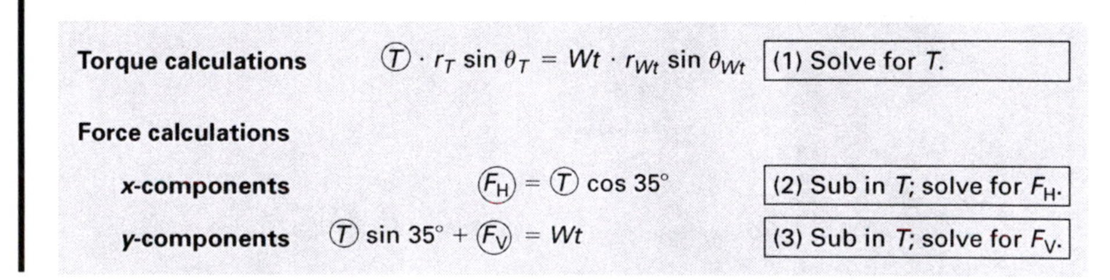

Answers

- $T \cong 282 \cong 280$ N
- $F_H \cong +231 \cong +230$ N (A *positive* magnitude means we guessed the *correct direction:* $\vec{F}_H$ is to the *right.*)
- $F_V \cong -122 \cong -120$ N (A *negative* magnitude means we guessed the *wrong direction:* We guessed it would be *up*, so $\vec{F}_V$ is actually *down*. We don't need to redo the problem, but just note this correct direction.) ■

11.6 HOW TO SET UP EQUILIBRIUM PROBLEMS

Here is a summary of what we did in the previous several exercises:

Equilibrium Problems—Mental and Written Steps

Mental →	**(1) Type of problem** Equilibrium for a rigid body: Equation 11.1 or 11.2, and Equations 11.3, 11.4a, and/or 11.4b.

Mental and written →	**(2) Sort by force vector and (3) Equations & unknowns**
	• **FBD**
	• **Torque calculations:**
	  • If 90° angles: Get Equation 11.1 for each force vector (also need *lever arm*).
	  • If non-90° angles: Get Equation 11.2 for each force vector (also need r and θ).
	  • Add and substitute these into Equation 11.3.
	• **Force calculations:** Use Equations 11.4a and/or 11.4b with force components from FBD.
	Circle the unknowns.

Mental →	**(4) Outline solution**

CHAPTER 12

MORE ANGULAR MOTION

IN THIS CHAPTER, we discuss angular motion problems not covered in Chapter 10, beginning with the relationship between torque and angular acceleration.

If a net torque acts on an object moving or rotating around an axis, then there is angular acceleration. The equation that relates these quantities is called *Newton's second law for rotation:*

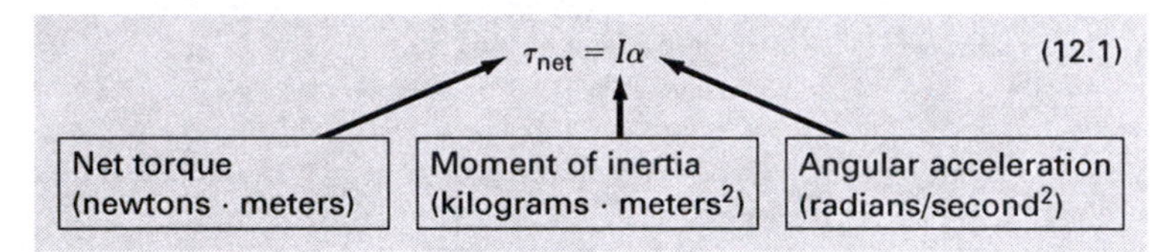

As usual, be consistent with the units!

We often need extra equations for one or more of these quantities. We already know how to get additional equations for *torque* (Equation 11.1 or 11.2) and *angular acceleration* (Equations 10.6a–d). Now we need equations for the *moment of inertia.*

12.1 MOMENT OF INERTIA

The *moment of inertia,* which is like "rotational mass" or "rotational inertia," depends on the *axis of rotation* and the *mass, size,* and *shape* of the object. Here is one example of an equation for the moment of inertia:

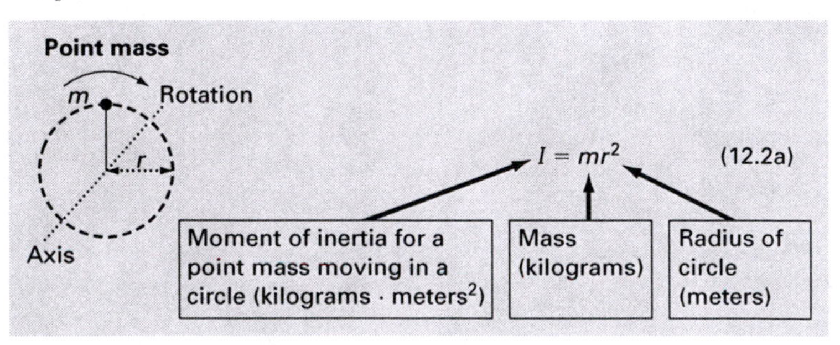

Other objects have different moment of inertia equations, but the *units are the same* as in Equation 12.2a. The equations for some *round* objects:

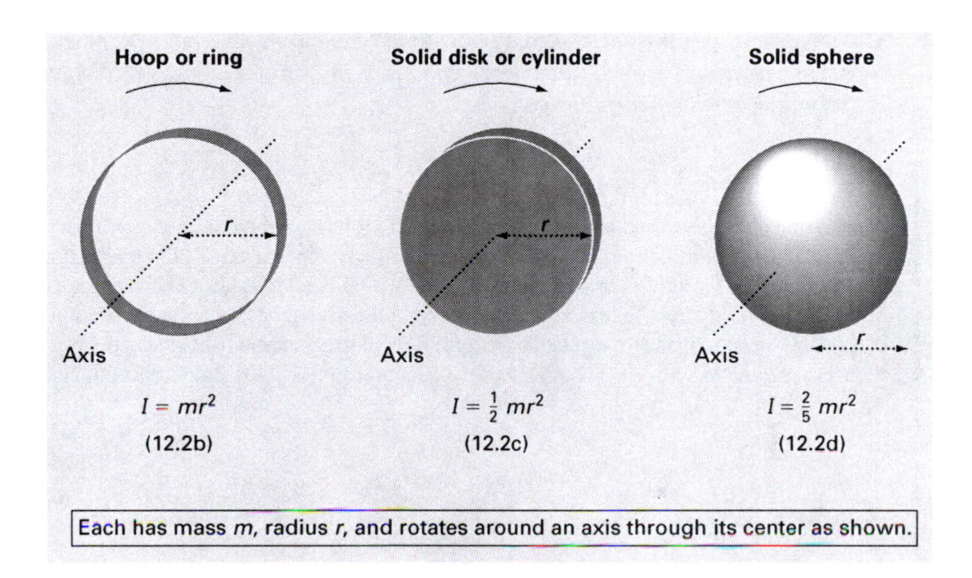

And some *nonround* objects:

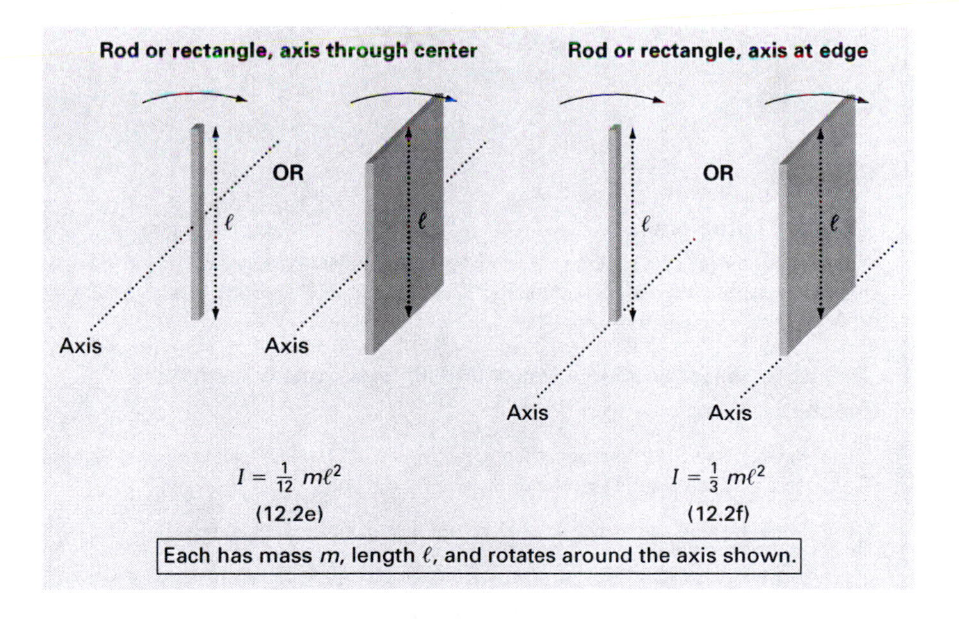

These are the most common shapes and equations, and we refer to them throughout this chapter. Your textbook may also have equations for other shapes.

12.2 TORQUE AND ANGULAR ACCELERATION PROBLEMS

The next exercise illustrates not only how to use Equation 12.1 but also how to get *separate equations* for each of the three quantities in it: τ_{net}, I, and α. As always, we show both *written* and *mental* steps.

EXERCISE 12.1

A uniform solid disk of radius 0.40 m and mass 2.5 kg rotates about a fixed, stationary axis, initially at 3.4 revolutions per second. A cord pulls with a tension of 12 N in a direction tangent to the rim of a light pulley of radius 0.10 m, which is attached at the center axis of the disk. The lower edge of the disk is in contact with a board that causes a kinetic friction force tangent to the rim, against the rotation. After 6.0 revolutions, the wheel slows to 1.8 revolutions per second. What is the magnitude of the friction force? (Ignore the mass of the pulley.)

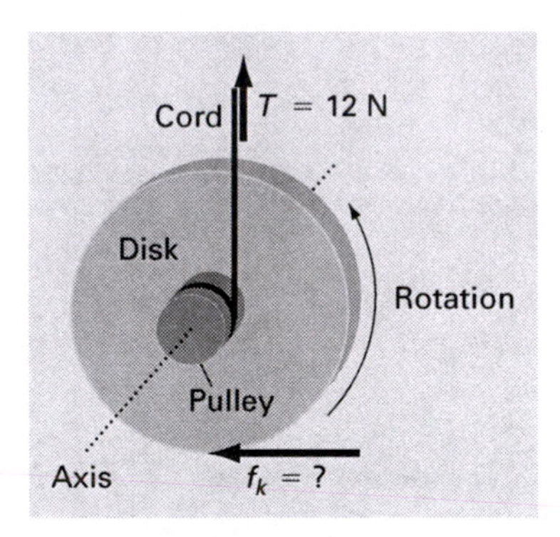

Solution

(1) Type of problem

Forces cause torque on a rotating object and give it angular acceleration (we know because the *angular velocity changes*): Use Equation 12.1 along with Equations 11.1 or 11.2, Equations 12.2a–f, and/or Equations 10.6a–d.

(2) Sort by object and force vector and (3) Equations & unknowns

The main, and simplest, part of the setup:

- **Newton's second law for rotation:** Equation 12.1 relates net torque, τ_{net}, moment of inertia, I, and angular acceleration, α.

We also may need to calculate one or more of these quantities separately:

- **Net torque:** Equation 11.1 or 11.2 for each force that causes torque.
- **Moment of inertia:** One or more of Equations 12.2a–f.
- **Angular acceleration:** Equations 10.6a–d.

We start with . . .

Newton's second law for rotation
Equation 12.1 for the disk/pulley together:

$$\tau_{net} = I\alpha$$

All three are unknown, so we try to get separate equations for *at least two.*

In this problem, we need all three!

Net torque
We don't need a full FBD because the forces and lengths (radii) are shown in the figure with the problem. Because all angles are 90°, we use Equation 11.1 to get the torque caused by each force.

First, the kinetic friction force, $\vec{f}_k$:

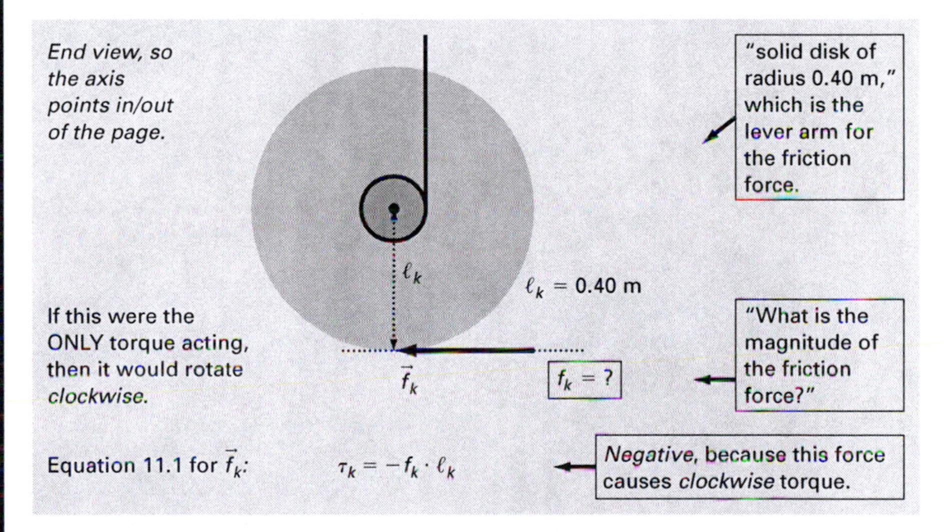

Now the tension force, $\vec{T}$:

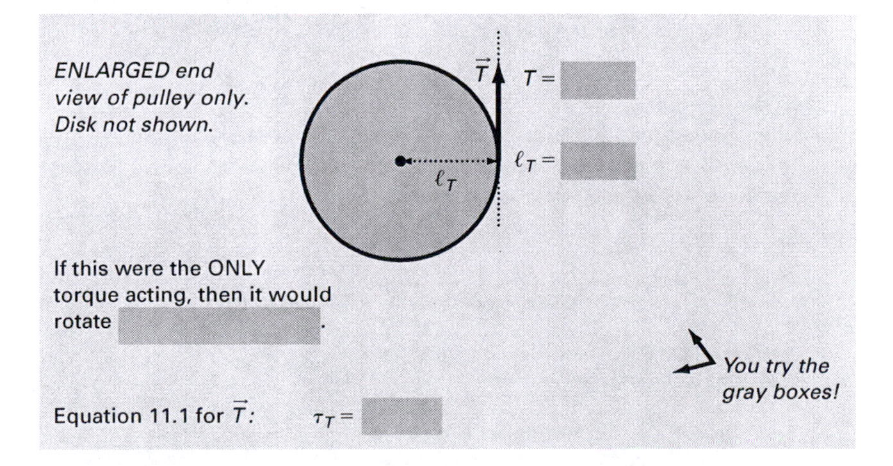

We have ignored two other forces: the *weight* and the *force caused by the axle* (at the axis) that keeps the disk in place. These both have lines of action *going through the axis* and so cause *zero torque.*

Next:

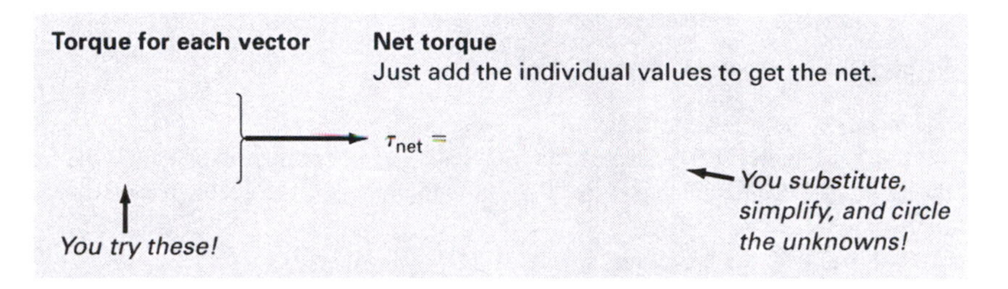

Moment of inertia

We pick from Equations 12.2a–f:

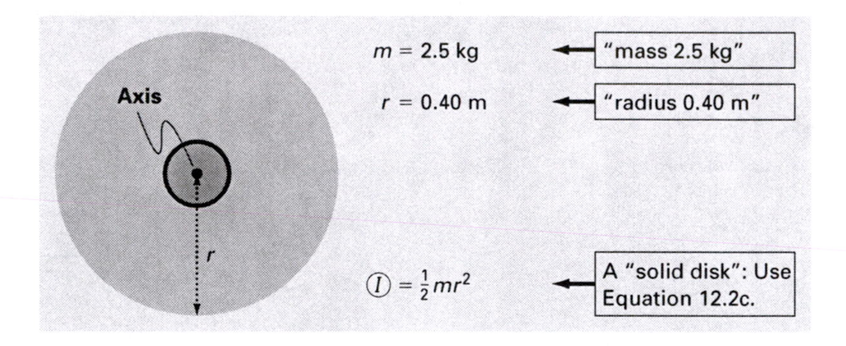

We "[i]gnore the mass of the pulley," so it has zero moment of inertia. If it had mass, we would calculate its moment of inertia and just add it to the disk's to get the total!

Angular acceleration

We set up an interval for Equations 10.6a–d to relate angular acceleration to the given angular motion quantities, which we recognize by the units (revolutions per second, revolutions, etc.). First, we convert to radians, seconds, and so on:

"6.0 *revolutions,*" which is an *angle!*	$\theta = 6.0 \text{ rev} \times \left(\frac{2\pi \text{ rad}}{1 \text{ rev}}\right) \cong 37.7 \text{ rad}$
"initially at 3.4 *revolutions per second,*" which is an *angular velocity!*	$\omega_0 = 3.4 \frac{\text{rev}}{\text{s}} \times \left(\frac{2\pi \text{ rad}}{1 \text{ rev}}\right) \cong 21.4 \frac{\text{rad}}{\text{s}}$
"slows to 1.8 *revolutions per second,*" which is an *angular velocity!*	$\omega = 1.8 \frac{\text{rev}}{\text{s}} \times \left(\frac{2\pi \text{ rad}}{1 \text{ rev}}\right) \cong 11.3 \frac{\text{rad}}{\text{s}}$

The interval and Equations 10.6a–d:

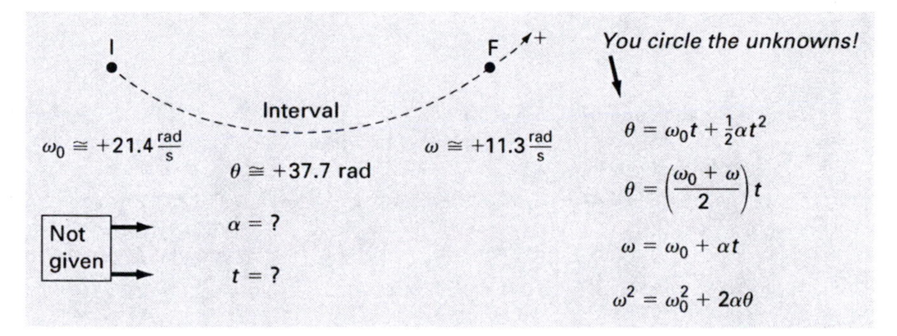

(4) Outline solution

We have done all the physics, and now we put the equations together:

You try an outline!

Newton's second law for rotation $\quad \tau_{net} = I\alpha$

Net torque $\quad \tau_{net} = T \cdot \ell_T - f_k \cdot \ell_k$

Moment of inertia $\quad I = \frac{1}{2}mr^2$

Angular acceleration

$$\theta = \omega_0 t + \frac{1}{2}\alpha t^2$$

$$\theta = \left(\frac{\omega_0 + \omega}{2}\right) t$$

$$\omega = \omega_0 + \alpha t$$

$$\omega^2 = \omega_0^2 + 2\alpha\theta$$

Answers for gray boxes

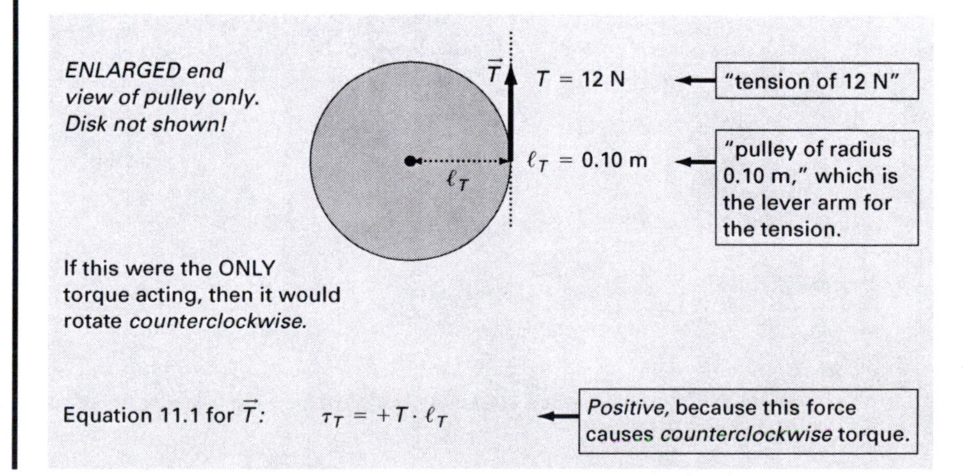

Answers for . . .

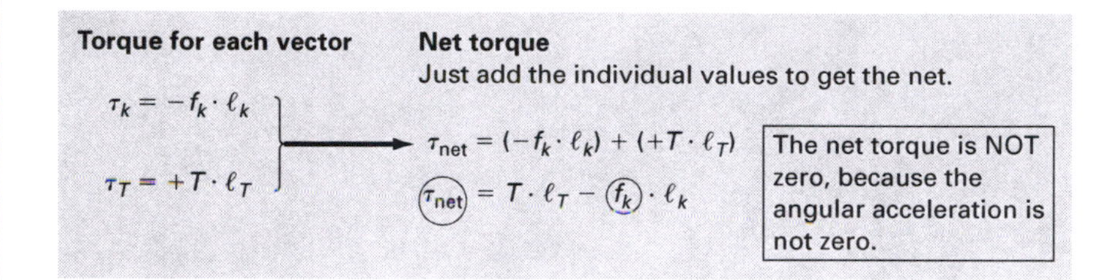

Answers for (4) Outline solution
One possible outline:

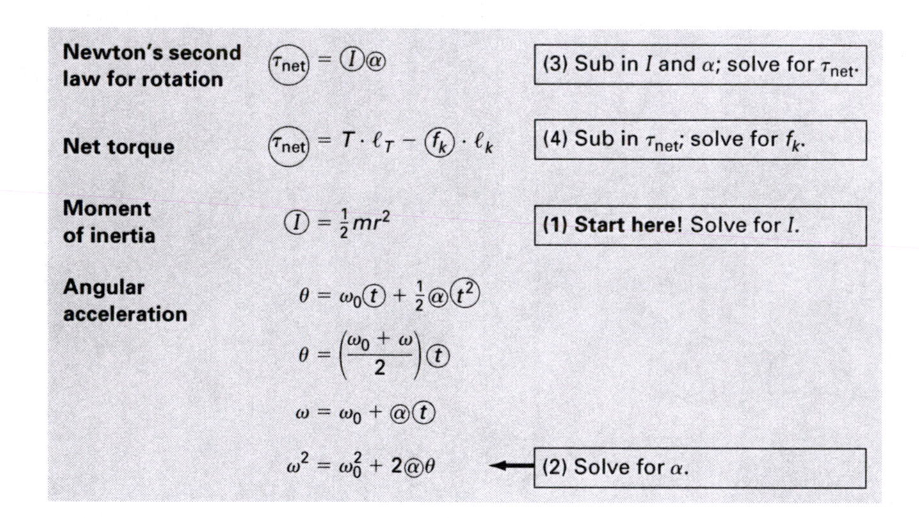

Answer $f_k \cong 5.19 \cong 5.2$ N (vector magnitude should be positive).

Intermediate answers:

- $\tau_{net} \cong -0.876$ N · m (*negative* means *clockwise,* or against the motion).
- $I = 0.200$ kg · m^2
- $\alpha \cong -4.38 \frac{\text{rad}}{\text{s}^2}$ (*negative* means *clockwise,* or against the motion, which agrees with the fact that it is slowing). ■

12.3 HOW TO SET UP TORQUE AND ANGULAR ACCELERATION PROBLEMS

Here is a summary of what we just did:

Torque and Angular Acceleration Problems—Mental and Written Steps

Mental →	**(1) Type of problem** Torque causes angular acceleration: Use Equation 12.1 along with Equations 11.1 or 11.2, Equations 12.2a–f, and/or Equations 10.6a–d.
Mental and written →	**(2) Sort by object and force vector and (3) Equations & unknowns** First: • **Newton's second law for rotation:** Sort quantities in Equation 12.1. We may need additional equations and quantities for: • **Net torque:** Equation 11.1 or 11.2 for each force that causes torque. • **Moment of inertia:** One or more of Equations 12.2a–f. • **Angular acceleration:** Interval for Equations 10.6a–d. Write equations as described above and circle the unknowns.
Mental →	**(4) Outline solution**

12.4 ROTATIONAL KINETIC ENERGY AND CONSERVATION OF ENERGY

To calculate kinetic energy of a *rotating* rigid body, we do NOT use $\frac{1}{2}mv^2$. The reason:

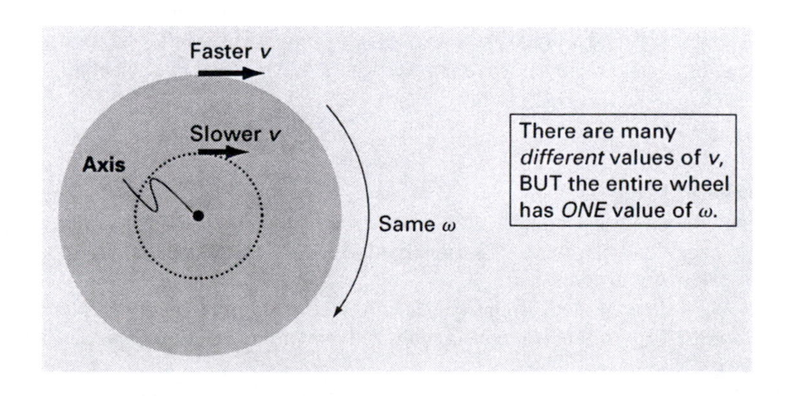

Instead of m and v, we use I and ω to calculate *rotational* kinetic energy:

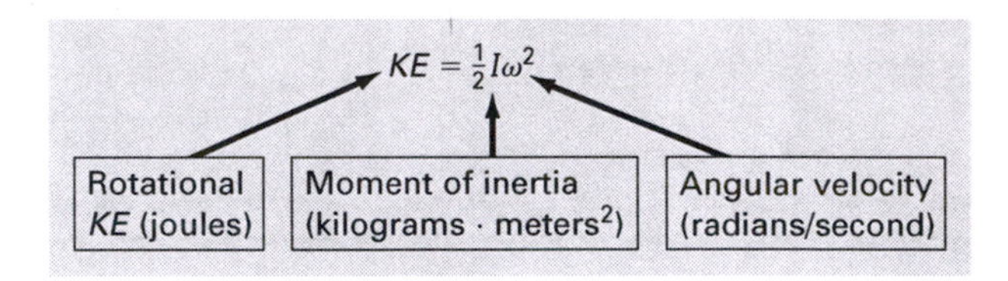

The moment of inertia, I, accounts for the *shape* and *mass* of the entire wheel, and it makes the units work out.

Now we modify any equation that has a *kinetic energy* term to allow for *three possibilities:*

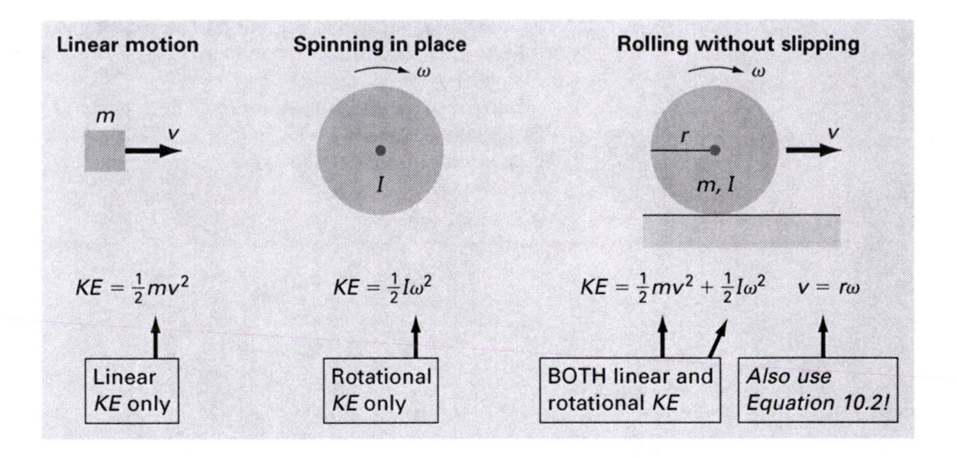

EXERCISE 12.2

The following three objects all have identical masses, m, and all have initial linear speed 5.0 m/s at the bottom of a hill. A block *slides without friction* up a hill, a hoop *rolls without slipping* up a hill, and a solid sphere *rolls without slipping* up a hill. The hoop and sphere each have radius r. Determine the maximum vertical height reached by each object.

Solution

(1) Type of problem

Motion interval for each object, and no nonconservative forces do work: Use conservation of energy, specifically Equation 8.4b (modified to allow for rotational kinetic energy if needed) since *speeds* are mentioned.

We follow the steps from Chapter 8 for conservation of energy problems, and also use Equation 10.2 for rolling without slipping, and one or more of Equations 12.2a–f for moment of inertia.

(2) Sort by interval, (3) Equations & unknowns, and (4) Outline solution

We do these steps all separately for each object.

Block

The quantities in Equation 8.4b for the block, which has only *linear* kinetic energy:

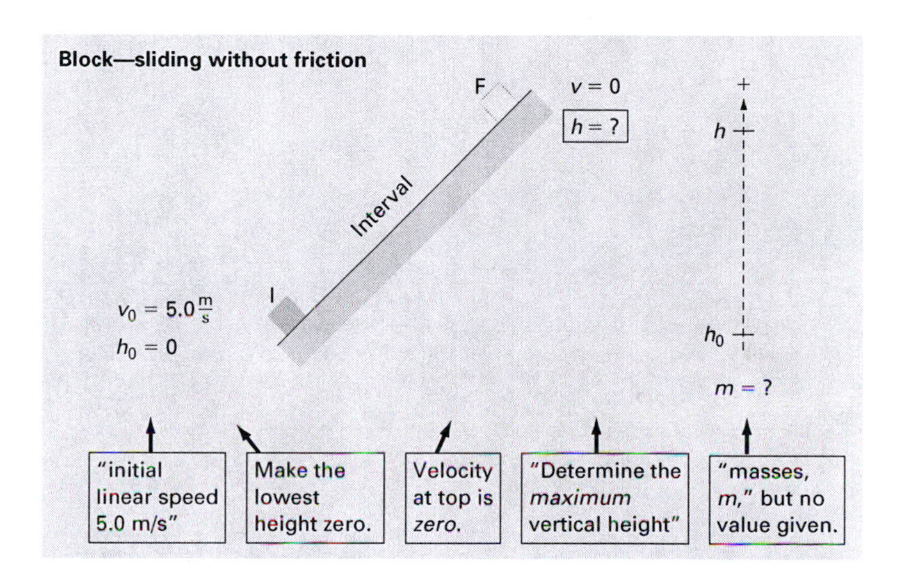

The equation and solution outline for the block:

Equation 8.4b $\frac{1}{2}ⓜv_0^2 + ⓜgh_0 = \frac{1}{2}ⓜv^2 + ⓜgⓗ$

(1) Divide each term by m, and sub in zero for h_0 and v.

$$\frac{1}{2}v_0^2 = gⓗ$$

(2) One unknown; solve.

Hoop

The hoop *rolls without slipping,* and so we modify Equation 8.4b to include both kinds of kinetic energy:

Equation 8.4b
With linear and rotational *KE*

$$\underbrace{\tfrac{1}{2}mv_0^2 + \tfrac{1}{2}I\omega_0^2}_{\text{Initial linear and rotational } KE} + mgh_0 = \underbrace{\tfrac{1}{2}mv^2 + \tfrac{1}{2}I\omega^2}_{\text{Final linear and rotational } KE} + mgh$$

The setup for this equation has a few additional quantities compared to that for the block:

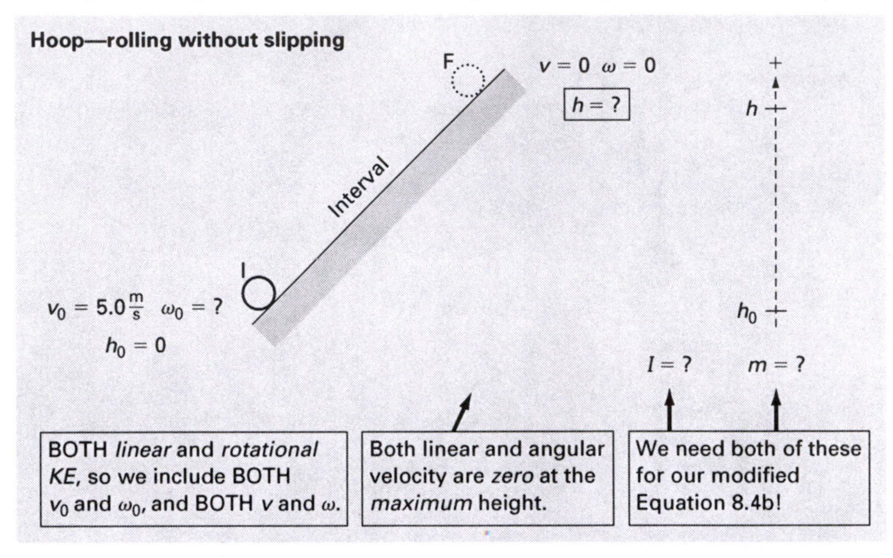

For the moment of inertia of the hoop, we pick from Equations 12.2a–f:

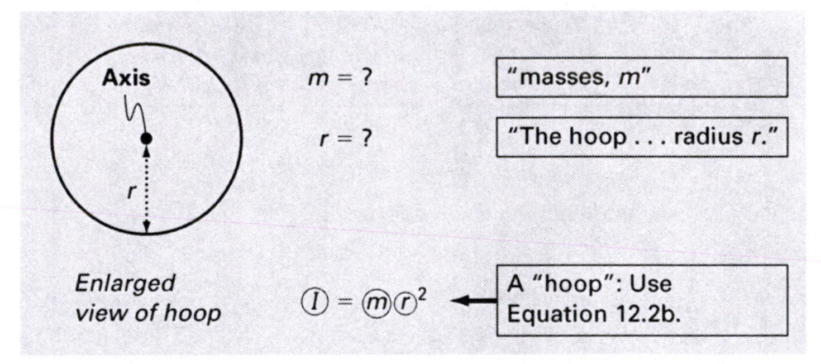

The previous two figures also have all the quantities needed for Equation 10.2, which we use for rolling without slipping. Now we put all the equations together:

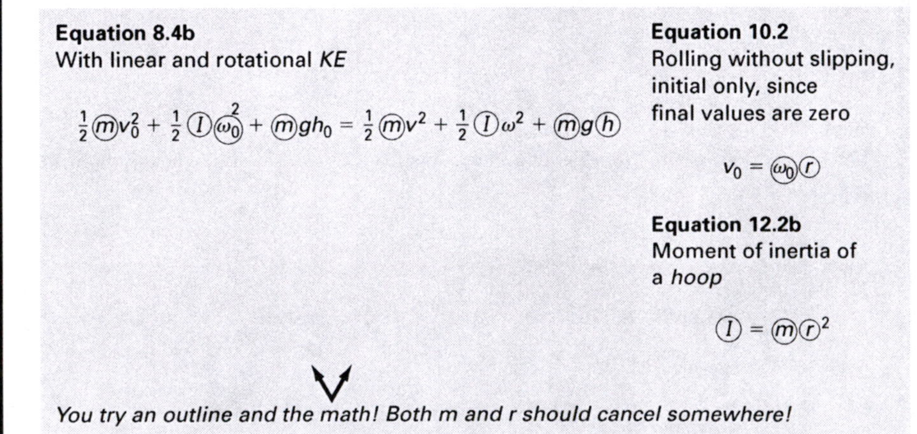

We will show the math for this part in the answers.

Solid sphere
The setup diagram for the sphere is identical to that for the hoop. The equations are nearly the same, *except* the moment of inertia equation is different:

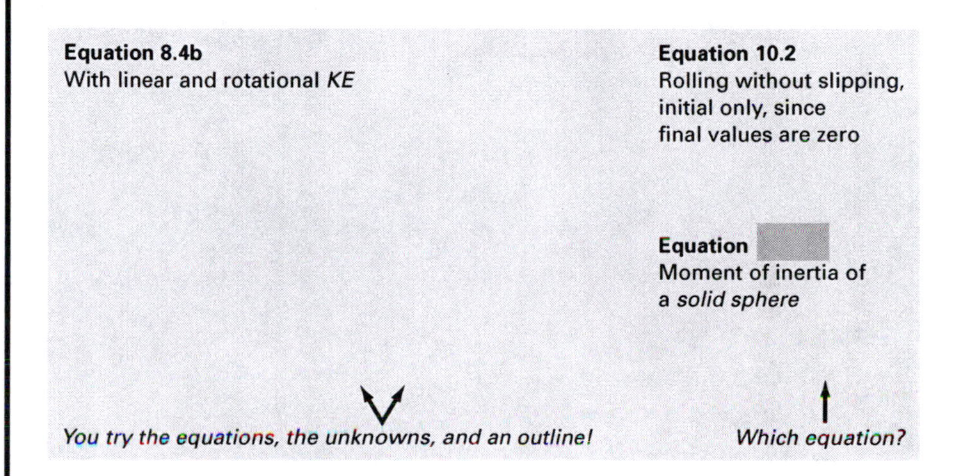

***Answers for (4) Outline solution and math for the* hoop**

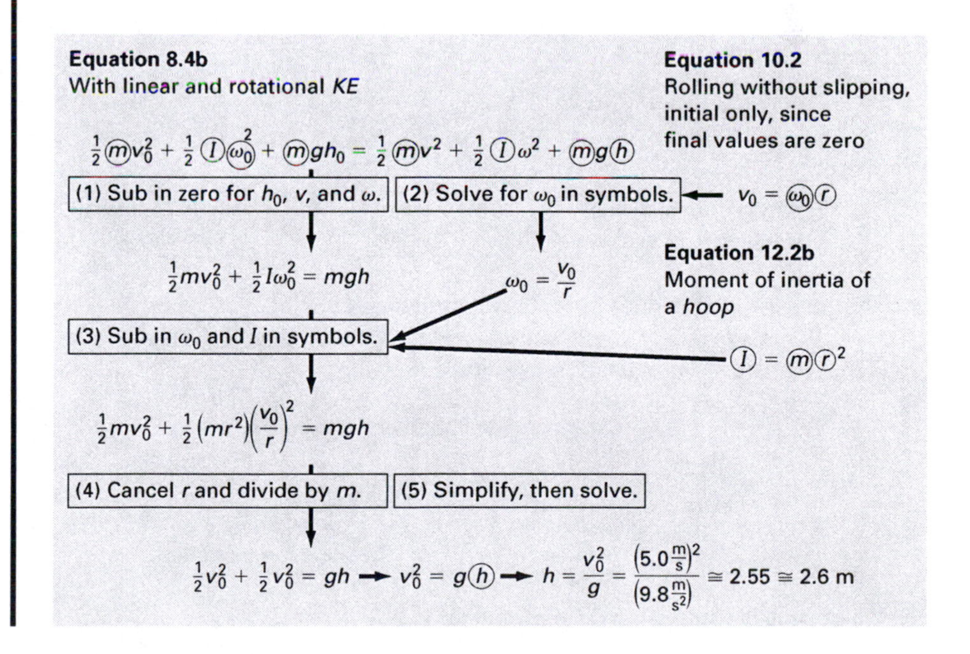

Answers for (3) Equations & unknowns, (4) Outline solution, and math for the sphere

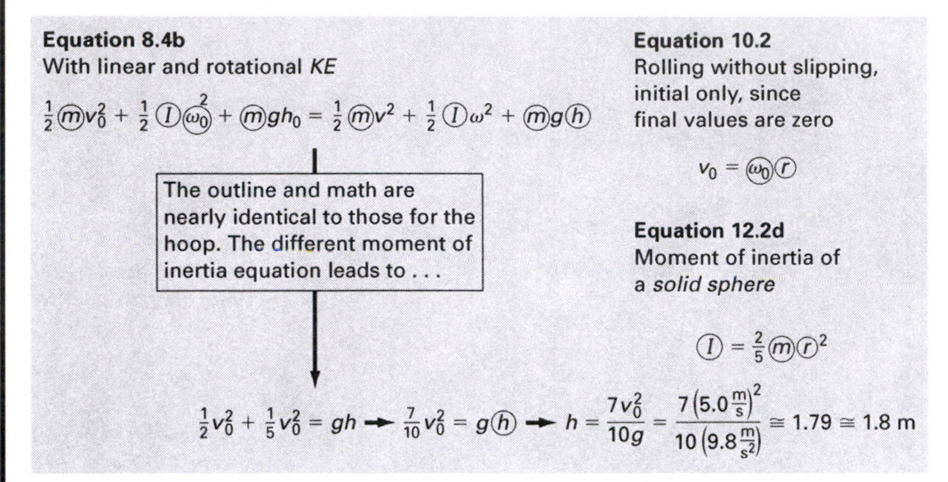

Answers

- Block: $h \cong 1.28 \cong 1.3$ m
- Hoop: $h \cong 2.55 \cong 2.6$ m
- Solid sphere: $h \cong 1.79 \cong 1.8$ m ■

12.5 CONSERVATION OF ANGULAR MOMENTUM

As with rotational kinetic energy, instead of m and v, we use I and ω to calculate *angular* momentum:

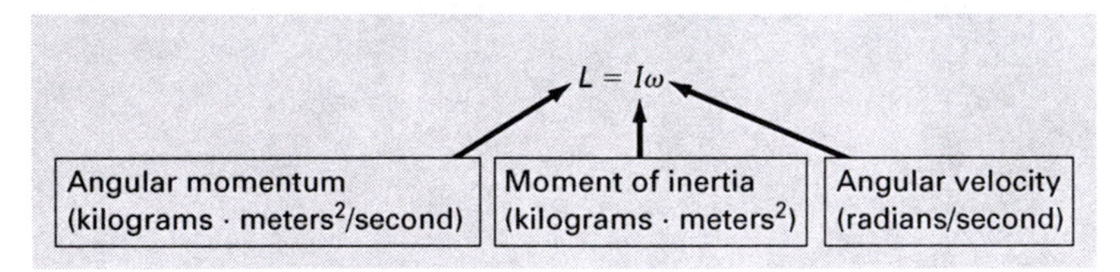

The units are *not* the same as for linear momentum.

When *moment of inertia* and/or *angular velocity* change during an interval, if there is *zero net torque* acting, we use the following:

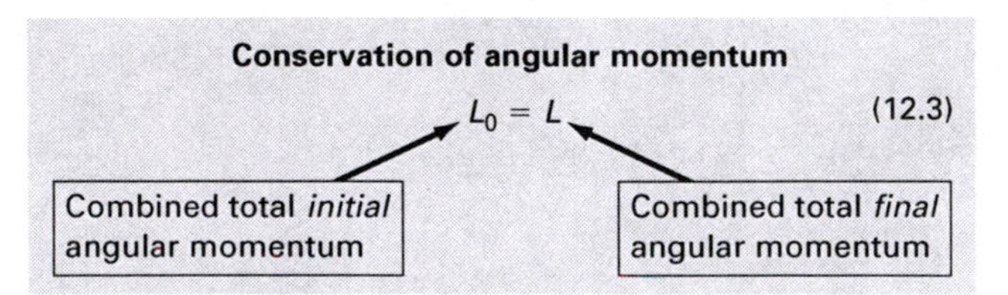

We see this in two main types of problem.

12.6 CONSERVATION OF ANGULAR MOMENTUM PROBLEMS—FIRST TYPE

This type of problem starts with a rotating object. Then its moment of inertia, I, somehow changes, which causes its angular velocity, ω, to also change. Think about these examples in terms of *instants* and *intervals:*

Example situation	Initial instant	Interval	Final instant
Ice skater spinning	Arms out, slow rotation	Pulls in her arms	Arms in, fast rotation
Boy on merry-go-round	Boy near center, fast rotation	Boy walks out toward edge	Boy near edge, slow rotation
Two kids at opposite ends of rope, spinning around common center	Long rope, slow rotation	Shorten rope	Short rope, fast rotation

EXERCISE 12.3

A 35-kg child is at the edge of a merry-go-round, which is a solid disk of radius 2.0 m and mass 160 kg. Initially the child and merry-go-round spin together at 4.0 rad/s. The child walks inward to a point 1.0 m from the center. What is the new angular velocity of the child and merry-go-round together?

Solution

(1) Type of problem

This is the first type of conservation of angular momentum problem: Use Equation 12.3.

(2) Sort by interval and object and (3) Equations & unknowns

We have two objects (child and merry-go-round) and one interval (the walk).

We need *moments of inertia* and *angular velocities* in order to calculate *angular momentum* and use Equation 12.3. The following shows how to mentally organize these quantities:

	Quantity	Initial (before walk)	Final (after walk)
Moment of inertia . . .	. . . of **child** (changes)	I_{0c}	I_c
	. . . of **merry-go-round** (constant)	I_m	I_m
Angular velocity of **both together**		ω_0	ω
Combined total **angular momentum** (Just add $I\omega$ for each object.)		$L_0 = (I_{0c} + I_m)\omega_0$	$L = (I_c + I_m)\omega$

This gives us an equation to use:

Conservation of angular momentum $L_0 = L$

Customized Equation 12.3 → $(I_{0c} + I_m)\omega_0 = (I_c + I_m)\omega$

The equation tells us which quantities to include in our setup:

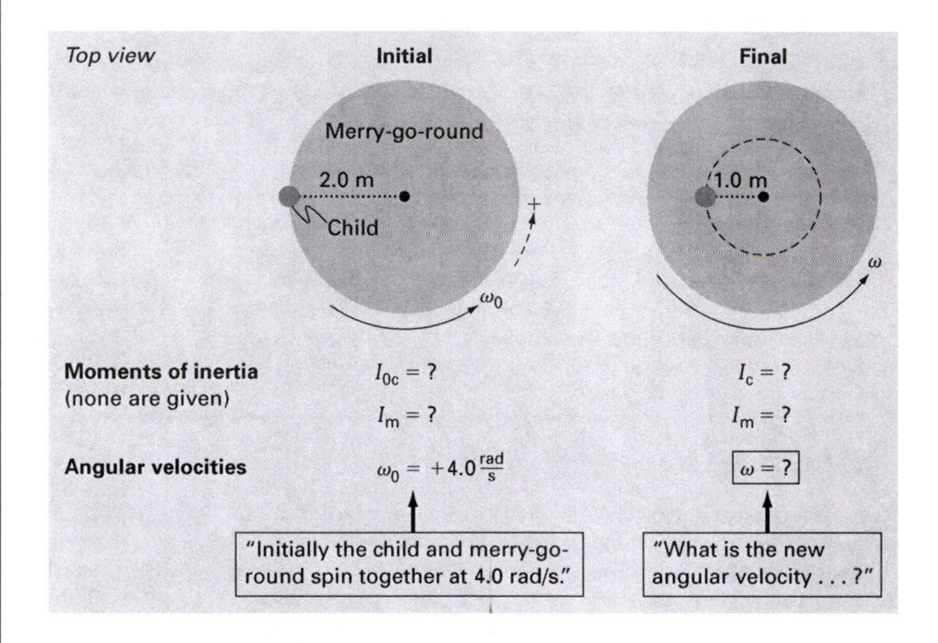

We circle the unknowns in our customized Equation 12.3:

Conservation of angular momentum $(I_{0c} + I_m)\omega_0 = (I_c + I_m)\omega$

Too many unknowns! We need equations for the moments of inertia in terms of the given masses and radii:

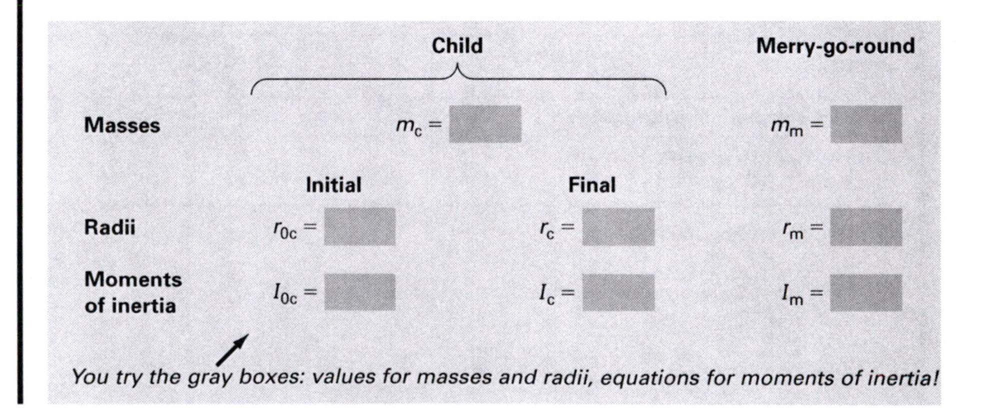

(4) Outline solution

At last, we put our equations together:

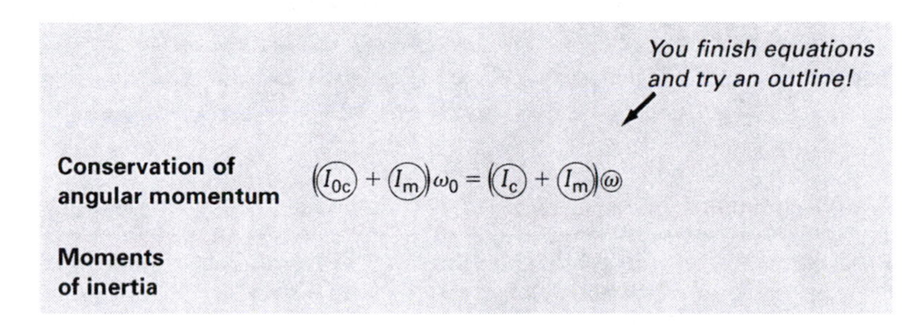

Answers for . . .

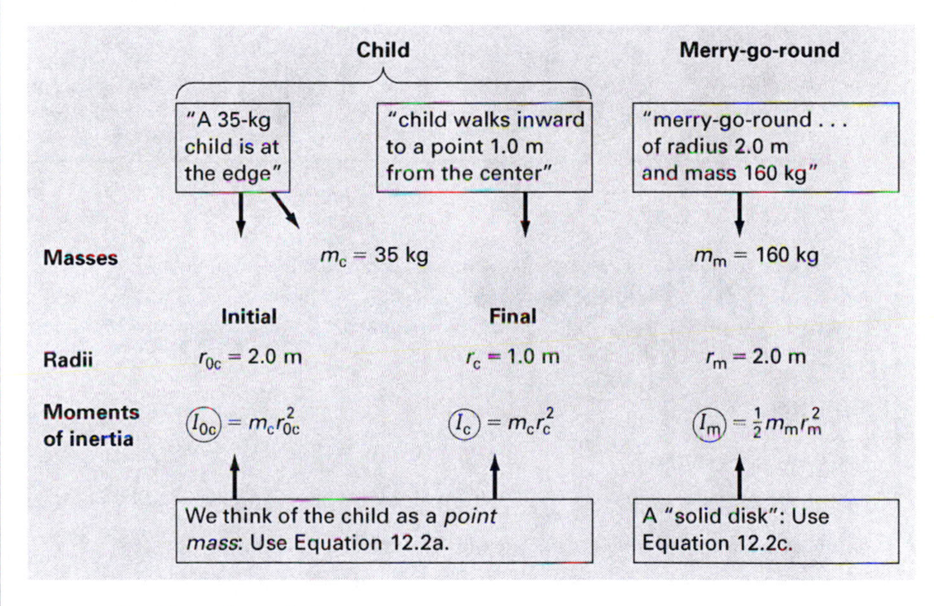

Answers for (4) Outline solution

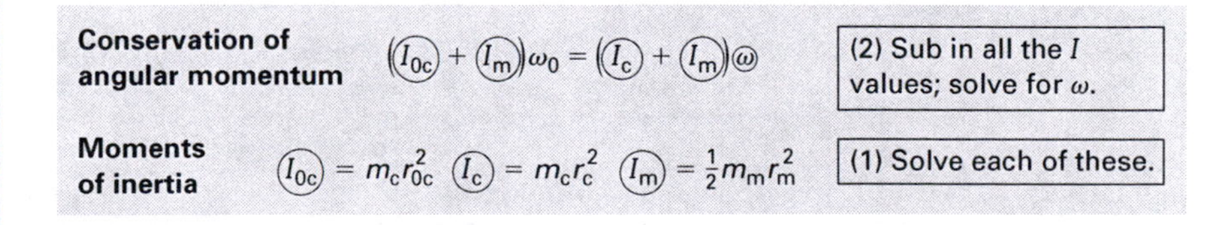

Answer $\omega \cong +5.18 \cong +5.2 \frac{\text{rad}}{\text{s}}$ (*positive* means *counterclockwise,* or in the original direction of rotation).

Intermediate answers: $I_{0c} = 140 \text{ kg} \cdot \text{m}^2$, $I_c = 35 \text{ kg} \cdot \text{m}^2$, $I_m = 320 \text{ kg} \cdot \text{m}^2$ ■

12.7 CONSERVATION OF ANGULAR MOMENTUM PROBLEMS—SECOND TYPE

This type of problem begins with two objects rotating with *different angular velocities.* Then they join together and rotate with the *same angular velocity.* Or it happens the other way around. After initially rotating together, the two objects rotate separately in the end. Some examples:

Example situation	Initial instant	Interval	Final instant
Two disks	Disk 1 is held *at rest* above disk 2, which *rotates* (different ω's).	Disk 1 is dropped onto disk 2.	They rotate *together* (same ω).
Girl and merry-go-round	Girl is *at rest* on the ground, but merry-go-round *rotates* (different ω's).	Girl jumps onto merry-go-round.	They rotate *together* (same ω).
Boy and merry-go-round	Boy is standing on edge of merry-go-round (same ω).	Boy starts walking.	Boy is walking around on edge of merry-go-round (different ω's).

EXERCISE 12.4

A 35-kg child is initially at rest on a low tree branch just above a merry-go-round, which is a solid disk of radius 2.0 m and mass 160 kg, initially spinning at 3.0 rpm. The child gently drops straight down onto the merry-go-round, to a point 1.0 m from the center. In the end, they spin together. What is the new angular velocity?

Solution

(1) Type of problem

This is the second type of conservation of angular momentum problem: Use Equation 12.3.

(2) Sort by interval and object and (3) Equations & unknowns

There are two objects (child and merry-go-round) and one interval (the drop).

Unlike in the previous problem, the two objects here have different initial angular velocities because the child is not initially rotating. Our mental organization:

	Quantity	Initial (before drop)	Final (after drop)
Moment of inertia . . .	. . . of **child**	(Not rotating)	I_c
	. . . of **merry-go-round**	I_m	I_m
Angular velocity . . .	. . . of **child**	(Not rotating)	ω
	. . . of **merry-go-round**	ω_{0m}	(Rotate together)
Combined total **angular momentum**		$L_0 =$ *You try!*	$L =$ *You try!*

This leads to our customized Equation 12.3:

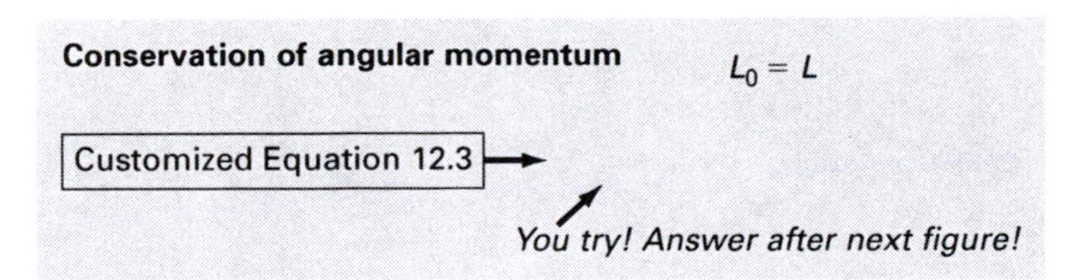

The equation tells us which quantities to put in our setup:

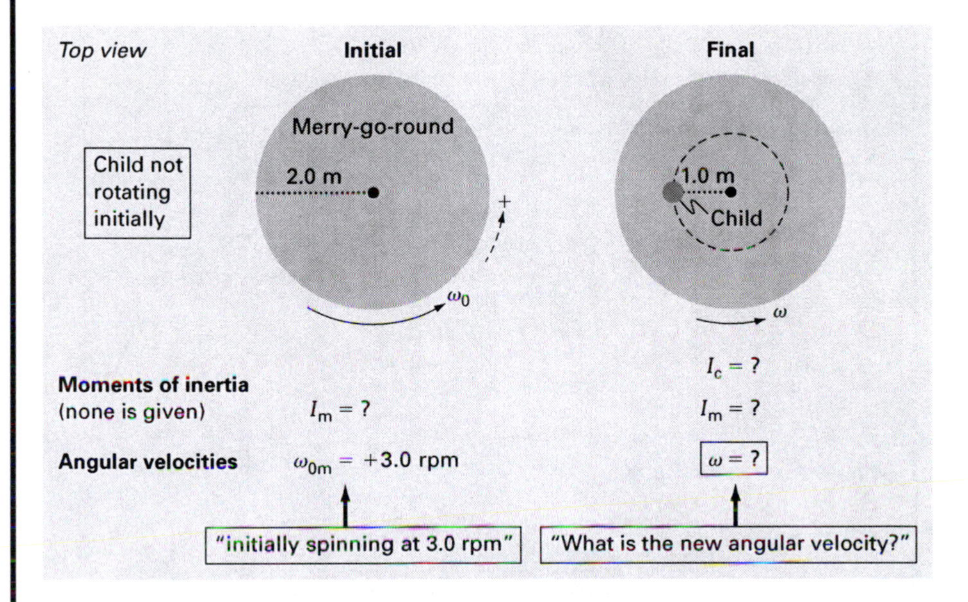

We circle the unknowns in our customized Equation 12.3:

Conservation of angular momentum $\quad \boxed{I_m}\omega_{0m} = (\boxed{I_c} + \boxed{I_m})\boxed{\omega}$

Now we need moment of inertia equations, which are identical to those in the previous exercise, except here the *child* has *no initial moment of inertia:*

	Child		Merry-go-round
Masses	$m_c = 35$ kg		$m_m = 160$ kg
	Initial	**Final**	
Radii	(None)	$r_c = 1.0$ m	$r_m = 2.0$ m
Moments of inertia	(None)	$\boxed{I_c} = m_c r_c^2$	$\boxed{I_m} = \frac{1}{2} m_m r_m^2$

(4) Outline solution

We put the equations together:

Conservation of angular momentum $I_m \omega_{0m} = (I_c + I_m)\omega$ ← *You try an outline!*

Moments of inertia $I_c = m_c r_c^2$ $I_m = \frac{1}{2} m_m r_m^2$

We will also show the calculation for ω, with units, in the answers.

Answers for (4) Outline solution

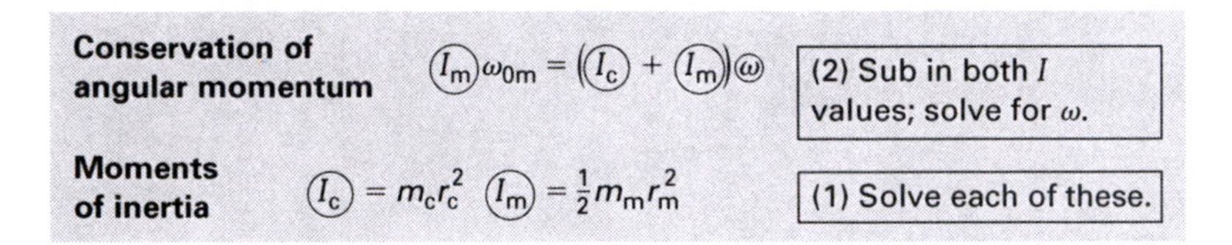

Answers for calculation of ω

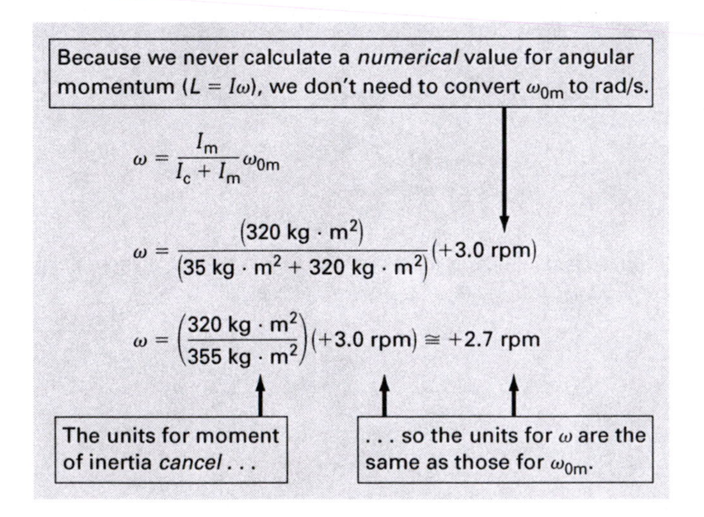

Answer $\omega \cong +2.7$ rpm

Intermediate answers: $I_c = 35\ \text{kg}\cdot\text{m}^2$, $I_m = 320\ \text{kg}\cdot\text{m}^2$ ■

12.8 HOW TO SET UP CONSERVATION OF ANGULAR MOMENTUM PROBLEMS

Here is a summary of what we did in the previous two exercises:

Conservation of Angular Momentum Problems—Mental and Written Steps

Mental→	**(1) Type of problem** Conservation of angular momentum, first or second type: Use Equation 12.3 and one or more of Equations 12.2a–f if needed.
Mental and written→	**(2) Sort by interval and object** • **Moment of inertia:** Initial and final for each object. • If moments of inertia are not given, sort quantities for one or more of Equations 12.2a–f. • **Angular velocity:** Initial and final for each object. Note: For two objects that rotate in opposite directions, like a kid walking clockwise on a merry-go-round that spins counterclockwise, make *counterclockwise* positive and *clockwise* negative.
Mental and written→	**(3) Equations & unknowns** • Write customized Equation 12.3 using *combined total angular momentum* (initial and final). • Write any needed moment of inertia equations from Equations 12.2a–f. • Circle the unknowns.
Mental→	**(4) Outline solution**

INDEX